Current Researches in Biodiversity

Current Researches in Biodiversity

Edited by **Jason Hendon**

R CALLISTO
REFERENCE

New York

Published by Callisto Reference,
106 Park Avenue, Suite 200,
New York, NY 10016, USA
www.callistoreference.com

Current Researches in Biodiversity
Edited by Jason Hendon

International Standard Book Number: 978-1-63239-143-8 (Hardback)

Printed in the United States of America.

Contents

Preface

Animals and plants of infinite shapes, sizes, colours, food habits and attitudes make this planet a mind-boggling place. There is so much diversity that it has given rise to its very own branch of science. Biodiversity is the branch of science that delves deep into the world of variation in various living beings.

Throughout history people have tried to understand these variations and make sense of them. Like why do rapid environmental changes cause mass extinctions? This branch of science explores terrestrial biodiversity, marine biodiversity and its finer nuances around the equator and how temperatures make a difference.

Biodiversity is why we now know that the first signs of life were seen in biogenic graphite found in Western Greenland. This entire branch of science is quite fascinating. It tracks history of life itself and how it has been changing over the eons. Biodiversity is also extremely relevant to the current state of various species and helps in determining their future.

The origins of the word 'biodiversity' can be traced to the wildlife scientist and conservationist Raymond F. Dasmann in his 1968 book. Biodiversity, as a discipline, essentially deals with amazing variety and infinite possibilities that such diversity opens up. It studies the degree of variation in human and plant life. From plants to animals, and even microbes, the diversity in the world is seemingly endless.

This book is the result of constant research happening all around the world. I would like to thank all the contributing authors for their efforts. I would also like to thank my family for their endless support.

Editor

Anthropomorphic Factors Influencing Spanish Conservation Policies of Vertebrates

Irene Martín-Forés,[1] **Berta Martín-López,**[2] **and Carlos Montes**[2]

[1] *Laboratory 8, Department of Ecology, Universidad Complutense de Madrid, c. José Antonio Novais,*
  28040 Madrid, Spain
[2] *Social-Ecological Systems Laboratory, Department of Ecology, Universidad Autónoma de Madrid, c. Darwin,*
  28049 Madrid, Spain

Correspondence should be addressed to Berta Martín-López; berta.martin@uam.es

Academic Editor: Rafael Riosmena-Rodríguez

National and international reports developed for the International Year of Biodiversity concluded that we have failed to meet the 2010 biodiversity target. There is an urgent need to analyze current policies for biodiversity conservation. We examined the anthropomorphic factors underlying the threatened species listings (both red lists and legal lists) and funding allocation for the conservation of vertebrates in Spain at different organizational levels, from the global to subnational level. Our results reveal a strong effect of anthropomorphic factors on conservation policies, mainly legal listings and species priority setting at national scale. Specifically, we found that those vertebrates that are phylogenetically close to humans or physically similar to human neonates tend to receive more conservation attention. Based on results, we suggest recommendations to improve conservation policies in Spain.

1. Introduction

Up to now, 193 countries endorsed through the Convention on Biological Diversity (CBD) a commitment to reduce the rates of biodiversity loss by 2010 [1]. For most nations, the 2010 biodiversity target has been their most important political commitment to conserve biodiversity [2, 3]. Although this target has stimulated considerable international and national interest, it is clear that we have failed to meet the 2010 biodiversity target [4, 5], especially in the case of vertebrates [6, 7]. One of the most important indicators developed for biodiversity is the International Union for Conservation of Nature (IUCN) Red List Index, which shows a net negative trend in the status of species [4]. This indicator uses information from the IUCN Red List (http://www.iucnredlist.org/) to trace trends in the comprehensive extinction risks of various sets of species [8]. The IUCN Red List is widely recognized as the most objective and authoritative listing of species at risk of global extinction (e.g., [9–14]). Approximately half of all countries worldwide have developed national and regional threatened species lists [15], establishing threatened

status as the most important indicator for conservation policies worldwide [16] and as an important tool in defining conservation priorities [17, 18].

Currently, there is an extensive debate on the use of the IUCN Red List in decision-making regarding conservation policies. Some authors argue that economic resources should not automatically be allocated to species according to their listing status because spending scarce conservation resources on species at the greatest risk of extinction are not an efficient way to minimize global extinction rates [19, 20], asking for using a broader range of criteria in the species priority setting [21, 22]. These criteria might include the probability of success of avoiding species extinction [20, 22], species' roles in ecosystem functioning [23], and social preferences [24–26]. However, increasingly governmental organizations rely on the IUCN Red List as well as on National Red Lists (NRLs; herein, national red lists and red data books) to influence conservation legislation inform priorities, and guide conservation investments [13, 18].

Some authors have recently identified a taxonomic bias in NRLs [14, 27, 28], which can influence legally threatened species listings [29–31] and conservation funding [32–35]. This phylogenetic bias indicates that mammals and birds are clearly overrepresented in both conservation legislation and conservation priorities [30, 34, 36].

In addition to taxonomic bias, previous studies have demonstrated that humans have an innate tendency to lavish attention and affection on individuals of nonhuman species with infantile physical features, such as large eyes, large rounded forehead, or short and narrow nose [37, 38]. Lorenz suggested that humans have a natural attraction to these neonates' features, that is, a baby schema or "Kindchenschema," which promotes in the last term a care behavior [39]. In fact, it seems that people feel a more positive affection towards animals which are phylogenetically close to humans or physically similar to human neonates than towards those which are phylogenetically distant or dissimilar to us [24, 39–42].

If anthropomorphic factors (i.e., phylogenetic distance from humans and neonatal morphological characteristics) influence human preferences towards species protection, then the question here is whether anthropomorphism influences vertebrates' conservation priority setting. In this context, we aim to explore the effect of anthropomorphic factors on the decision-making process regarding the conservation listing (both Red lists and legal listing) and conservation priority setting of vertebrates in Spain. We specifically examined the effect of vertebrates' phylogenetic distance from humans (using the taxonomic classification of species) and the effect of species' morphological characteristics on the vertebrates conservation priority setting at different organizational levels, from international to subnational. At the international level, we used the global IUCN Red List and examined European legislation regarding vertebrate conservation and conservation funding allocation. At the national level, we focused on Spain, which is considered a Mediterranean biodiversity hotspot [43]. Finally, on a subnational level, we selected Andalucía, which contains more terrestrial vertebrate species than larger areas such as the United Kingdom or Sweden [44].

2. Methods

2.1. Data Sources. We developed a data matrix of all species of vertebrates (specifically, 679 species) that are present in Spain according to the National Inventory of Biodiversity (http://www.mma.es/portal/secciones/biodiversidad/inventarios/inb/inventario_vertebrados/index.htm), which are distributed among different classes of vertebrates as follows: 48 fishes, 33 amphibians, 73 reptiles, 406 birds, and 119 mammals. In order to analyze the effect of anthropomorphic factors on vertebrates conservation priority setting, we considered three dependent variables (Table 1): (1) the threatened species category in red lists, (2) the threatened species category in conservation legal listings, and (3) funding allocation for the conservation of vertebrates.

The conservation of Spanish vertebrates is regulated by different laws at the international, national, and subnational levels. At the European level, the Habitats Directive (Council Directive 92/43/EEC) and the Birds Directive (Council

Directive 79/409/EEC) are the two most important legal tools for protecting Europe's threatened species, and both were transposed into national law. Additionally, the Spanish government has listed threatened vertebrates in the National Catalogue of Threatened Species (NCTS) through the Royal Decree 439/90. The NCTS considers four threatened categories: endangered (EN), sensitive to habitat change (SHC), vulnerable (VU), and of special interest (SI). These categories are similar but not identical to those of the IUCN, which are: extinct (EX), extinct in the wild (EW), critically endangered (CR), endangered (EN), vulnerable (VU), near threatened (NT), least concern (LC), and data deficient (DD) (for more details, see [29]). At the subnational level, legislation based on the NCTS categorization system for the conservation of vertebrates has been developed in Andalucía. Here, we searched species listings in each taxonomic group of vertebrates in the global IUCN Red List, NRLs and Red list of Andalucía, as well as conservation legislation in Europe, Spain, and Andalucía, and we recorded the number of Spanish vertebrates for each vertebrate class regarding their threatened status at all organizational levels (Table 2).

Data sources for funding allocation at European and national levels were obtained from 2003 to 2007. On the European level, we looked up the Life Project's database, which is the most important European financial instrument supporting biodiversity conservation (http://ec.europa.eu/environment/life/project/Projects/index.cfm), and at national level we consulted three different sources: (1) the Official Spanish Gazette, (2) annual reports of the activities of seven Spanish national parks, and (3) the projects database of the Biodiversity Foundation, which is a nonprofit making nature foundational organization. At subnational level, we obtained data on conservation expenditures for Doñana Protected Area, which is one of the most important natural areas of the European Union. We consulted annual activity reports for Doñana National and Natural Parks and we carried out personal interviews with environmental managers responsible for endangered species programs (for more details, see [32]). More information about data sources regarding these variables at different organizational levels is presented in Table 1.

As variables (Table 3), we considered (1) those that measured the species phylogenetic distance from humans, recording the vertebrates' class and order of each species, and (2) those that can measure the effect of "Kindchenschema" phenomenon (in the sense of [39]) using morphological traits, such as the length and relative measures of the weight and the eye size of vertebrates, which were calculated employing the quotient between the weight and the length and between the eye size and the length, respectively. We explored the effect of phylogenetic distance from humans on the conservation of vertebrates at two taxonomic levels: (1) class (i.e., if the species is fish, amphibian, reptile, bird, or mammal) and (2) order, incorporating 44 different orders.

There are few caveats regarding our data sources of conservation funding that must be taken into account in the interpretation of our results. The first one is that at the subnational level, we only considered funding allocation in the Doñana Protected Area due to the difficulty of obtaining information regarding funding allocation in other protected

TABLE 1: Description and data source of the dependent variables.

Variable type	Attributes	Organizational level	Data source
		Red lists[a]	
Ordinal	Vertebrates' threatened category: 7: EX; RE; 6: EW; 5: CR; 4: EN; 3: VU; 2: NT; 1: LC; 0: DD and non	International	IUCN [69] (http://www.iucnredlist.org/)
		National	Atlas and Red Book of fish in Spain [70] Atlas and Red Book of amphibians and reptiles in Spain [71] The breeding bird Atlas in Spain [72] Atlas and Red Book of terrestrial mammals in Spain [73]
		Subnational	Red Book of the Vertebrates Threatened in Andalucía [74]
		Legal listing[b]	
Ordinal	1: included 0: excluded	International	Habitats Directive (Council Directive 92/43/EEC) Birds Directive (Council Directive 79/409/EEC)
	Vertebrates' threatened category: 4: EN; 3: SHC; 2: VU; 1: SI; 0: non	National	National Catalogue of Endangered Species (Law 4/1989) and Royal Decree (439/1990).
		Subnational	*Law 8/2003* of Wild Flora and Fauna of Andalucía
		Conservation budget	
Continuous	Ln (Funding allocation to vertebrates' conservation).	International	Life projects database (2003–2007 years) (http://ec.europa.eu/environment/lif/project/Projects/index.cfm)
		National	Environmental projects published in the Official Spanish Gazette (2003–2007 years) Annual Reports of National Parks Organization for Spanish National Parks (2003–2007 years) The Biodiversity Foundation database (2004–2007 years)

[a](EX) extinct; (RE) at subnational level, regionally extinct; (EW) extinct in the wild; (CR) critically endangered; (EN) endangered; (VU) vulnerable; (NT) near threatened; (LC) least concern; (DD) data deficient; (Non) nonlisted.
[b](EN) endangered; (SHC) sensitive to habitat change; (VU) vulnerable; (SI), special interest; (Non) nonlisted.

areas in Andalucía. We considered our approximation valid because the Doñana Protected Area is a highly emblematic protected area in Spain and receives a great majority of the national conservation budget [45]. The second caveat is that no database of conservation budgets is available at the international level. We thus decided to use LIFE conservation funds.

2.2. Data Analysis. We used nonparametric statistics (i.e., Kruskal-Wallis test) to compare the threatened category of Red lists and legal listing, as well as on funding allocation for vertebrates, among different taxonomic groups. Additionally, we explored the relationship between morphological characteristics (i.e., length, weight, and eye size) and the threatened category of vertebrates (both in red lists and legal listing) as well as their funding allocation, using Spearman and Pearson correlation tests. To avoid problems with heteroscedasticity, we transformed the continuous variables by their natural logs.

3. Results

3.1. Effect of Phylogenetic Distance from Humans on Conservation Policies. Table 4 shows the effect of phylogenetic distance from humans on the vertebrates red lists, legal listing, and funding allocation for their conservation.

While fish was the most threatened vertebrate class in the Red lists at the international, national, and subnational levels ($\chi^2 = 46.76$, df = 4, $P < 0.0001$; $\chi^2 = 126.57$, df = 4, $P < 0.0001$; and $\chi^2 = 35.33$, df = 4, $P < 0.0001$, resp.), reptiles were the most threatened class in legal listing at the national and the subnational levels ($\chi^2 = 17.86$, df = 4, $P < 0.001$; $\chi^2 = 17.63$, df = 4, $P = 0.001$). Phylogeny also had an effect on funding allocation. While amphibians was the most favored vertebrate class at the international level ($\chi^2 = 106.33$, df = 4, $P < 0.0001$), funds at the national level were mainly directed to mammals ($\chi^2 = 66.03$, df = 4, $P < 0.0001$).

Among the different orders of fish the only order with a significantly higher threatened status in the red lists at all

TABLE 2: Number of Spanish vertebrates for each vertebrate class regarding both, their red list threatened status and their legal listing categories, at international, national, and subnational levels.

			Fish	Amphibians	Reptiles	Birds	Mammals
Red lists[a]	IUCN Red List	EX	—	—	—	1	1
		EW	—	—	—	—	—
		CR	5	—	7	3	2
		EN	6	1	7	5	6
		VU	10	4	2	9	7
		NT	2	7	10	21	10
		LC	22	21	39	364	83
		DD and non	3	—	8	3	10
	National	EX		—	—	7	1
		EW		—	—	—	—
		CR	3	1	5	15	2
		EN	11	2	6	30	4
		VU	24	8	11	44	13
		NT	7	7	11	30	14
		LC	—	14	34	—	—
		DD and non	3	1	6	280	85
	Subnational	RE	1	—	—	3	—
		EW	—	—	—	—	—
		CR	3	—	1	14	6
		EN	6	—	6	13	7
		VU	7	2	4	23	19
		NT	2	3	2	18	4
		LC	—	—	—	—	—
		DD and non	29	28	60	335	83
Legal lists[b]	International	Included	19	16	34	171	47
		Excluded	29	17	39	235	72
	National	EN	5	1	5	17	7
		SHC	—	—	3	3	2
		VU	6	1	1	14	23
		SI	1	20	42	252	23
		Non	36	11	21	120	64
	Subnational	EN	5	1	6	19	7
		SHC	—	—	3	3	2
		VU	6	1	1	13	23
		SI	1	22	42	251	24
		Non	12	8	7	120	41

[a](EX) extinct; (RE) at subnational level, regionally extinct; (EW) extinct in the wild; (CR) critically endangered; (EN) endangered; (VU) vulnerable; (NT) near threatened; (LC) least concern; (DD) data deficient; (Non) nonlisted.
[b](EN) endangered; (SHC) sensitive to habitat change; (VU) vulnerable; (SI), special interest; (Non) nonlisted.

organizational levels was Acipenseriformes (χ^2 = 147.95, df = 45, P < 0.0001 at the international level; χ^2 = 208.43, df = 45, P < 0.0001 at the national level; and χ^2 = 118.71, df = 45, P < 0.0001 at the subnational level). Both Acipenseriformes and Cyprinodontiformes had a higher threatened category at the international, national, and subnational levels in legal listings

(χ^2 = 174.52, df = 45, P < 0.0001; χ^2 = 193.04, df = 45, P < 0.0001; and χ^2 = 199.46, df = 45, P < 0.0001). In contrast, the orders that received significantly more conservation funding were Procellariiformes, Gaviiformes, and Pelecaniformes at the international level (χ^2 = 336.21, df = 45, P < 0.0001), Cetaceaat the national level (χ^2 = 360.18, df = 45, P < 0.0001),

TABLE 3: Description and data source of the explanatory variables used in the study.

Variable type	Variables	Data source
	Phylogeny	
Nominal	Vertebrates' class	The Iberian Fauna Project (http://iberfauna.mncn.csic.es/)
	Vertebrate's order	Species 2000 (http://www.sp2000.org/)
	Morphology	
Continuous	Length Relative weight (quotient between weight and length)	Data source for fish' morphology was obtained from FishBase (http://www.fishbase.org/search.php) and [70]. For amphibians' morphology, we used García-París et al. [75] and Pleguezuelos et al. [71]. For reptiles' morphology we used Salvador [76]. For birds' morphology we used Díaz et al. [77], Martí and del Moral [72], and Tellería et al. [78]. For mammals' morphology, we used, Palomo et al. [73], and Rodríguez [79]. Additionally, we based Cetacea data in Kiefner [80].
	Relative eye size (quotient between eye size and length)	[81–83]

TABLE 4: Taxonomic groups at class and order levels highly considered as threatened in red and legal listing as well as in conservation priority setting.

		Taxonomic bias	
		Class level	Order level
Red lists	International	Fish	Acipenseriformes, Anguilliformes (fish)
	National	Fish	Acipenseriformes, Accipitriformes (fish)
	Subnational	Fish	Acipenseriformes, Cyprinodontiformes (fish)
Legal lists	International	Amphibians	Cetacea (mammals), Falconiformes (birds)
	National	Reptiles	Acipenseriformes, Cyprinodontiformes (fish)
	Subnational	Reptiles	Acipenseriformes, Cyprinodontiformes (fish)
Funding allocation to vertebrates' conservation	International	Amphibians	Procellariiformes, Gaviiformes, Pelecaniformes (birds)
	National	Mammals	Cetacea (mammals)
	Subnational	Mammals	Lagomorpha (mammals)

and Lagomorpha at the subnational level ($\chi^2 = 101.42$, df = 45, $P < 0.0001$).

3.2. Effect of Morphology on Conservation Policies. At the national level, we found relationships between the relative weight of vertebrates and the threatened status of NRLs ($\rho = 0.14$, $P = 0.060$) and between relative eye size and the threatened status of NRLs ($\rho = 0.15$, $P = 0.038$). Finally, at subnational level, we found significant relationships between the threatened status considered in red list of Andalucía and physical variables, that is, length ($\rho = 0.26$, $P < 0.0001$), relative weight ($\rho = 0.42$, $P < 0.0001$), and relative eye size ($\rho = 0.32$, $P < 0.001$).

In legal listing of threatened species at the international level, we found relationships between threatened status and physical variables, that is, length (Spearman's rho = 0.23, $P < 0.0001$), relative weight ($\rho = 0.49$, $P < 0.0001$), and relative eye size ($\rho = 0.47$, $P < 0.0001$). At the national and subnational levels, there also were a significant correlation

between relative eye size and threatened status ($\rho = 0.16$, $P = 0.027$; and $\rho = 0.16$, $P = 0.003$, resp.).

Our analysis yielded significant correlations between funding allocation and relative eye size at the national and subnational levels (national: $r = 0.46$, $P < 0.0001$; subnational: $r = 0.22$, $P = 0.002$). At the national level, there also existed a significant correlation between vertebrates' length and funding allocation ($r = 0.49$, $P < 0.0001$) and relative weight and funding allocation ($r = 0.74$, $P < 0.0001$).

Similar results were obtained for vertebrates classes (Table 5). However, we found that the effect of morphological characteristics was higher in those vertebrate classes phylogenetically close to humans (i.e., birds and mammals).

4. Discussion

4.1. Effect of Phylogenetic Distance from Humans on Conservation Policies. Previous studies have demonstrated that humans prefer species that are phylogenetically close to us;

TABLE 5: Significant correlations between morphological characteristics and both listing and funding allocations, for each vertebrate class.

Vertebrate class	Physical characteristics	Independent variables	Organizationallevel	Correlation results
Amphibians	Length	Funding allocation	International	$r = 0.472, P = 0.010$
Reptiles	Length	Red lists	Subnational	$\rho = 0.359, P = 0.048$
		Funding allocation	National	$r = 0.417, P = 0.004$
	Relative weight	Legal listing	International	$\rho = 0.535, P = 0.042$
		Funding allocation	National	$r = 0.829, P < 0.0001$
	Relative eye size	Legal listing	International	$\rho = 0.535, P = 0.042$
		Funding allocation	National	$r = 0.737, p = 0.002$
Birds	Length	Red lists	International	$\rho = 0.170, P = 0.002$
			National	$\rho = 0.183, P = 0.001$
			Subnational	$\rho = 0.298, P < 0.0001$
		Legal listing	International	$\rho = 0.308, P < 0.0001$
		Funding allocation	International	$r = 0.264, P < 0.0001$
			National	$r = 0.217, p < 0.0001$
			Subnational	$r = 0.155, P = 0.004$
	Relative weight	Legal listing	International	$\rho = 0.245, P = 0.031$
			National	$\rho = -0.293, P = 0.009$
		Funding allocation	Subnational	$r = 0.229, P = 0.004$
	Relative eye size	Legal listing	International	$\rho = 0.267, P = 0.019$
		Funding allocation	Subnational	$r = 0.401, P < 0.0001$
Mammals	Length	Legal listing	International	$\rho = 0.608, P < 0.0001$
			National	$\rho = 0.556, P < 0.0001$
		Funding allocation	National	$r = 0.674, P < 0.0001$
	Relative weight	Legal listing	International	$\rho = 0.604, P < 0.0001$
			National	$\rho = 0.540, P < 0.0001$
		Funding allocation	National	$r = 0.705, P < 0.0001$
	Relative eye size	Legal listing	International	$\rho = 0.576, P < 0.0001$
			National	$\rho = 0.491, P < 0.0001$
		Funding allocation	National	$r = 0.589, P < 0.0001$

these species tend to evoke a more positive affect than those that are phylogenetically distant from humans or physically dissimilar to human features [24, 25, 41, 42, 46]. Similarly, structural complexity, as an indicator of phylogenetic distance, is positively related to the amount of scientific output on different species, and this relationship underlies the high existence values and societal popularity of complex organisms [33]. In this sense, recent studies have shown that both conservation biology research and public support are skewed significantly towards birds and mammals [32, 35, 46–50]. Both results suggest that phylogenetic distance from humans underlies scientific and social preferences.

Our results also show that funding allocation for vertebrate conservation mostly favors the protection of those species phylogenetically close to humans [30, 32, 51]. As in [25], we found that different groups of vertebrates vary in the amount of political attention they receive. On the one hand, few species of amphibians; bird orders such as *Procellariiformes*, *Gaviiformes*, and *Pelecaniformes*; and mammalian orders such as *Cetacea* receive relatively high amounts of political attention, as measured by their conservation budget (Table 4). On the other hand, fish, reptiles,

and also some orders of small mammals (*Rodentia* and *Chiroptera*) have low political power and receive fewer funds for their conservation, despite the important roles they play in ecosystem function [52]. Similarly, although many other nocturnal creatures are classified as threatened species in the Red lists, they do not receive conservation funds at national and subnational levels. These animals include most of amphibians, which often evoke feelings of disgust [53], and bats, which inspire primal fears related to the vampire myth [54].

This overall taxonomic bias is stronger at national and subnational levels than at international scale. Taxonomic bias in Spanish conservation projects was evident in the overrepresentation of mammals and birds. This bias can occur because Spanish conservation efforts are based on available scientific information, and this information is biased towards species phylogenetically close to humans [29, 32, 55].

4.2. Effect of Morphology on Conservation Policies. Previous studies have demonstrated that humans' preferences for animals are significantly influenced by physical characteristics of the species (e.g., [39, 40, 56–58]). In fact, people are more inclined to protect species that are large, aesthetically attractive, and regarded as possessing the capacities for feeling, thought, and pain [59].

We found that there is a strong bias in both conservation legal listing and funding allocation towards species that are large, have a large relative eye size, and have a high relative weight (both in relation to their length). Our results agree with Lorenz's "Kindchenschema" phenomenon because both listing and funding are biased towards those vertebrates with relative higher eye size and weight, especially in those vertebrates classes close to humans, that is, mammals and birds.

We also found a stronger correlation for those taxonomic classes phylogenetically close to humans. Thus, we described a correlation between physical characteristics and funding allocation at the national and subnational levels for birds, and mammals. In fact, at lower organizational levels, especially at national level, mammals are the focus of conservation because of their charismatic appeal. They are more likely to receive conservation funds if they are charismatic, well-known, and large bodied [60]. Fortunately, some large charismatic vertebrate predators that are easily recognized, such as carnivores or raptors, can be used as flagship or umbrella species when the area under protection is small sized [61].

5. Conclusions

Understanding which factors motivate species conservation legislation and species priority setting is essential for redefining criteria for future conservation initiatives [62]. In this context, our results suggest that many conservation choices are made on subjective grounds, that is, anthropomorphic factors. Consistent with the conclusions of Metrick and Weitzman [30], we showed that likeability factors or "visceral" characteristics, including physical size, relative weight, and relative eye size, as well as whether the animals were higher life forms, play a more important role in setting priorities for vertebrate conservation. This effect was especially pronounced in legislation and funding allocation for vertebrate conservation at national organizational level. In this sense, according to Bottrill et al. [63], we highlight the need to improve management at the national level with greater connectivity among state and international agencies, having in account both expert opinions and conservation policies assessments.

Although anthropomorphism could be a conservation tool because it has the potential to promote public participation in conservation actions [64], the legal bias towards charismatic species could reduce the probability of achieving the 2020 biodiversity target as policy attention is focused towards few taxonomic groups [25], obscuring those key taxonomic groups essential for maintaining ecological properties as well as a diverse flow of ecosystem services to society [65, 66]. Moreover, funding concentrated on just few charismatic species with neonatal features perpetuates the dearth of social, scientific, and political attention of many less visible species, promoting a sort of pit-fall trap in which few charismatic and cute species, mainly better-known species,

tend to receive most of the conservation funds and policy attention [29].

Therefore, it is essential to rethink the vertebrate conservation priority setting process in Spain because most of the social, scientific, and policy attention are allocated towards few charismatic species [7, 67]. Here, we should abandon the automatic allocation of resources to species based on these anthropomorphic factors and take into account a broader range of factors in funding decisions, such as the degree of taxonomic uniqueness of a species, the level of endemicity, the role of biodiversity in maintaining the resilience of ecosystems to disturbance, and the capacity to deliver a set of ecosystem services to society. Decisions could also consider cultural and spiritual values, which must be recognized to involve different groups of stakeholders in conservation decision making. In order to raise awareness of the value of biodiversity, including the value of less attractive species, it is essential to intensify efforts in providing information to the whole society about less cute and charismatic species through adequate environmental education programs. In order to create an environmentally responsible population that contributes to biodiversity conservation, we need to develop programs of environmental education beyond aesthetic appealing that address the ethical and instrumental values of the whole species diversity [68].

Funding concentrated on just a few species with neonatal features perpetuate the dearth of knowledge of many less visible and cute species but essential in the maintenance of ecological functioning and therefore, in the delivery of ecosystem services for human wellbeing.

References

[1] A. Balmford, P. Crane, A. Dobson, R. E. Green, and G. M. Mace, "The 2010 challenge: data availability, information needs and extraterrestrial insights," *Philosophical Transactions of the Royal Society B*, vol. 360, no. 1454, pp. 221–228, 2005.

[2] G. M. Mace, W. Cramer, S. Díaz et al., "Biodiversity targets after 2010," *Current Opinion in Environmental Sustainability*, vol. 2, no. 1-2, pp. 3–8, 2010.

[3] M. R. W. Rands, W. M. Adams, L. Bennun et al., "Biodiversity conservation: challenges beyond 2010," *Science*, vol. 329, no. 5997, pp. 1298–1303, 2010.

[4] S. H. M. Butchart, M. Walpole, B. Collen et al., "Global biodiversity: indicators of recent declines," *Science*, vol. 328, no. 5982, pp. 1164–1168, 2010.

[5] I. J. Gordon, N. Pettorelli, T. Katzner et al., "International year of biodiversity: missed targets and the need for better monitoring, real action and global policy," *Animal Conservation*, vol. 13, no. 2, pp. 113–114, 2010.

[6] M. Hoffmann, C. Hilton-Taylor, A. Angulo et al., "The impact of conservation on the status of the world's vertebrates," *Science*, vol. 330, no. 6010, pp. 1503–1509, 2010.

[7] EME-Evaluación de los Ecosistemas del Milenio de España, "La Evaluación de los Ecosistemas del Milenio de España. Síntesis de resultados. Madrid: fundación Biodiversidad. Ministerio de Medio Ambiente, y Medio Rural y Marino," 2011, http://www.ecomilenio.es/informe-sintesis-eme/2321.

[8] S. H. M. Butchart, H. R. Akçakaya, J. Chanson et al., "Improvements to the red list index," *PLoS One*, vol. 2, no. 1, article e140, 2007.

[9] T. J. Regan, M. A. Burgman, M. McCarthy et al., "The consistency of extinction risk classification protocols," *Conservation Biology*, vol. 19, no. 6, pp. 1969–1977, 2005.

[10] P. C. de Grammont and A. D. Cuarón, "An evaluation of threatened species categorization systems used on the american continent," *Conservation Biology*, vol. 20, no. 1, pp. 14–27, 2006.

[11] G. M. Mace, N. J. Collar, K. J. Gaston et al., "Quantification of extinction risk: IUCN's system for classifying threatened species," *Conservation Biology*, vol. 22, no. 6, pp. 1424–1442, 2008.

[12] R. M. Miller, J. P. Rodríguez, T. Aniskowicz-Fowler et al., "National threatened species listing based on IUCN criteria and regional guidelines: current status and future perspectives," *Conservation Biology*, vol. 21, no. 3, pp. 684–696, 2007.

[13] A. S. L. Rodrigues, J. D. Pilgrim, J. F. Lamoreux, M. Hoffmann, and T. M. Brooks, "The value of the IUCN red list for conservation," *Trends in Ecology and Evolution*, vol. 21, no. 2, pp. 71–76, 2006.

[14] T. J. Zamin, J. E. M. Baillie, R. M. Miller, J. P. Rodríguez, A. Ardid, and B. Collen, "National red listing beyond the 2010 target," *Conservation Biology*, vol. 24, no. 4, pp. 1012–1020, 2010.

[15] J. P. Rodríguez, "National red lists: the largest global market for IUCN red list categories and criteria," *Endangered Species Research*, vol. 6, no. 2, pp. 193–198, 2008.

[16] J. C. Vié, C. Hilton-Taylor, C. M. Pollock et al., "The IUCN red list: a key conservation tool," in *Wildlife in A Changing World—An Analysis of the 2008 IUCN Red List of Threatened Species*, J. C. Vié, C. Hilton-Taylor, and S. N. Stuart, Eds., pp. 1–13, IUCN, Gland, Switzerland, 2009.

[17] D. S. Schmeller, B. Gruber, E. Budrys, E. Framsted, S. Lengyel, and K. Henle, "National responsibilities in European species conservation: a methodological review," *Conservation Biology*, vol. 22, no. 3, pp. 593–601, 2008.

[18] M. Hoffmann, T. M. Brooks, G. A. B. da Fonseca et al., "Conservation planning and the IUCN red list," *Endangered Species Research*, vol. 6, no. 2, pp. 113–125, 2008.

[19] H. P. Possingham, S. J. Andelman, M. A. Burgman, R. A. Medellín, L. L. Master, and D. A. Keith, "Limits to the use of threatened species lists," *Trends in Ecology and Evolution*, vol. 17, no. 11, pp. 503–507, 2002.

[20] T. M. Rout, D. Heinze, and M. A. McCarthy, "Optimal allocation of conservation resources to species that may be extinct," *Conservation Biology*, vol. 24, no. 4, pp. 1111–1118, 2010.

[21] D. Farrier, R. Whelan, and C. Mooney, "Threatened species listing as a trigger for conservation action," *Environmental Science and Policy*, vol. 10, no. 3, pp. 219–229, 2007.

[22] L. N. Joseph, R. F. Maloney, and H. P. Possingham, "Optimal allocation of resources among threatened species: a project prioritization protocol," *Conservation Biology*, vol. 23, no. 2, pp. 328–338, 2009.

[23] J. Muñoz, "Biodiversity conservation including uncharismatic species," *Biodiversity and Conservation*, vol. 16, no. 7, pp. 2233–2235, 2007.

[24] S. R. Kellert and J. K. Berry, *Phase III: Knowledge, Affection and Basic Attitudes Toward Animals in American Society*, United States Government Printing Office, Washington, DC, USA, 1980.

[25] B. Czech, P. R. Krausman, and R. Borkhataria, "Social construction, political power, and the allocation of benefits to endangered species," *Conservation Biology*, vol. 12, no. 5, pp. 1103–1112, 1998.

[26] E. Meuser, H. W. Harshaw, and A. Ø. Mooers, "Public preference for endemism over other conservation-related species attributes," *Conservation Biology*, vol. 23, no. 4, pp. 1041–1046, 2009.

[27] P. Hutchings, "Invertebrates and threatened species legislation," in *Threatened Species Legislation—Is It Just An Act?* P. Hutchings, D. Lunney, and C. Dickman, Eds., pp. 88–93, Royal Zoological Society of NSW, Mosman, Australia, 2004.

[28] M. Burgman, "Expert frailties in conservation risk assessment and listing decisions," in *Threatened Species Legislation—Is It Just An Act?* P. Hutchings, D. Lunney, and C. Dickman, Eds., pp. 20–29, Royal Zoological Society of NSW, Mosman, Australia, 2004.

[29] B. Martín-López, J. A. González, and C. Montes, "The pit-fall trap of species conservation priority setting," *Biodiversity and Conservation*, vol. 20, no. 11, pp. 663–682, 2011.

[30] A. Metrick and M. L. Weitzman, "Patterns of behavior in endangered species preservation," *Land Economics*, vol. 72, no. 1, pp. 1–16, 1996.

[31] A. Ø. Mooers, L. R. Prugh, M. Festa-Bianchet, and J. A. Hutchings, "Biases in legal listing under canadian endangered species legislation," *Conservation Biology*, vol. 21, no. 3, pp. 572–575, 2007.

[32] B. Martín-López, C. Montes, L. Ramírez, and J. Benayas, "What drives policy decision-making related to species conservation?" *Biological Conservation*, vol. 142, no. 7, pp. 1370–1380, 2009.

[33] V. M. Proença, H. M. Pereira, and L. Vicente, "Organismal complexity is an indicator of species existence value," *Frontiers in Ecology and the Environment*, vol. 6, no. 6, pp. 298–299, 2008.

[34] M. Restani and J. M. Marzluff, "Funding extinction? Biological needs and political realities in the allocation of resources to endangered species recovery," *Bioscience*, vol. 52, no. 2, pp. 169–177, 2002.

[35] J. R. U. Wilson, S. Proches, B. Braschler, E. S. Dixon, and D. M. Richardson, "The (bio)diversity of science reflects the interests of society," *Frontiers in Ecology and the Environment*, vol. 5, no. 8, pp. 409–414, 2007.

[36] P. J. Seddon, P. S. Soorae, and F. Launay, "Taxonomic bias in reintroduction projects," *Animal Conservation*, vol. 8, no. 1, pp. 51–58, 2005.

[37] T. R. Alley, "Head shape and the perception of cuteness," *Developmental Psychology*, vol. 17, no. 5, pp. 650–654, 1981.

[38] T. R. Alley, "Infantile head shape as an elicitor of adult protection," *Merrill-Palmer Quarterly*, vol. 29, no. 4, pp. 411–427, 1983.

[39] K. Lorenz, *Studies in Animal and Human Behavior*, vol. 2, Harvard University Press, Cambridge, Mass, USA, 1971.

[40] K. Lorenz, *The Foundations of Ethology*, Springer, New York, NY, USA, 1978.

[41] S. Plous, "Psychological mechanisms in the human use of animals," *Journal of Social Issues*, vol. 49, no. 1, pp. 11–52, 1993.

[42] G. M. Burghardt and H. A. Herzog, "Animals, evolution and ethics," in *Perceptions of Animals in American Culture*, R. J. Hoage, Ed., pp. 129–151, Smithsonian Institution Press, Washington, DC, USA, 1989.

[43] N. Myers, R. A. Mittermeier, C. G. Mittermeier, G. A. B. da Fonseca, and J. Kent, "Biodiversity hotspots for conservation priorities," *Nature*, vol. 403, pp. 853–858, 2000.

[44] A. Estrada, R. Real, and J. M. Vargas, "Assessing coincidence between priority conservation areas for vertebrate groups in a Mediterranean hotspot," *Biological Conservation*, vol. 144, no. 3, pp. 1120–1129, 2011.

[45] M. Múgica, C. Martínez, J. Gómez-Limón, J. Puertas, J. A. Atauri, and J. V. de Lucio, *Anuario EUROPARC-España del estado de los espacios naturales protegidos 2009*, FUNGOBE, Madrid, Spain, 2010.

[46] B. Martín-López, C. Montes, and J. Benayas, "The non-economic motives behind the willingness to pay for biodiversity conservation," *Biological Conservation*, vol. 139, no. 1-2, pp. 67–82, 2007.

[47] I. Fazey, J. Fischer, and D. B. Lindenmayer, "What do conservation biologists publish?" *Biological Conservation*, vol. 124, no. 1, pp. 63–73, 2005.

[48] B. Bajomi, A. S. Pullin, G. B. Stewart, and A. Takács-Sánta, "Bias and dispersal in the animal reintroduction literature," *Oryx*, vol. 44, no. 3, pp. 358–365, 2010.

[49] J. A. Clark and R. M. May, "Taxonomic bias in conservation research," *Science*, vol. 297, no. 5579, pp. 191–192, 2002.

[50] J. Schlegel and R. Rupf, "Attitudes towards potential animal flagship species in nature conservation: a survey among students of different educational institutions," *Journal for Nature Conservation*, vol. 18, no. 4, pp. 278–290, 2010.

[51] T. D. Male and M. J. Bean, "Measuring progress in US endangered species conservation," *Ecology Letters*, vol. 8, no. 9, pp. 986–992, 2005.

[52] B. Clucas, K. McHugh, and T. Caro, "Flagship species on covers of US conservation and nature magazines," *Biodiversity and Conservation*, vol. 17, no. 6, pp. 1517–1528, 2008.

[53] P. Muris, B. Mayer, J. Huijding, and T. Konings, "A dirty animal is a scary animal! Effects of disgust-related information on fear beliefs in children," *Behaviour Research and Therapy*, vol. 46, no. 1, pp. 137–144, 2008.

[54] P. Prokop, J. Fančovičová, and M. Kubiatko, "Vampires are still alive: Slovakian students' attitudes toward bats," *Anthrozoos*, vol. 22, no. 1, pp. 19–30, 2009.

[55] L. M. Bautista and J. C. Pantoja, "What animal species should we study next?" *Bulletin of the British Ecological Society*, vol. 36, pp. 27–28, 2005.

[56] S. Bitgood, D. Patterson, and A. Benefield, "Exhibit design and visitor behavior: empirical relationships," *Environment and Behavior*, vol. 20, no. 4, pp. 474–491, 1988.

[57] A. Gunnthorsdottir, "Physical attractiveness of an animal species as a decision factor for its preservation," *Anthrozoös*, vol. 14, no. 4, pp. 204–215, 2001.

[58] B. Martín-López, C. Montes, and J. Benayas, "Economic valuation of biodiversity conservation: the meaning of numbers," *Conservation Biology*, vol. 22, no. 3, pp. 624–635, 2008.

[59] S. R. Kellert, "Attitudes toward animals: age-related development among children," *Journal of Environmental Education*, vol. 16, no. 3, pp. 29–39, 1985.

[60] N. Sitas, J. E. M. Baillie, and N. J. B. Isaac, "What are we saving? Developing a standardized approach for conservation action," *Animal Conservation*, vol. 12, no. 3, pp. 231–237, 2009.

[61] R. Home, C. Keller, P. Nagel, N. Bauer, and M. Hunziker, "Selection criteria for flagship species by conservation organizations," *Environmental Conservation*, vol. 36, no. 2, pp. 139–148, 2009.

[62] K. H. Redford, P. Coppolillo, E. W. Sanderson et al., "Mapping the conservation landscape," *Conservation Biology*, vol. 17, no. 1, pp. 116–131, 2003.

[63] M. C. Bottrill, J. C. Walsh, J. E. M. Watson, L. N. Joseph, A. Ortega-Argueta, and H. P. Possingham, "Does recovery planning improve the status of threatened species?" *Biological Conservation*, vol. 144, no. 5, pp. 1595–1601, 2011.

[64] A. Chan, "Anthropomorphism as a conservation tool," *Biodiversity and Conservation*, vol. 21, no. 7, pp. 1889–1892, 2012.

[65] B. J. Cardinale, J. E. Duffy, A. Gonzalez et al., "Biodiversity loss and its impact on humanity," *Nature*, vol. 486, pp. 59–67, 2012.

[66] C. Kremen, "Managing ecosystem services: what do we need to know about their ecology?" *Ecology Letters*, vol. 8, no. 5, pp. 468–479, 2005.

[67] B. Martín-López, I. Martín-Forés, J. A. González, and C. Montes, "La conservación de biodiversidad en España: atención científica, construcción social e interés político," *Ecosistemas*, vol. 20, pp. 104–113, 2011.

[68] M. Kassas, "Environmental education: biodiversity," *The Environmentalist*, vol. 22, no. 4, pp. 345–351, 2002.

[69] IUCN, *IUCN Red List of Threatened Species*, IUCN, Cambridge, UK, 2010, http://www.iucnredlist.org.

[70] I. Doadrio, *Atlas y Libro Rojo de los Peces Continentales de España*, National Museum of Natural Science, General Direction of Nature Conservation, Madrid, Spain, 2001.

[71] J. M. Pleguezuelos, R. Márquez, and M. Lizana, Eds., *Atlas y Libro rojo de Anfibios y reptiles de España*, The Spanish Ministry of the Environment, Autonomous Organism of Nature Reserves, Madrid, Spain, 2004.

[72] R. Martí and J. C. del Moral, *Atlas de las Aves Reproductoras de España*, The Spanish Ministry of the Environment (Autonomous Organism of Nature Reserves) and Spanish Ortnithologist Society, Madrid, Spain, 2003.

[73] J. L. Palomo, J. Gisbert, and J. C. Blanco, *Atlas y Libro Rojo de los mamíferos terrestres de España*, The Spanish Ministry of the Environment, Autonomous Organism of Nature Reserves, Madrid, Spain, 2007.

[74] A. Franco and M. Rodríguez de los Santos, *Libro Rojo de los vertebrados amenazados de Andalucía*, Regional Environment Ministry of Andalusian Board, Sevilla, Spain, 2001.

[75] M. García-París, A. Montori, and P. Herrero, *Fauna Ibérica: Amphibia: Lissamphibia*, vol. 24, National Museum of Natural Science, CSIC, Madrid, Spain, 2004.

[76] A. Salvador, *Fauna Ibérica: Reptiles*, vol. 10, National Museum of Natural Science, CSIC, Madrid, Spain, 1998.

[77] M. Díaz, B. Asensio, and J. L. Tellería, *Aves Ibéricas I. No Paseriformes*, J. M. Reyero, Madrid, Spain, 1996.

[78] J. L. Tellería, B. Asensio, and M. Díaz, *Aves Ibéricas II. Paseriformes*, J. M. Reyero, Madrid, Spain, 1999.

[79] J. L. Rodríguez, *Guía de Mamíferos Ibéricos*, Natural Content, Ávila, Spain, 1999.

[80] R. Kiefner, *Guía de los cetáceos del Mundo: Océano Pacífico, Océano Índico, Mar Rojo, Océano Atlántico, Caribe, Océano Ártico, Océano Antártico*, Editorial Group M&G Difusión, D. L., Elche, Spain, 2002.

[81] M. D. L. Brooke, S. Hanley, and S. B. Laughlin, "The scaling of eye size with body mass in birds," *Proceedings of the Royal Society B*, vol. 266, no. 1417, pp. 405–412, 1999.

[82] H. C. Howland, S. Merola, and J. R. Basarab, "The allometry and scaling of the size of vertebrate eyes," *Vision Research*, vol. 44, no. 17, pp. 2043–2065, 2004.

[83] R. J. Thomas, T. Székely, I. C. Cuthill et al., "Eye size in birds and the timing of song at dawn," *Proceedings of the Royal Society B*, vol. 269, no. 1493, pp. 831–837, 2002.

2

Marine Nematodes from the Shallow Subtidal Coast of the Adriatic Sea: Species List and Distribution

F. Semprucci

Dipartimento di Scienze della Terra, della Vita e dell'Ambiente (DiSTEVA), Università degli Studi di Urbino "Carlo Bo", Località Crocicchia, 61029 Urbino, Italy

Correspondence should be addressed to F. Semprucci, federica.semprucci@uniurb.it

Academic Editor: Rafael Riosmena-Rodríguez

This study is the first attempt aiming to assess the composition and number of free-living marine nematode species on the coasts of the Marches region, Italy. A high number of putative species of nematodes were recognized (84), these belonging to 52 genera in 22 families. Fifty-one taxa have been identified to the species level increasing the number of known nematode species for the Adriatic Sea from 283 to 310 and for the Mediterranean Sea from 700 to 723. The highest diversity and abundance were registered for the nematode families typically of intertidal zones characterized as medium-fine sands. The majority of the species found in the present study are known to occur in the North European coasts or the North Atlantic Sea, the best known regions for nematode distribution. Inferences on the biogeography of marine nematodes are preliminary since most Biodiversity literature concerning the Mediterranean of the basin is very out of date. Considering the great importance of nematodes in the assessment of ecosystem health conditions, an intensification of sampling efforts should be pursued in other regions in order to improve our current knowledge of the distribution pattern of marine nematode species as well as clarify their biogeographical patterns.

1. Introduction

Marine nematodes are the most abundant metazoans in marine sediments, reaching densities as high as 20 million individuals per square meter [1]. They generally are the dominant component of the meiofauna in any aquatic habitat, often one order of magnitude higher than any other major taxon [1]. Estimates of global nematode species diversity have significantly varied over the past 15 years, but with a growing agreement suggesting about one million species [2, 3]. Their great importance in marine ecology as bioindicators is well recognized [4, 5]. The study of nematode communities offers several advantages for assessing the quality of freshwater, marine, and terrestrial ecosystems. These organisms show great diversity and abundance which make easier their sampling; they present different trophic groups as well as being intimately related with their environment, the sediment. Therefore, they can provide important information on the ecological state of a given area once they are exposed to pollutants. Furthermore, they have been recently proposed within the Water Framework Directive (WFD, Directive 2000/60/EC) [6] as a tool for evaluating the ecological quality status of marine vulnerable ecosystems.

The taxonomy of marine nematodes is known to be very difficult (see for review [7]), in part due to their high diversity. Identification of nematodes at specific level is often hampered by the fact that a significant proportion of the specimens collected are juveniles or females, which often lack the diagnostic features required for an accurate identification. When males are encountered, it may be necessary to examine a high number of specimens before a final decision can be made. Owing to these limitations, biodiversity studies focusing on marine nematodes have been frequently replaced by ecological studies where identification is mostly limited to the genus level (or more rarely morphotypes). Unfortunately, this kind of information may be hardly used to create a map of distribution of the free-living marine nematodes.

A recent review of the state of knowledge of the meiofauna in the Adriatic sea has been carried out by Balsamo et al. [8]. Marine nematodes known in the Italian seas represent 443 species which are distributed into 262 genera and 46 families [9]; these are a quarter of the species reported for the

European sea waters. Although these data are an important base line for the area, most of these records are the result of old surveys, mainly carried out in the first part of the XX century [9]. In particular, the Adriatic coast poses a high number of nematode species: 283 species in the entire Adriatic Basin and 263 in the Italian coasts (see [8] for review). Extensive investigations in this area were carried out by Allgén [10], Schuurmans-Stekhoven [11], Gerlach [12], and Travizi and Vidaković [13]. Most of the species were recorded in the area of the river Po, Venice lagoon, and Gulf of Trieste [13–15]. A significant contribution to the knowledge of marine nematodes for the southern Adriatic Basin, in particular the Apulian littoral, was provided by De Zio [16, 17] and Grimaldi-De Zio [18–20]. Therefore, most of the investigations on nematodes have been limited to the North Adriatic sector whereas from the coast of Emilia-Romagna to Apulia few studies were carried out [9, 13].

Despite their great ecological importance [6], the knowledge of the nematodes in the Adriatic has changed a little, except for data reported by some ecological studies where nematode identification is mostly limited to the genus level [21–26].

In this context, a nematode survey in the coastline between Fiorenzuola and Pesaro (Marches region) has been carried out. Currently, there is no data available on nematode diversity for this area as well as in the Marches region. Furthermore, this coast includes the Natural Regional Park of Monte San Bartolo, an area of high naturalistic value. Thus, the goal of the present study was to assess nematode biodiversity along this coastal area as well as to identify patterns of species distribution.

2. Materials and Methods

2.1. Study Area and Sampling Sites. The Adriatic sea is an elongated NW/SE oriented basin located in the central Mediterranean Sea and characterized by the most extensive development of continental shelf in the Mediterranean. It is a relatively shallow basin, especially in the northern sector, with an average water depth of 35 m. The general circulation is cyclonic, with a flow towards the northwest along the eastern side and a return flow towards the southeast along the western side [27]. The central and southern regions of the Adriatic Sea are characterized by a low primary productivity, with the continental inputs and the benthic-pelagic interactions being of minor importance in comparison to the northern area [28].

The investigated area is located in the northern part of the Marches region (Italy) and extends from Fiorenzuola di Focara to Pesaro (Figure 1).

The nematode species reported in the present study were collected in three different sites from the sublittoral zone of the Marches coast: Fiorenzuola di Focara (FI) (43°57′N–12°49′E), Monte Brisighella (BR) (43°56′N–12°50′E), and Pesaro (P) (43°55′N–12°53′E) (Figure 1). In each site, two stations were established (1 and 3 m of depth) and sediment samples collected in May and November 2003. In addition, a third station was established at FI site (12 m in depth) which

was sampled in May, July, and October 2002. The total of the samples studied for the taxonomical study of the nematodes were 15.

2.2. Sample Processing. Bulk sublittoral samples were taken using a 1.5 L plastic scoop digging it directly into the sediment at shallow sites and using a Van Veen grab at the deeper sites (3 and 12 m). Nematode samples were first fixed in a 7% magnesium chloride aqueous solution and then posteriorly using a 4% formaldehyde solution (in a buffered sea-water). Two additional samples were taken to determinate the sedimentological features as well as total organic content of each site, depth, and period.

2.3. Abiotic Parameters. Grain size analysis was performed on the collected samples using a vibrosiever for fractions larger than 63 μm and an X-ray analyzer for those smaller than 63 μm. Sediments were classified following the Wentworth scale [29].

Total organic matter content was gravimetrically determined after loss on ignition [30]. The sediment samples were first dried at 60°C for 6 h and weighed on a Scaltec SBC21 microbalance (accuracy 0.1 mg) to obtain the dry weight. Samples were then combusted in a muffle furnace (550°C for 4 h) and reweighed to determine the ash fraction. The organic fraction content was calculated by subtracting the ash weight from the total weight.

2.4. Nematode Study. Nematode specimens were extracted by decantation or by centrifugation through a silica gel gradient (Ludox HS 30, density 1.18 g cm^{-3}) [31]. About 100 nematode specimens from each sample were picked out under a stereomicroscope, transferred into glycerine, and mounted on permanent slides. Nematode species identification was performed using a light microscope equipped with Nomarski optics (Optiphot-2 Nikon) and aided by the pictorial keys in Platt and Warwick [32, 33], Warwick et al. [34], and other relevant literature available on Deprez et al. [35]. Genera and species identified in this study are reported in alphabetical order (Table 1). The validity of scientific names was verified based on Deprez et al. [35] and WoRMS (http://www.marinespecies.org/). Synonymies species were also consulted in Gerlach and Riemann [36]. The habitat features (i.e., depth and sediment type) of the described taxa are indicated by symbols.

Current data on the distribution of Italian marine fauna as well as nematodes species [9] cover the nine biogeographical zones of the Italian sea as suggested by Bianchi [37]. In this sense, the area sampled in the present study is also part of the North Adriatic sector.

3. Results

The granulometric analysis of the sediment samples revealed that all the shallow subtidal sites (1.5–3 m in depth) are characterized as medium-fine sand (Mz 250 μm). These samples contained a high percentage of sand (96%) very low contents of mud (3%) and gravel (1%). On the other

TABLE 1: List of nematode species found along the Marches coast and their distribution in the Mediterranean and Adriatic Seas. Sampling site and habitat/sediment type are reported for each species.

Species	Sampling site	Habitat/sediment	New records[1,2,3]	Distribution in the Mediterranean Sea[2,3]
Order Enoplida				
Family Anoplostomatidae				
Chaetonema riemanni Platt 1973	BR	SSL/MS	● (A and M)	—
Family Anticomidae				
Anticoma eberthi Bastian 1865	FI	SSL/MS		North Adriatic Sea
Family Enchelidiidae				
Eurystomina ornata (Eberth 1863)	BR	SSL/MS		North Adriatic, Ligurian and South Tyrrhenian Sea
Family Oncholaimidae				
Metoncholaimus albidus (Bastian 1865)	FI, BR, P	SSL/MS	● (A)	South Tyrrhenian Sea
Oncholaimellus calvadosicus De Man 1890	FI, BR	SSL/MS	● (A and M)	—
Oncholaimus campylocercoides De Coninck & Schuurmans Stekhoven 1933	FI, BR, P	SSL/MS		North and South Adriatic Sea, Ligurian and South Tyrrhenian Sea
Oncholaimus dujardinii de Man 1876	FI, BR	FO, SSL/MS		North Adriatic, Ligurian and South Tyrrhenian Sea
Oncholaimus oxyuris Ditlevsen 1911	FI, BR, P	FO, SSL/MS	● (A and M)	—
Viscosia abyssorum (Allgén 1933)	BR	SSL/MS		North Adriatic Sea
Viscosia elegans (Kreis 1924)	FI, BR, P	SSL/MS	● (A)	French Coast
Family Oxystominidae				
Thalassoalaimus tardus de Man 1893	FI	SL/MU	● (A and M)	—
Family Thoracostomopsidae				
Mesacanthion diplechma (Southern 1914)	FI, BR	SSL/MS		North Adriatic Sea
Family Tripyloididae				
Bathylaimus tenuicaudatus (Allgén 1933)	FI, BR	SSL/MS	● (A and M)	—
Order Trefusiida				
Family Trefusiidae				
Trefusia filicauda Allgén 1933	BR	SSL/MS		North Adriatic Sea and French Coast
Order Chromadorida				
Family Chromadoridae				
Chromadorella duopapillata Platt 1973	FI, BR	SSL/MS		North Adriatic Sea
Hypodontolaimus cf. *dimorphus* Wieser 1954	FI, BR	SSL/MS	● (A and M)	—
Prochromadorella ditlevseni (de Man 1922)	FI, BR	SSL/MS		North Adriatic Sea
Prochromadorella septempapillata Platt 1973	FI, BR, P	SSL/MS		North Adriatic Sea
Ptycholaimellus pandispiculatus (Hopper 1961)	FI, BR	SSL/MS	● (A and M)	—
Family Cyatholaimidae				
Cyatholaimus gracilis (Eberth 1863)	FI, BR	SSL/MS		North Adriatic, Ligurian and South Tyrrhenian Sea, and French coast
Pomponema sedecima Platt 1973	FI, BR	SSL/MS	● (A and M)	—
Praeacanthonchus opheliae (Warwick 1970)	BR	SSL/MS	● (A and M)	—
Family Desmodoridae				
Chromaspirina inglisi Warwick 1970	FI	SSL/MS	● (A and M)	—

TABLE 1: Continued.

Species	Sampling site	Habitat/sediment	New records[1,2,3]	Distribution in the Mediterranean Sea[2,3]
Family Neotonchidae				
Filitonchus filiformis (Warwick 1971)	FI	SSL/MS	● (A and M)	—
Family Leptolaimidae				
Stephanolaimus jayasreei Platt 1983	FI	SSL/MS	● (A and M)	—
Family Microlaimidae				
Microlaimus teutonicus Riemann 1966	FI	SSL/MS	● (A and M)	—
Family Monoposthiidae				
Monoposthia mirabilis Schulz 1932	FI	SSL/MS		North Adriatic and South Tyrrhenian Sea
Monoposthia costata (Bastian 1865)	BR, FI	SSL/MS		North Adriatic, Ligurian and Tyrrhenian Sea
Family Selachinematidae				
Halichoanolaimus dolichurus Ssaweljev 1912	FI	SL/MU		North Adriatic and South Tyrrhenian Sea, French coast
Synonchiella riemanni Warwick 1970	FI	SSL/MS	● (A and M)	—
Order Monhysterida				
Family Axonolaimidae				
Ascolaimus elongatus (Bütschli 1874)	FI	SL/MU	● (A)	Ligurian and South Tyrrhenian Sea
Odontophora rectangula Lorenzen 1972	FI, BR, P	SSL, SL/MS, MU		North Adriatic Sea
Odontophora wieseri Luc & De Coninck 1959	FI	SSL/MS	● (A and M)	—
Family Comesomatidae				
Dorylaimopsis mediterranea Grimaldi-De Zio 1968	FI	SL/MU		North and South Adriatic Sea and French coast
Sabatieria breviseta Stekhoven 1935	FI, BR	SL/MU		North Adriatic Sea
Sabatieria ornata (Ditlevsen 1918)	FI, BR	SSL, SL/MS, MU		North Adriatic Sea
Sabatieria pulchra (Schneider 1906)	BR, FI, P	SSL, SL/MS, MU		North Adriatic Sea and French Coast
Setosabatieria fibulata (Wieser 1949)	FI	SSL, SL/MS, MU	● (A and M)	—
Setosabatieria hilarula (de Man 1922)	FI	SSL, SL/MS, MU		North and South Adriatic and Ligurian Sea
Family Linhomoeidae				
Metalinhomoeus longiseta Kreis 1929	FI	SSL/MS	● (A and M)	—
Terschellingia longicaudata de Man 1907	FI	SL/MU		North Adriatic Sea and French Coast
Family Xyalidae				
Cobbia caledonia Warwick & Platt 1973	FI	SSL/MS	● (A and M)	—
Cobbia trefusiaeformis de Man 1907	FI	SSL/MS	● (A and M)	—
Daptonema curvatus Gerlach 1956	FI, BR	SSL/MS	● (A and M)	—
Daptonema fistulatum (Wieser & Hopper 1967)	FI, BR	SSL/MS	● (A and M)	—
Daptonema furcatum (Juario 1974)	FI, BR	SSL/MS		North Adriatic Sea
Paramonhystera buetschlii (Bresslau & Stekhoven in Stekhoven 1935)	FI, BR	SSL/MS	● (A)	Mediterranean Sea
Theristus acer Bastian 1865	FI, BR	SSL/MS		Spanish coast, North Adriatic, and South Tyrrhenian Sea

TABLE 1: Continued.

Species	Sampling site	Habitat/sediment	New records[1,2,3]	Distribution in the Mediterranean Sea[2,3]
Theristus denticulatus Warwick 1970	FI	SSL/MS	• (A and M)	—
Theristus flevensis Stekhoven 1935	FI	SSL/MS		North Adriatic Sea
Trichotheristus floridanus (Wieser & Hopper 1967)	FI, BR	SSL/MS	• (A and M)	—
Trichotheristus mirabilis (Stekhoven & De Coninck 1933)	FI	SSL/MS		North Adriatic, Ligurian and South Tyrrhenian Sea

[•: new record in the Adriatic Sea (A) and in the Mediterranean Sea (M); FO: foreshore; MS: medium sands; MU: muddy sediments; SL: sublittoral; SSL: shallow sublittoral].
Note: [1]Travizi and Vidaković [13]; [2]Semprucci et al. [9]; [3]http://www.marinespecies.org.

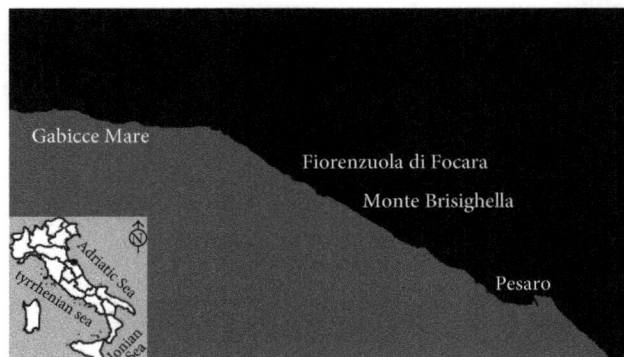

FIGURE 1: Study area and study sites.

hand, the sediments of the subtidal site (12 m in depth) were characterized as muddy sediments (Mz 25 μm) with a higher mud content (64%), followed by sand (35%) and gravel (1%). The total organic content (TOM) was lower in the shallow subtidal sites (2%) than that in the deeper sites (5%).

Overall, 84 putative species of nematodes were recognized. These belonged to 52 genera and 22 families, 51 of which have been identified to species level (Table 1). The highest number of species belonged to the families Xyalidae (16), Chromadoridae (10), Axonolaimidae (7), Comesomatidae (7), Cyatholaimidae (7), and Oncholaimidae (7). The highest number of species was recorded at the shallow subtidal sites. The most common families were the Anticomidae, Axonolaimidae, Chromadoridae, Comesomatidae, Cyatholaimidae, Oncholaimidae, and Xyalidae, while Anoplostomatidae, Ceramonematidae, Enchelidiidae, Leptolaimidae, Neotonchidae, Oxystominidae, Sphaerolaimidae, and Trefusiidae were very rare.

The Xyalidae family, one of the most specious families, had 16 putative species (11 identified to species level) distributed into 6 genera: *Cobbia*, *Daptonema*, *Gonionchus*, *Paramonhystera*, *Theristus*, and *Trichotheristus*. Most species were identified in the genera *Daptonema* (3 species and 2 putative species) and *Theristus* (3 species) (Table 1). Chromadoridae showed 11 putative species (6 identified to species) in 7 genera: *Chromadora*, *Chromadorella*, *Dichromadora*, *Hypodontolaimus*, *Neochromadora*, *Prochromadorella* and *Ptycholaimellus*. *Dichromadora* and *Prochromadorella* presented the higher number of putative species (*Dichromadora* with 2 putative species, and *Prochromadorella* with 2 species *P. ditlevseni* and *P. septempapillata*).

Comesomatidae were represented by 7 putative species in 4 genera (*Dorylaimopsis*, *Paramesonchium*, *Sabatieria*, and *Setosabatieria*). Axonolaimidae presented 7 putative species in 4 genera: *Ascolaimus*, *Axonolaimus*, *Odontophora*, and *Parodontophora*. *Odontophora* was the more frequent and specious genus with 2 species (*O. rectangula* and *O. wieseri*) and two putative species.

Cyatholaimidae were represented by 7 putative species (3 only identified to species) which belong to the following genera: *Cyatholaimus*, *Pomponema*, and *Praeacanthonchus*.

Oncholaimidae were represented by 4 genera: *Metoncholaimus*, *Oncholaimellus*, *Oncholaimus*, *Viscosia*. *Oncholaimus* was the richer genus, with 3 species (*O. campylocercoides*, *O. dujardinii*, and *O. oxyuris*), followed by *Viscosia* (*V. elegans* and *V. abyssorum*).

A comparison with the data on species distribution from Gerlach and Riemann [36], Deprez et al. [35], WoRMS [38], and Travizi and Vidaković [13] revealed that 27 of the 51 species found in the present study were new for the Adriatic nematofauna: *Chaetonema riemanni* Platt, 1973; *Metoncholaimus albidus* (Bastian 1865); *Oncholaimellus calvadosicus* De Man, 1890; *Oncholaimus oxyuris* Ditlevsen 1911; *Viscosia elegans* (Kreis 1924); *Thalassoalaimus tardus* De Man 1893; *Bathylaimus tenuicaudatus* (Allgén 1933); *Hypodontolaimus* cf. *dimorphus* Wieser 1954, *Ptycholaimellus pandispiculatus* (Hopper 1961), *Setosabatieria fibulata* (Wieser 1949); *Pomponema sedecima* Platt 1973; *Praeacanthonchus opheliae* (Warwick 1970); *Chromaspirina inglisi* Warwick 1970; *Filitonchus filiformis* (Warwick 1971); *Stephanolaimus jayasreei* Platt 1983; *Microlaimus teutonicus* Riemann 1966; *Synonchiella riemanni* Warwick 1970; *Ascolaimus elongatus*

(Bütschli 1874); *Odontophora wieseri* Luc & De Coninck 1959; *Metalinhomoeus longiseta* Kreis 1929; *Cobbia caledonia* Warwick & Platt 1973; *Cobbia trefusiaeformis* de Man 1907; *Daptonema curvatus* Gerlach 1956; *Daptonema fistulatum* (Wieser & Hopper 1967); *Paramonhystera buetschlii* (Bresslau & Stekhoven in Stekhoven 1935); *Trichotheristus floridanus* (Wieser & Hopper 1967). Only four of these species were already known in Mediterranean Sea: *Metoncholaimus albidus* (Bastian 1865) recorded in the South Tyrrhenian Sea, *Viscosia elegans* (Kreis 1924) present in the French Coasts, *Ascolaimus elongatus* (Bütschli 1874) in the Ligurian and South Tyrrhenian Sea, and *Paramonhystera buetschlii* (Bresslau & Stekhoven in Stekhoven 1935) in the Mediterranean Sea. The species identified during this survey have been only been recorded in the northern Atlantic and/or North Sea (Table 1) and in some cases also in the Black Sea, Canary Islands, and Antarctic Ocean [35, 36, 38].

4. Discussion

The general composition of the nematode fauna found in this study seems related to the habitat and sediment types (i.e., medium-fine sands) [13]. Most of the recorded species are in the shallow sublittoral zones: this may be related both to a more extensive investigation carried out in the shallow waters and/or to a lower nematode richness and diversity found offshore [26]. Indeed, the increase in the biodiversity of nematodes in parallel with the size of the sediment is well known [4, 39, 40]. In fact, the sediment grain size appears to be one of the main environmental variables influencing the horizontal distribution of nematodes. Different sediment types are characterized by different nematode assemblages and by a predominance of specific families, so-called iso-communities or parallel communities (see [4, 41, 42]).

Although Xyalidae is a family usually predominant in fine sediments, it may occur (with different ratio) in a wide range of sediment types, from muddy to coarse sands [4, 26, 40, 43]. All the Xyalidae specimens identified at specific level were found in shallow sites and are typical of intertidal habitats (see [34]) (Table 1). The highest diversity of this family in the shallow waters proves the importance of genera like *Theristus* and *Daptonema* in this environment, which includes sheltered habitats with medium-to-fine sands. On the other hand, Comesomatidae representatives are commonly related to muddy sediments, often characterized by anoxic conditions [4, 26, 44]. Indeed, these species were more diverse and abundant in the offshore station of FI, characterized by sediments with a high content of mud and organic matter. In muddy sediments, this family can be dominant, representing up to 34% of the total nematode community [26]. Comesomatidae were mostly represented by species of the genus *Sabatieria*: *S. breviseta, S. pulchra, S. ornata*, as well as by *Dorylaimopsis mediterranea, S. fibulata*, and *S. hilarula*. Among these species, *S. pulchra* is known to be physiologically well-adapted to stress conditions (see [39]), being able to survive in de-oxygenated sediments as a facultative anaerobic species [45]. However, *S. ornata* seems to be confined to the oxidized sediment layers. Among the

species found offshore, *Terschellingia longicaudata, T. tardus, Halichoanolaimus dolichurus, O. rectangula*, and *A. elongatus* are typical of fine sediments rich in silt fraction [34, 46–48]. However, some of the offshore species were also found in the coastal stations (e.g., *O. rectangula, S. breviseta, S. pulchra, S. ornata*, and *S. fibulata*). The specious Oncholaimidae was one of the most widespread and abundant nematode family in the shallow waters, with some species typical of intertidal zone and coarse sands [4, 34], while being less representative offshore [26]. This might be related to the life strategy of some species within this family, which might be sensible to anoxic conditions found offshore [49]. Species, such as *O. campylocercoides*, might be an exception. Frequently in anoxic sediments, *O. campylocercoides* and *T. longicaudata* present sub-epidermic grains that probably represent a detoxification mechanism from the high sulfide concentrations [50, 51]. As Comesomatidae, Linhomoeidae and Xyalidae are dominant and characteristic of muddy sediments, whereas the importance of Chromadoridae increases with the increasing of median grain size [4, 40, 48]. In fact, Chromadoridae were more frequent and abundant in the shallow sublittoral samples, with species typical of the intertidal zone and sandy substrates.

Biogeographic hypotheses for most of the inconspicuous meiofaunal organisms, and thus also for nematodes, are often very difficult. This is mainly due to the "taxonomic crisis" that concerns most meiofaunal taxa and also noticeable by the significant decrease of biodiversity surveys in the last decades [8, 52]. Costello et al. [53], updating the European marine biodiversity inventory, have highlighted that the majority of the nematode reports in the Mediterranean area comes from out of date literature as well as total absence of identification guides for the Mediterranean basin. Furthermore, the guides often only report the widespread, common, and/or ecologically significant species. In contrast, many of the rarer or taxonomically ambiguous species (often because of the very poor quality of the original descriptions) are not covered in any guides. A consequence is that the first species are much more frequently reported in the faunistic studies and thus present wider distribution, while the latter species appear to be locally distributed.

Overall, the families and genera reported in this study show a wide global distribution, while the species detected are typical of the North Western Europe or the Northern Atlantic; see Deprez et al. [35] and WoRMS [38]. This is in part due to the presence of most marine nematode taxonomists in the North Western Europe. In addition, most of the marine nematode literature is concentrated in this region. All these elements must be considered when inferences on nematode distribution and biogeography are discussed. Furthermore, recently Fontaneto and Hortal [54] have recently brought up the question "if microorganisms—freshwater animals lower than 2 mm—can have a biogeography" to the attention of the scientific community. The authors argue about the "Everything is Everywhere" (EisE) hypothesis, suggesting that the geographical distribution of these organisms may be strongly influenced by their animal size. Indeed, for inconspicuous freshwater organisms, the high potential for dispersal might contribute more in determining a wider distribution than that in the large ones [55]. The occurrence of morphologically identical nematode species in completely divergent habitats ("meiofauna paradox" *sensu* Giere, [56]) is also common in marine meiofauna. However, the existence of cryptic species (a complex of morphologically identical species which can only be differentiated by DNA) or a parallel evolution in response to similar microhabitats seems to be more plausible explanation to this trend than the dispersal capacities (see for review [41]). Therefore, theories on the biogeography of nematodes as well as other meiofaunal groups should be proposed with caution.

The present study represents a first attempt to increase the knowledge of free-living marine nematodes in the Adriatic basin and the Mediterranean Sea. It aims to establish the basis for a future discussion of the biogeography of marine nematodes in the Mediterranean Sea, currently biased by also dated and fragmentary information.

5. Conclusions

Biodiversity investigations aim to integrate species checklists and the compilation of databases that represent a regional and global benefit for researchers worldwide. Furthermore, the monitoring of biodiversity over time is of great importance for planning conservation actions, which seems to be more urgent these days, especially in vulnerable coastal systems. This study represents the first survey of the marine nematodes in the Marches coasts. The families with the highest abundance and diversity found in this study are typical of intertidal habitats, especially in sediments characterized by medium-fine sands. Most nematode species found in this study were previously reported only in the North Western Europe and the North Atlantic which are the most known regions regarding marine nematodes. This shows the necessity of intensifying sampling efforts in other regions to improve the current knowledge of marine nematode distribution as well as to clarify their biogeographic patterns.

Acknowledgments

The author warmly thanks the anonymous reviewers for their helpful comments that were material in improving the paper. A special thank is due to Professor Susanna Grimaldi-De Zio for her encouragement and precious theoretical and practical help in the study of marine nematodes.

References

[1] H. M. Platt and R. M. Warwick, "Chapter 10, The significance of free-living nematodes to the littoral ecosystem," in *The Shore Environment—2. Ecosystems*, J. H. Price and D. E. G. Irvine, Eds., pp. 729–759, London, UK, 1980.

[2] P. J. D. Lambshead, in *Nematology: Advances and Perspectives—Vol 1: Nematode Morphology, Physiology and Ecology*, Z. X. Chen, S. Y. Chen, and D. W. Dickson, Eds., Marine Nematode Biodiversity, pp. 436–467, CABI Publishing, Wallingford, UK, 2004.

[3] I. Andrássy, "A short census of free-living nematodes," *Fundamental and Applied Nematologie*, vol. 15, pp. 187–188, 1992.

[4] C. Heip, M. Vincx, and G. Vranken, "The ecology of marine nematodes," *Oceanography and Marine Biology an Annual Review*, vol. 23, pp. 399–489, 1985.

[5] O. Giere, *Meiobenthology: The Microscopic Motile Fauna of Aquatic Sediments*, Springer, Berlin, Germany, 2nd edition, 2009.

[6] M. Moreno, F. Semprucci, L. Vezzulli, M. Balsamo, M. Fabiano, and G. Albertelli, "The use of nematodes in assessing ecological quality status in the Mediterranean coastal ecosystems," *Ecological Indicators*, vol. 11, no. 2, pp. 328–336, 2011.

[7] E. Abebe, W. Decraemer, and P. De Ley, "Global diversity of nematodes (Nematoda) in freshwater," *Hydrobiologia*, vol. 595, no. 1, pp. 67–78, 2008.

[8] M. Balsamo, G. Albertelli, V. U. Ceccherelli et al., "Meiofauna of the adriatic sea: present knowledge and future perspectives," *Chemistry and Ecology*, vol. 26, no. 1, pp. 45–63, 2010.

[9] F. Semprucci, R. Sandulli, and S. De Zio-Grimaldi, "Adenophorea, nematodi marini," in *Checklist della Flora e della Fauna dei Mari Italiani*, G. Relini, Ed., vol. 15 of *Biologia Marina Mediterranea*, pp. 184–209, 2008.

[10] C. Allgén, "Die freilebenden Nematoden des Mittelmeeres," *Zoologische Jahrbücher*, vol. 76, pp. 1–102, 1942.

[11] J. H. Schuurmans-Stekhoven, "The free-living marine nemas of the Mediterranean—I. The Bay of Villefranche," *Institut Royal Des Sciences Naturelles De Belgique Memoires*, vol. 37, pp. 1–220, 1950.

[12] S. A. Gerlach, "Die Nematodenbesiedlung des Sandstrandes und des Küstengrundwassers an der italienischen Küste—I. Systematischer Teil," *Archivio Zoologico Italiano*, vol. 37, pp. 517–640, 1953.

[13] A. Travizi and J. Vidaković, "Nematofauna in the Adriatic Sea: review and check-list of free-living nematode species," *Helgolander Meeresuntersuchungen*, vol. 51, no. 4, pp. 503–519, 1997.

[14] V. U. Ceccherelli and F. Cevidalli, "Osservazioni preliminari sulla bionomia dei popolamenti meiobentonici della Sacca di Scardovari (Delta del Po), con particolare riferimento ai Nematodi ed ai Copepodi," Quaderni del laboratorio di Tecnologia della Pesca. Atti del Congresso S.I.B.M., 1981.

[15] A. Guerrini, M. A. Colangelo, and V. U. Ceccherelli, "Recolonization patterns of meiobenthic communities in brackish vegetated and unvegetated habitats after induced hypoxia/anoxia," *Hydrobiologia*, vol. 375-376, pp. 73–87, 1998.

[16] S. De Zio, "Distribuzione dei nematodi in spiagge pugliesi," *Bollettino Di Zoologia*, vol. 31, no. 2, pp. 907–920, 1964.

[17] S. De Zio, "Nematodi marini del litorale pugliese," *Bollettino di Zoologia*, vol. 33, p. 182, 1966.

[18] S. Grimaldi-De Zio, "Il popolamento di Nematodi di una spiaggia pugliese in rapporto al ritmo di marea," *Bollettino di Zoologia*, vol. 34, p. 126, 1967.

[19] S. Grimaldi-De Zio, "Una nuova specie di nematodi Comesomatidae: dorylaimopsis mediterraneus," *Bollettino di Zoologia*, vol. 35, pp. 137–141, 1968.

[20] S. Grimaldi-De Zio, "Confronto fra Nematodi del fango coralligeno di piattaforma e una comunità del fango dello stesso distretto adriatico," *Bollettino di Zoologia*, vol. 35, p. 347, 1968.

[21] R. Sandulli, S. De Zio-Grimaldi, M. Gallo D'Addabbo, L. Caló, and E. Bressan, "Aspetti della biodiversità della meiofauna lungo il litorale pugliese," *Biologia Marina Mediterranea*, vol. 9, no. 1, pp. 484–493, 2001.

[22] R. Sandulli, C. de Leonardis, and J. Vanaverbeke, "Meiobenthic communities in the shallow subtidal of three Italian Marine Protected Areas," *Italian Journal of Zoology*, vol. 77, no. 2, pp. 186–196, 2010.

[23] J. Vidaković, A. Travizi, and G. Boucher, "Two new species of the genus *Metacyatholaimus* (Nematoda, Cyatholaimidae) from the Adriatic Sea with a key to the species," *Cahiers de Biologie Marine*, vol. 44, no. 2, pp. 111–120, 2003.

[24] C. de Leonardis, R. Sandulli, J. Vanaverbeke, M. Vincx, and S. De Zio, "Meiofauna and nematode diversity in some Mediterranean subtidal areas of the Adriatic and Ionian Sea," *Scientia Marina*, vol. 72, no. 1, pp. 5–13, 2008.

[25] A. Travizi, "The nematode fauna of the northern Adriatic offshore sediments: community structure and biodiversity," *Acta Adriatica*, vol. 51, no. 2, pp. 169–180, 2010.

[26] F. Semprucci, P. Boi, A. Manti et al., "Benthic communities along a littoral of the Central Adriatic Sea (Italy)," *Helgoland Marine Research*, vol. 64, no. 2, pp. 101–115, 2010.

[27] P. Franco, L. Jeftic, P. Malanotte-Rizzoli, A. Michelato, and M. Orlic, "Descriptive model of the northern Adriatic," *Oceanologica Acta*, vol. 5, pp. 379–389, 1982.

[28] M. Zavatarelli, J. W. Baretta, J. G. Baretta-Bekker, and N. Pinardi, "The dynamics of the Adriatic Sea ecosystem. An idealized model study," *Deep-Sea Research Part I*, vol. 47, no. 5, pp. 937–970, 2000.

[29] J. B. Buchanan, "Sediment analysis," in *Methods for the Study of Marine Benthos*, N. A. Holme and A. D. McIntyre, Eds., pp. 41–65, Blackwell Scientific Publications, Oxford, UK, 1984.

[30] J. B. Buchanan, "Measurement of the physical and chemical environment," in *Methods for the Study of Marine Benthos*, N. A. Holme and A. D. McIntyre, Eds., pp. 30–52, Blackwell Scientific Publications, Oxford, UK, 1971.

[31] O. Pfannkuche and H. Thiel, "Sample processing," in *Introduction to the Study of Meiofauna*, R. P. Higgins and H. Thiel, Eds., pp. 134–145, Smithsonian Institute, Washington, DC, USA, 1988.

[32] H. M. Platt and R. M. Warwick, "Free-living marine nematodes—part I. British Enoplids," in *Synopses of the British Fauna*, vol. 28, Cambridge University Press, Cambridge, UK, 1983.

[33] H. M. Platt and R. M. Warwick, "Free-living marine nematodes—part II. British Chromadorids," in *Synopses of the British Fauna*, vol. 38, Brill, Leiden, Holland, 1988.

[34] R. M. Warwick, H. M. Platt, and P. J. Somerfield, "Free-living marine nematodes—part III. British monhysterids," in *Synopses of the British Fauna*, vol. 53, Field Studies Council, Shrewsbury, UK, 1998.

[35] T. Deprez, E. Vanden Berghe, and M. Vincx, World Wide Web electronic publication, 2005, http://www.nemys.ugent.be.

[36] S. A. Gerlach and F. Riemann, The Bremerhaven Checklist of Aquatic Nematodes. A catalogue of Nematoda Adenophorea excluding the Dorylaimida. Veröffentlichungen des Instituts für Meeresforschung in Bremerhaven, 4 Part I and Part II, 1973-1974.

[37] C. N. Bianchi, "Proposta di suddivisione dei mari italiani in settori biogeografici," *Notiziario SIBM*, vol. 46, pp. 57–59, 2004.

[38] World Register of Marine Species (WoRMS), 2009, http://www.marinespecies.org.

[39] M. Steyaert, N. Garner, D. Van Gansbeke, and M. Vincx, "Nematode communities from the North Sea: environmental controls on species diversity and vertical distribution within the sediment," *Journal of the Marine Biological Association of the United Kingdom*, vol. 79, no. 2, pp. 253–264, 1999.

[40] F. Semprucci, P. Colantoni, G. Baldelli, M. Rocchi, and M. Balsamo, "The distribution of meiofauna on back-reef sandy platforms in the Maldives (Indian Ocean)," *Marine Ecology*, vol. 31, no. 4, pp. 592–607, 2010.

[41] M. Raes, W. Decraemer, and A. Vanreusel, "Walking with worms: coral-associated epifaunal nematodes," *Journal of Biogeography*, vol. 35, no. 12, pp. 2207–2222, 2008.

[42] T. Gheskiere, M. Vincx, B. Urban-Malinga, C. Rossano, F. Scapini, and S. Degraer, "Nematodes from wave-dominated sandy beaches: diversity, zonation patterns and testing of the isocommunities concept," *Estuarine, Coastal and Shelf Science*, vol. 62, no. 1-2, pp. 365–375, 2005.

[43] F. Semprucci, P. Colantoni, C. Sbrocca, G. Baldelli, M. Rocchi, and M. Balsamo, "Meiofauna in sandy back-reef platforms differently exposed to the monsoons in the Maldives (Indian Ocean)," *Journal of Marine Systems*, vol. 87, no. 3-4, pp. 208–215, 2011.

[44] M. Moreno, G. Albertelli, and M. Fabiano, "Nematode response to metal, PAHs and organic enrichment in tourist marinas of the mediterranean sea," *Marine Pollution Bulletin*, vol. 58, no. 8, pp. 1192–1201, 2009.

[45] P. Jensen, "Ecology of benthic and epiphytic nematodes in brackish waters," *Hydrobiologia*, vol. 108, no. 3, pp. 201–217, 1984.

[46] C. Heip, R. Huys, M. Vincx et al., "Composition, distribution, biomass and production of North Sea meiofauna," *Netherlands Journal of Sea Research*, vol. 26, no. 2–4, pp. 333–342, 1990.

[47] A. Vanreusel, "Ecology of free-living marine nematodes in the Voordelta (Southern Bight of the North Sea)—II. Habitat preferences of the dominant species," *Nematologica*, vol. 37, no. 3, pp. 343–359, 1991.

[48] M. Schratzberger, K. Warr, and S. I. Rogers, "Patterns of nematode populations in the southwestern North Sea and their link to other components of the benthic fauna," *Journal of Sea Research*, vol. 55, no. 2, pp. 113–127, 2006.

[49] T. Bongers, R. Alkemade, and G. W. Yeates, "Interpretation of disturbance-induced maturity decrease in marine nematode assemblages by means of the Maturity Index," *Marine Ecology Progress Series*, vol. 76, no. 2, pp. 135–142, 1991.

[50] L. J. Jacobs and J. Heyns, "Notes brèves. An ecological strategy in the genus Monhystera—an hypothesis," *Revue de Nématologie*, vol. 13, pp. 109–111, 1990.

[51] F. Thiermann, B. Vismann, and O. Giere, "Sulphide tolerance of the marine nematode *Oncholaimus campylocercoides*—a result of internal sulphur formation?" *Marine Ecology Progress Series*, vol. 193, pp. 251–259, 2000.

[52] M. Curini-Galletti, "Platyhelminthes. La checklist della fauna marina Italiana," in *Checklist della Flora e della Fauna dei Mari Italiani*, G. Relini, Ed., vol. 15 of *Biologia Marina Mediterranea*, pp. 110–124, 2008.

[53] M. J. Costello, P. Bouchet, C. S. Emblow, and A. Legakis, "European marine biodiversity inventory and taxonomic resources: state of the art and gaps in knowledge," *Marine Ecology Progress Series*, vol. 316, pp. 257–268, 2006.

[54] D. Fontaneto and J. Hortal, "Do microorganisms have biogeography?" *The International Biogeography Society*, vol. 6, no. 1, pp. 3–7, 2008.

[55] C. A. Kellogg and D. W. Griffin, "Aerobiology and the global transport of desert dust," *Trends in Ecology and Evolution*, vol. 21, no. 11, pp. 638–644, 2006.

[56] O. Giere edition, *Meiobenthology, The Microscopic Motile Fauna of Aquatic Sediments*, Springer, Berlin, Germany, 1st edition, 1993.

3

Seasonal Diversity of Arbuscular Mycorrhizal Fungi in Mangroves of Goa, India

James D'Souza and Bernard Felinov Rodrigues

Department of Botany, Goa University, Taleigao, Goa 403 206, India

Correspondence should be addressed to James D'Souza; james2442@gmail.com

Academic Editor: Curtis C. Daehler

Seasonal dynamics of arbuscular mycorrhizal (AM) fungal community composition in three common mangrove plant species, namely, *Acanthus ilicifolius, Excoecaria agallocha,* and *Rhizophora mucronata,* from two sites in Goa, India, were investigated. In all three species variation in AM fungal spore density was observed. Maximum spore density and AM species richness were recorded in the premonsoon season, while minimum spore density and richness were observed during monsoon season at both sites. A total of 11 AM fungal species representing five genera were recorded. *Acaulospora laevis* was recorded in all seasons at both sites. Multivariate analysis revealed that season and host coaffected AM spore density and species richness with the former having greater influence than the latter.

1. Introduction

Mangroves are a type of coastal woody vegetation that fringes muddy saline shores and estuaries in tropical and subtropical regions [1]. They are characterized by high levels of productivity and fulfill essential ecological functions, harbouring precious natural resources [2]. Mangroves have become the center of many conservation and environmental issues because of the beneficial effects they have on the coastal environment. Recent evidence suggests that growth of mangroves is limited primarily by phosphorus (P) availability as it is adsorbed and coprecipitated within carbonate-dominated environments [3]. Phosphate solubilizers, N fixers, and AM fungi are known to interact in the rhizosphere soils [4] where hyphae of AM fungi assist in accessing nutrients by extending beyond the root depletion zone [5]. These fungi also alleviate salt stress and aid physiological processes such as osmotic adjustment *via* accumulation of soluble sugars in root cells [6] and contribute to the nutritional status of plants [7]. They play a crucial role in determining plant diversity, production, and species composition [8]. The seasons are a result of the tilt of Earth's axis that causes variation in environmental conditions and spore density, and community composition of AM fungi are influenced by these changes.

To understand the ecology and function of plant-fungus associations in natural ecosystems, it is necessary to clarify seasonal diversity of AM fungi, providing insight into the factors and processes regulating ecosystem development [9]. Studies on the occurrence and diversity of AM fungi from different mangrove plants have been documented [2, 10]. However, no studies have been reported on the seasonal dynamics of AM fungi in mangroves. In this paper, spore density and species richness pattern of AM fungi in relation to seasons have been elucidated.

2. Methodology

2.1. Study Sites and Sample Collection. Two study sites, namely, Terekhol ($15°72'28''$ N and $73°72'99''$ E) with a stretch of 28 Km and Zuari ($15°32'56''$ N and $73°89'71''$ E) having 67 Km, were selected for the study. Three dominant plant species, namely, *Acanthus ilicifolius* L. (Acanthaceae), *Excoecaria agallocha* L. (Euphorbiaceae), and *Rhizophora mucronata* Poir. (Rhizophoraceae), common to the two sites, were undertaken for the study. The tropical environment at both sites is warm and humid, with marshy soils. Mean temperature range is $22-35°$C and average annual rainfall

2500 mm. Mangroves species were identified using the local floras [11]. Rhizosphere soil samples were randomly collected in the premonsoon (March 2009–May 2009), monsoon (July 2009–Sep 2009), and postmonsoon (Oct 2009–Feb 2010) seasons from the two sites. The samples were placed in polyethylene bags, transported to the laboratory, and stored at 4°C until processed. Rhizosphere soil of individual plants was air dried at room temperature, sieved (mesh size 720 μ), and divided into two parts, one for isolation, enumeration, and identification of AM spores and the other for preparation of trap cultures.

2.2. Soil Analyses. Three soil samples from each of the study site were separately collected in polyethylene bags from a depth of 0–25 cm, air dried in the laboratory before passing through a 2 mm sieve, and mixed thoroughly to obtain a composite sample. Soil pH was measured in soil water (1 : 2) suspension using a pH meter (LI 120 Elico, India). Electrical Conductivity (EC) was measured at room temperature in 1 : 5 soil suspension, using a conductivity meter (CM-180 Elico, India). Standard soil analysis techniques, namely, Walkley and Black rapid titration method [12] and Bray and Kurtz method [13], were employed for determination of organic carbon and available P, respectively. Available potassium was estimated by ammonium acetate method [14] using a flame photometer (Systronic 3292). Available zinc (Zn), copper (Cu), manganese (Mn), and iron (Fe) were quantified by DTPA-CaCl$_2$-TEA method [15] using an Atomic Absorption Spectrophotometer (AAS 4139).

2.3. Trap Cultures, Isolation, and Taxonomic Identification of AM Fungal Spores. For identification of AM species, trap cultures were prepared in pots using field soil and sterile sand (1 : 1). *Solenostemon scutellarioides* (L.) Codd was used as the catch plant and the pots were maintained in a polyhouse at 27°C. All cultures were provided a 14 h day/10 h night photoperiod for six months. The pots were watered when required, and Hoagland's solution minus P was added fortnightly. Intact and nonparasitized spores used for identification, obtained from both rhizosphere soil samples and trap cultures, were isolated using wet sieving and decanting technique [16]. Identification was based on spore morphology and subcellular characteristics [17, 18].

2.4. Diversity Studies and Statistical Analysis. Diversity studies were conducted for each site separately by calculating Simpson's Index of Diversity $1 - D$ [19] $D = 1 - \Sigma(P_i)^2$, where $P_i = n_i/N$ (n_i, the relative abundance (RA) of the species calculated as the proportion of individuals of a given species to the total number of individuals in a community, N). Shannon diversity index (H') is commonly used to characterize species diversity in a community, accounting for both abundance and evenness of the species present, $H = -\Sigma(P_i \ln(P_i))$ [20]. Species richness (SR) is the number of species present. Species evenness (E), which indicates the distribution of individuals within species, was calculated by using the following formula: $\Sigma(H') = H'/H'_{\max}$, where $H'_{\max} = \ln(\text{SR})$.

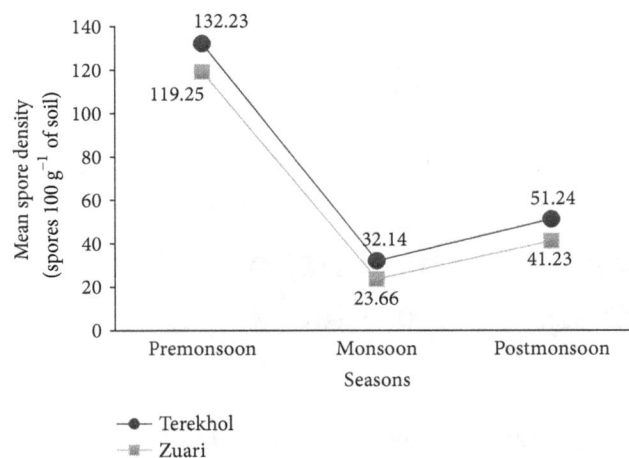

FIGURE 1: Mean spore density in rhizosphere soils of mangrove plant species in different seasons at the two study sites.

Pearson's correlation coefficient was calculated to assess the relationship between spore density and species richness at each site, using WASP software (Web Based Agricultural package) 2.0 ($P \leq 0.05$). Relative abundance of AM fungal species common to all seasons was correlated with soil pH, P, and EC ($P \leq 0.05$). Data on seasons and host coaffect the AM fungal spore density, species richness, and Shannon-Weiner diversity index was analyzed using multivariate analysis of variance (MANOVA). The statistically significant difference was determined at $P \leq 0.05$.

3. Results

3.1. Soil Analyses. Results of the soil physicochemical analyses are depicted in Table 1. The study revealed acidic soils (pH range 5.5–6.8) at both sites. Electrical Conductivity (EC) ranged from 4.03 to 8.49 dSm^{-1}. Organic carbon content was higher in the premonsoon season at both study sites. Soils at both sites were deficient in P. Levels of micronutrients such as Cu, Zn, Mn, and Fe varied between the two study sites.

3.2. AM Fungal Spore Density. Sitewise results of the seasonal variations in spore density of AM fungi in the three mangrove plant species undertaken for the study are depicted in Table 2. Spore density varied significantly between the seasons and mangrove species. At both sites the mean spore density was significantly higher in the premonsoon season compared to monsoon and postmonsoon seasons (Figure 1). Maximum spore density was recorded for *A. ilicifolius* (230 spores 100 g^{-1}) at the Terekhol site, while maximum spore density was recorded for *E. agallocha* (186 spores 100 g^{-1}) at the Zuari site. Minimum spore density was recorded for *R. mucronata* at both sites.

3.3. Distribution and Relative Abundance (RA) of AM Species. A total of 11 AM fungal species representing five genera were recorded. *Glomus* was the dominant genus followed by *Acaulospora, Rhizophagus, Funneliformis,* and *Racocetra.* In

TABLE 1: Soil physicochemical analyses of the study sites.

Soil characteristics	Premonsoon		Monsoon		Postmonsoon	
	Terekhol	Zuari	Terekhol	Zuari	Terekhol	Zuari
pH	6.5 ± 0.12	5.9 ± 0.08	6.8 ± 0.07	6.2 ± 0.02	6.7 ± 0.19	5.5 ± 0.10
EC (dSm^{-1})	4.12 ± 1.02	8.19 ± 1.64	4.03 ± 1.79	8.11 ± 1.24	4.30 ± 1.12	8.49 ± 1.24
OC (%)	4.79 ± 1.12	3.61 ± 1.29	4.01 ± 1.37	3.01 ± 1.02	4.45 ± 1.41	3.12 ± 1.06
P (g/kg)	0.17 ± 0.02	0.13 ± 0.01	0.16 ± 0.04	0.14 ± 0.06	0.13 ± 0.01	0.11 ± 0.01
K (g/kg)	63.23 ± 2.12	87.96 ± 4.12	68.14 ± 1.96	81.28 ± 2.12	70.34 ± 3.16	85.12 ± 2.12
N (g/kg)	0.54 ± 0.06	0.73 ± 0.03	0.57 ± 0.04	0.79 ± 0.03	0.61 ± 0.02	0.81 ± 0.05
Zn (g/kg)	0.049 ± 0.01	0.030 ± 0.01	0.031 ± 0.01	0.029 ± 0.04	0.043 ± 0.01	0.032 ± 0.01
Mn (g/kg)	0.042 ± 0.04	0.029 ± 0.04	0.033 ± 0.04	0.021 ± 0.04	0.037 ± 0.04	0.026 ± 0.04
Cu (g/kg)	0.069 ± 0.02	0.051 ± 0.01	0.041 ± 0.01	0.032 ± 0.01	0.052 ± 0.01	0.040 ± 0.01
Fe (g/kg)	0.189 ± 0.06	0.621 ± 0.02	0.112 ± 0.04	0.511 ± 0.07	0.143 ± 0.08	0.563 ± 0.03

Data presented are means of three readings at each season; ± indicates S.E.

TABLE 2: Seasonal variation in spore density of AM fungi in selected mangrove plant species at two study sites.

Sr. no.	Plant species	Spore density (spores100 g^{-1} of soil)	
		Terekhol site	Zuari site
1	Acanthus ilicifolius L.[a]		
	Premonsoon season	230.00 ± 9.21^{a}	149.00 ± 5.61^{a}
	Monsoon season	40.00 ± 4.41^{c}	35.00 ± 3.52^{c}
	Postmonsoon season	76.00 ± 3.56^{b}	67.00 ± 5.45^{b}
2	Excoecaria agallocha L.[b]		
	Premonsoon season	128.00 ± 8.57^{a}	186.00 ± 9.32^{a}
	Monsoon season	20.00 ± 2.35^{c}	27.00 ± 2.41^{c}
	Postmonsoon season	49.00 ± 4.25^{b}	56.00 ± 3.24^{b}
3	Rhizophora mucronata Poir.[c]		
	Premonsoon season	38.00 ± 5.24^{a}	22.00 ± 2.98^{b}
	Monsoon season	7.00 ± 1.20^{c}	21.00 ± 7.37^{b}
	Postmonsoon season	17.45 ± 2.56^{a}	24.00 ± 8.23^{a}

Means in a column for a mangrove species followed by a different superscripts indicate significant differences ($n = 6$ for each season, $n = 54$ overall, $P \leq 0.05$).

the present study two sporocarp species were recovered, *G. aggregatum* and *G. rubiforme*. Identification of AM fungal species was confirmed by trap culture method where no additional AM fungal species were recovered. Within AM species the highest RA was recorded for *R. intraradices* followed by *A. scrobiculata*, *A. laevis*, and *A. bireticulata*, and the lowest RA was recorded for *R. gregaria*. *Acaulospora laevis* was recovered in all three seasons, *R. intraradices*, and *R. gregaria* were recorded in two seasons from both sites, while *G. nanolumen*, *R. fasciculatus*, and *G. multicaule* were recorded in only one season from either site (Table 3). In the present study RA of dominant AM fungal species showed no significant correlation with soil pH, P, or EC values ($P \geq 0.05$) (Table 4).

3.4. AM Fungal Species Richness and Species Evenness. Species richness in combined sites was maximum in premonsoon (9 species) and minimum in the monsoon season (4 species). Correlation analysis indicated that the spore density in premonsoon was significantly correlated with species richness at both sites (Terekhol $r = 0.726$; Zuari $r = 0.645$;

$P \leq 0.05$) while no significant correlation was observed in either postmonsoon or monsoon season in either site (Table 5). Species evenness was maximum in postmonsoon season in both sites, and Shannon-Weiner (H) and Simpson's indices showed variation between the different seasons (Table 6). Multivariate analysis revealed that seasons ($F_{2,26} = 2.346$; $P < 0.001$) and host ($F_{2,18} = 1.854$; $P < 0.001$) coaffected AM fungal spore density and species richness. The seasons had a greater influence than host species as evidenced by higher F values; however the interaction was found to be nonsignificant.

4. Discussion

Variation in pH and EC levels in the mangrove soils observed in the present study may be attributed to the constant flushing of tidal water, leading to deposition of salts [21]. Soils at both of the study sites were deficient in available P. It is reported that nearly 80–85% of P is made unavailable to plants due to fixation and immobilization [22]. Degradation of litter in mangrove ecosystems is active and continuous, resulting

TABLE 3: Seasonal variation in relative abundance of AM fungi in the selected study sites.

AM fungal species	Premonsoon		Monsoon		Postmonsoon	
	Terekhol	Zuari	Terekhol	Zuari	Terekhol	Zuari
Rhizophagus intraradices (Schenck and Smith) Walker and Schüβler	18.68	15.71	58.82	42.16	—	—
Glomus multicaule Gerdemann and Bakshi	14.14	—	—	—	—	—
Glomus aggregatum Schenck and Smith	—	10.72	10.29	—	—	15.64
Rhizophagus fasciculatus (Thaxter) Walker and Schüβler	—	—	—	—	20.42	—
Funneliformis geosporum (Nicol. and Gerd.) Walker and Schüβler	—	—	—	—	7.04	16.32
Glomus nanolumen Koske and Gemma	9.84	—	—		—	—
Glomus rubiforme Gerdemann and Trappe	—	13.24	—	25.30	—	—
Acaulospora bireticulata Rothwell and Trappe	8.83	31.67		—	—	—
Acaulospora laevis Gerdemann and Trappe	9.84	33.41	30.88	32.53	23.94	25.85
Acaulospora scrobiculata Trappe	33.83	—	—	—	35.91	48.97
Racocetra gregaria (Schenck and Nicolson) Oehl, Souza, and Sieverd	4.79	5.23	—	—	12.67	8.84

All values are means of the composite sample of three plant species for each season.

TABLE 4: Pearson's correlation coefficient (*r* value) between relative abundance of dominant AM fungal species and soil pH, P, and EC at the two study sites.

Ecological parameters	Terekhol site			Zuari site		
	Premonsoon	Monsoon	Postmonsoon	Premonsoon	Monsoon	Postmonsoon
RA versus soil pH	0.535	0.071	0.393	0.070	0.475	0.327
RA versus P	0.423	0.475	0.417	0.429	0.563	0.311
RA versus EC	0.512	0.326	0.418	0.419	0.375	0.426

RA: relative abundance; P: soil phosphorus; EC: electrical conductivity. ($P \leq 0.05$); all values are means of the composite sample of three plant species for each season.

in the release of various acids during hydrolysis of tannins [23] and the oxidation of iron sulfide (pyrite) that releases dissolved ferrous iron [24], that are known to shift soils towards more acidic conditions. The present study revealed a high organic carbon content in the mangrove sediments, and its degradation resulting in low pH [23]. Concentration of Fe was higher at Zuari than at Terekhol. At the Zuari site, maximum concentration of Fe was recorded in the premonsoon season. Earlier study suggested that increased Fe content in mangrove soils could be attributed to the precipitation of the respective metal sulfide compounds in anaerobic sediments [25].

The selected mangrove plant species showed variation in AM spore density. An earlier study reported that spore density patterns may not reflect the activity of AM fungi in roots, but rather the tendency to sporulate along wide environmental conditions [26]. Other studies suggested that P availability [27], plant physiology, and turnover of plant roots [28] are among the drivers of AM fungal seasonality. Spore density showed variation, maximum in premonsoon and minimum in the monsoon season at both sites. Similar observations have been reported in earlier studies [29, 30]. Higher spore density in premonsoon season is thought to be an indication of root senescence and available nutrients, stimulating fungal sporulation as plant nutrient requirement is reduced [31]. Higher AM spore density in premonsoon may also be attributed to soil temperature, as previous studies [32, 33] suggested that high soil temperature favours AM fungal sporulation.

In the present study no significant correlation between relative abundance (RA) of dominant AM fungal species, and soil pH, P, or EC values suggests that AM fungi have a specific multidimensional niche determined by host plant species. This may affect variation between and within sites in AM fungal community composition [34]. Some reports suggest that AM fungi are obligately aerobic, flooding reduces sporulation [35], and total spore density correlates negatively with soil moisture [36]. Others have found higher values of total spore density in wet soils than in dry soils [37] and have suggested that high sporulation is a stress response to adverse or extreme environmental conditions.

Glomus species are known to be widely distributed and are commonly found in different geographical regions [38]. Furthermore, *Glomus* species are more adaptable to adjustment of sporulation patterns in varied environmental conditions [39] resulting in dominance. In our study, *Acaulospora laevis* was recorded in all of the three seasons at both study sites. The acidic nature of mangrove soils explains the presence of *A. laevis* in all the seasons as *Acaulospora* species are known to occur in acidic soils [40]. Based on RA, the dominant species, namely, *A. laevis* and *R. intraradices*, showed different patterns of sporulation and distribution. These differences in sporulation pattern may be attributed to plant phenological events including new root growth [41], flowering, and fruiting [42]. They may also suggest differences in AM functionality.

Multivariate analysis revealed that seasons and host coaffected AM fungal spore density and species richness, and

TABLE 5: Pearson correlation coefficient (r value) between spore density and species richness at the two study sites.

Ecological parameters	Terekhol site			Zuari site		
	Premonsoon	Monsoon	Postmonsoon	Premonsoon	Monsoon	Postmonsoon
SD versus SR	*0.726	0.675	0.513	*0.645	0.326	0.342

SD: spore density; SR: species richness; *: significant at $P \leq 0.05$.

TABLE 6: Diversity measurements of AM fungal communities in different seasons at the two study sites.

Study sites and ecological parameters	Premonsoon	Monsoon	Postmonsoon
Terekhol site			
Shannon-Weiner index (H)	0.795	0.358	0.852
Simpsons index of dominance (D)	0.93	0.79	0.97
Evenness (E)	0.40	0.32	0.52
Zuari site			
Shannon-Weiner index (H)	0.927	0.361	0.774
Simpsons index of dominance (D)	0.99	0.784	0.99
Evenness (E)	0.48	0.32	0.51

the seasons have a greater influence than host by analyses of F values. The seasons and host are important factors influencing AM fungal spore density and species richness in natural ecosystems because the host plant can regulate carbon allocation to roots, produces secondary metabolites, and changes the soil environment during different seasons [9]. Beside seasons other factors such as disturbance [43], sporulation efficiency [44], and dormancy [45] are known to affect the abundance of AM fungal species. Seasonal diversity observed in the present study is higher than that in an earlier mangrove study in South China, where only six AM fungal species were reported [2]. Similarly, only four AM fungal species belonging to two genera in 16 aquatic and marshy plant species were reported from Goa, India [46]. In general the AM fungal diversity in wetland ecosystems is lower than in terrestrial ecosystems [47, 48]. Preference of different host plants and dormancy may be factors attributing to lower diversity of AM fungi in wetlands [49].

5. Conclusion

Mangrove plant communities interact with rhizosphere soil and can modify edaphic properties. Similarly edaphic factors interact with plant communities and modify their composition. Consequently in this study there was no clearly observed separation between the plant and soil factors influence on AM fungal sporulation. Seasonal studies of AM fungi help to predict the conditions crucial for development of AM fungi. Further targeted ecological studies are needed to consider the combined effect of occurrence of AM fungi in the different phenological stages of mangroves to provide an accurate picture of AM fungal development and function prevailing in the given ecosystem. Our study suggests that the uneven spatial distribution of AM fungal spores and the complex structure of a mangrove ecosystem should be considered as major factors affecting the spore density of AM fungi in different seasons.

Acknowledgment

The authors gratefully acknowledge the financial assistance received from Planning Commission, Government of India, New Delhi, India, to carry out this study.

References

[1] B. Gopal and M. Chauhan, "Biodiversity and its conservation in the sundarban mangrove ecosystem," *Aquatic Sciences*, vol. 68, no. 3, pp. 338–354, 2006.

[2] Y. Wang, Q. Qiu, Z. Yang, Z. Hu, N. F. Y. Tam, and G. Xin, "Arbuscular mycorrhizal fungi in two mangroves in South China," *Plant and Soil*, vol. 33, no. 1, pp. 181–191, 2010.

[3] C. E. Lovelock, I. C. Feller, K. L. McKee, B. M. J. Engelbrecht, and M. C. Ball, "The effect of nutrient enrichment on growth, photosynthesis and hydraulic conductance of dwarf mangroves in Panama," *Functional Ecology*, vol. 18, no. 1, pp. 25–33, 2004.

[4] D. M. Alongi, "Present state and future of the world's mangrove forests," *Environmental Conservation*, vol. 29, no. 3, pp. 331–349, 2002.

[5] M. Cui and M. M. Caldwell, "Facilitation of plant phosphate acquisition by arbuscular mycorrhizas from enriched soil patches II. Hyphae exploiting root-free soil," *New Phytologist*, vol. 133, no. 3, pp. 461–467, 1996.

[6] G. Feng, F. S. Zhang, X. L. Li, C. Y. Tian, C. Tang, and Z. Rengel, "Improved tolerance of maize plants to salt stress by arbuscular mycorrhiza is related to higher accumulation of soluble sugars in roots," *Mycorrhiza*, vol. 12, no. 4, pp. 185–190, 2002.

[7] R. B. Zandavalli, L. R. Dillenburg, and V. D. Paulo, "Growth responses of *Araucaria angustifolia* (Araucariaceae) to inoculation with the mycorrhizal fungus *Glomus clarum*," *Applied Soil Ecology*, vol. 25, no. 3, pp. 245–255, 2004.

[8] M. G. A. van der Heijden, J. N. Klironomos, M. Ursic et al., "Mycorrhizal fungal diversity determines plant biodiversity, ecosystem variability and productivity," *Nature*, vol. 396, no. 6706, pp. 69–72, 1998.

[9] Y. Y. Su, X. Sun, and L. D. Guo, "Seasonality and host preference of arbuscular mycorrhizal fungi of five plant species in the inner

mongolia steppe, China," *Brazilian Journal of Microbiology*, vol. 42, no. 1, pp. 57–65, 2011.

[10] T. Kumar and M. Ghose, "Status of arbuscular mycorrhizal fungi (AMF) in the Sundarbans of India in relation to tidal inundation and chemical properties of soil," *Wetlands Ecology and Management*, vol. 16, no. 6, pp. 471–483, 2008.

[11] R. S. Rao, *Flora of Goa, Daman, Dadra and Nagar Haveli Volume I & II.Botanical Survey of India*, Deep Printers, New Delhi, India, 1985.

[12] A. Walkley and J. A. Black, "An examination of the Degtjareff method for determining soil organic matter and a proposed modification of the chromic titration method," *Soil Science*, vol. 37, pp. 29–38, 1934.

[13] R. H. Bray and L. T. Kurtz, "Determination of total, organic and available forms of phosphorus in soils," *Soil Science*, vol. 59, pp. 39–45, 1945.

[14] J. J. Hanway and H. Heidel, "Soil analysis method as used in Iowa State College soil testing laboratory," *Iowa State College of Agriculture*, vol. 57, pp. 1–31, 1952.

[15] W. L. Lindsay and W. A. Norvell, "Development of DTPA soil test for zinc, iron, manganese and copper," *Soil Science Society of America Journal*, vol. 42, pp. 421–428, 1978.

[16] J. W. Gerdemann and T. H. Nicolson, "Spore density of the *Endogone* species extracted from soil by wet sieving and decanting," *Transactions of British Mycological Society*, vol. 46, pp. 235–244, 1963.

[17] B. F. Rodrigues and T. Muthukumar, *Arbuscular Mycorrhizae of Goa—A Manual of Identification Protocols*, Goa University, Goa, India, 2009.

[18] N. C. Schenck and Y. Perez, *Manual for the Identification of VA Mycorrhizal Fungi*, University of Florida, Gainesville, Fla, USA, 1990.

[19] E. H. Simpson, "Measurement of diversity," *Nature*, vol. 163, no. 4148, p. 688, 1949.

[20] C. E. Shannon and W. Weaver, *The Mathematical Theory of Communication*, The University of Illinois Press, Urbana, Ill, USA, 1949.

[21] B. F. Rodrigues and N. Anuradha, "Arbuscular mycorrhizal fungi in Khazan land agro-ecosystem," in *Frontiers in Fungal Ecology, Diversity and Metabolites*, K. R. Sridhar, Ed., pp. 141–150, I.K. International, New Delhi, India, 2009.

[22] R. T. M. Padma and D. Kandaswamy, "Effect of interactions between VA mycorrhizae and graded levels of phosphorus on growth of papaya (*Carica papaya*)," in *Current Trends in Mycorrhizal Research*, B. L. Jalai and H. Chand, Eds., pp. 133–134, Haryana Agricultural University, Hisar, India, 1990.

[23] J. F. Liao, "The chemical properties of the mangrove Solonchak in the northeast part of Hainan Island," *Acta Scientiarum Naturalium Universitatis*, vol. 9, pp. 67–72, 1990.

[24] M. Stumm and J. J. Morgan, *Aquatic Chemistry*, John Wiley and Sons, New York, NY, USA, 3rd edition, 1996.

[25] R. W. Howarth, "Pyrite: its rapid formation in a salt marsh and its importance in ecosystem metabolism," *Science*, vol. 203, no. 4375, pp. 49–51, 1979.

[26] S. P. Miller and J. D. Bever, "Distribution of arbuscular mycorrhizal fungi in stands of the wetland grass *Panicum hemitomon* along a wide hydrologic gradient," *Oecologia*, vol. 119, no. 4, pp. 586–592, 1999.

[27] A. L. Ruotsalainen, H. Väre, and M. Vestberg, "Seasonality of root fungal colonization in low-alpine herbs," *Mycorrhiza*, vol. 12, no. 1, pp. 29–36, 2002.

[28] M. A. Lugo, M. E. G. Maza, and M. N. Cabello, "Arbuscular mycorrhizal fungi in a mountain grassland II: seasonal variation of colonization studied, along with its relation to grazing and metabolic host type," *Mycologia*, vol. 95, no. 3, pp. 407–415, 2003.

[29] S. S. Dhillion and R. C. Anderson, "Seasonal dynamics of dominant species of arbuscular mycorrhizae in burned and unburned sand prairies," *Canadian Journal of Botany*, vol. 71, no. 12, pp. 1625–1630, 1993.

[30] J. N. Gemma, R. E. Koske, and M. Carreiro, "Seasonal dynamics of selected species of V-A mycorrhizal fungi in a sand dune," *Mycological Research*, vol. 92, no. 3, pp. 317–321, 1989.

[31] S. P. Bentivenga and B. A. D. Hetrick, "Seasonal and temperature effects on mycorrhizal activity and dependence of cool- and warm-season tallgrass prairie grasses," *Canadian Journal of Botany*, vol. 70, no. 8, pp. 1596–1602, 1992.

[32] D. S. Hayman, "Plant growth responses to vesicular-arbuscular mycorrhiza. VI. Effect of light and temperature," *New Phytologist*, vol. 73, pp. 71–78, 1970.

[33] A. Saravanakumar, M. Rajkumar, S. J. Serebiah, and G. A. Thivakaran, "Seasonal variations in physico-chemical characteristics of water, sediment and soil texture in arid zone mangroves of Kachchh-Gujarat," *Journal of Environmental Biology*, vol. 29, no. 5, pp. 725–732, 2008.

[34] E. M. Ahulu, A. Gollotte, V. Gianinazzi-Pearson, and M. Nonaka, "Cooccurring plants forming distinct arbuscular mycorrhizal morphologies harbor similar AM fungal species," *Mycorrhiza*, vol. 17, no. 1, pp. 37–49, 2006.

[35] T. Aziz and D. M. Sylvia, "Activity and species composition of arbuscular mycorrhizal fungi following soil removal," *Ecological Applications*, vol. 5, no. 3, pp. 776–784, 1995.

[36] R. C. Anderson, A. E. Liberta, and L. A. Dickman, "Interaction of vascular plants and vesicular-arbuscular mycorrhizal fungi across a soil moisture-nutrient gradient," *Oecologia*, vol. 64, no. 1, pp. 111–117, 1984.

[37] D. H. Rickerl, F. O. Sancho, and S. Ananth, "Vesicular-arbuscular endomycorrhizal colonization of wetland plants," *Journal of Environmental Quality*, vol. 23, no. 5, pp. 913–916, 1994.

[38] J. C. Stutz, R. Copeman, C. A. Martin, and J. B. Morton, "Patterns of species composition and distribution of arbuscular mycorrhizal fungi in arid regions of southwestern North America and Namibia, Africa," *Canadian Journal of Botany*, vol. 78, no. 2, pp. 237–245, 2000.

[39] J. C. Stutz and J. B. Morton, "Successive pot cultures reveal high species richness of vesicular-arbuscular mycorrhizal fungi across a soil moisture nutrient gradient," *Oecologia*, vol. 64, pp. 111–117, 1996.

[40] L. K. Abbott and A. D. Robson, "Factors influencing the occurrence of vesicular-arbuscular mycorrhizas," *Agriculture, Ecosystems and Environment*, vol. 35, no. 2-3, pp. 121–150, 1991.

[41] W. E. van Duin, J. Rozema, and W. H. O. Ernst, "Seasonal and spatial variation in the occurrence of vesicular-arbuscular (VA) mycorrhiza in salt marsh plants," *Agriculture, Ecosystems and Environment*, vol. 29, no. 1-4, pp. 107–110, 1989.

[42] D. L. Stenlund and I. D. Charvat, "Vesicular arbuscular mycorrhizae in floating wetland mat communities dominated by *Typha*," *Mycorrhiza*, vol. 4, no. 3, pp. 131–137, 1994.

[43] P. Guadarrama and F. J. Alvarez-Sanchez, "Abundance of arbuscular mycorrhizal fungi spores in different environments in a tropical rain forest, Veracruz, Mexico," *Mycorrhiza*, vol. 8, no. 5, pp. 267–270, 1991.

[44] Z. W. Zhao, "Population composition and seasonal variation of VA mycorrhizal fungi spores in the rhizosphere soil of four pteridophytes," *Acta Botanica Yunnanica*, vol. 21, pp. 437–441, 1999.

[45] C. Walker, C. W. Mize, and H. S. McNabb, "Populations of endogonaceous fungi at two locations in central Iowa," *Canadian Journal of Botany*, vol. 60, no. 12, pp. 2518–2529, 1982.

[46] K. P. Radhika and B. F. Rodrigues, "Arbuscular mycorrhizae in association with aquatic and marshy plant species in Goa, India," *Aquatic Botany*, vol. 86, no. 3, pp. 291–294, 2007.

[47] K. P. Radhika and B. F. Rodrigues, "Arbuscular mycorrhizal fungal diversity in some commonly occurring medicinal plants of Western Ghats, Goa region," *Journal of Forestry Research*, vol. 21, no. 1, pp. 45–52, 2010.

[48] D. Zhao and Z. Zhao, "Biodiversity of arbuscular mycorrhizal fungi in the hot-dry valley of the Jinsha River, Southwest China," *Applied Soil Ecology*, vol. 37, no. 1-2, pp. 118–128, 2007.

[49] X. He, S. Mouratov, and Y. Steinberger, "Spatial distribution and colonization of arbuscular mycorrhizal fungi under the canopies of desert halophytes," *Arid Land Research and Management*, vol. 16, no. 2, pp. 149–160, 2002.

4

Diversity of Mercury Resistant *Escherichia coli* Strains Isolated from Aquatic Systems in Rio de Janeiro, Brazil

Raquel Costa de Luca Rebello,[1] **Karen Machado Gomes,**[2]
Rafael Silva Duarte,[2] **Caio Tavora Coelho da Costa Rachid,**[3] **Alexandre Soares Rosado,**[3]
and Adriana Hamond Regua-Mangia[1]

[1] *Departamento de Ciências Biológicas, Escola Nacional de Saúde Pública Sergio Arouca (Ensp), Fundação Oswaldo Cruz (FIOCRUZ), Rua Leopoldo Bulhões 1480 Manguinhos, 21041-210 Rio de Janeiro, RJ, Brazil*
[2] *Departamento de Microbiologia Médica, Instituto de Microbiologia Paulo de Góes, Universidade Federal do Rio de Janeiro (UFRJ), Rio de Janeiro, RJ, Brazil*
[3] *Departamento de Microbiologia Geral, Instituto de Microbiologia Paulo de Góes, Universidade Federal do Rio de Janeiro (UFRJ), Rio de Janeiro, RJ, Brazil*

Correspondence should be addressed to Adriana Hamond Regua-Mangia; regua@ensp.fiocruz.br

Academic Editor: Steven Lee Stephenson

Escherichia coli may harbor genetic mercury resistance markers which makes this bacterial species a promising alternative for bioremediation processes. The objective of this study was to investigate phenotypic and genetic characteristics related to diversity and mercury resistance among 178 *Escherichia coli* strains isolated from residential, industrial, agricultural, and hospital wastewaters and recreational waters at Rio de Janeiro city. Genetic and conventional methods were carried out in order to determine mercury resistance. Random amplification of polymorphic DNA (RAPD-PCR) and denaturing gradient gel electrophoresis (DGGE) were used to investigate genetic variability. RAPD data revealed a high degree of polymorphism among *E. coli* mercury resistant strains and showed reproducibility and good discriminative results. DGGE typing detected diversity within the *merA* gene fragment. Our findings represent an improvement in epidemiological studies of Hg^R *E. coli* and support the evidence of nonclonal nature of mercury resistant *E. coli* strains circulating in rural and urban aquatic systems in Rio de Janeiro city.

1. Introduction

Chemical contamination of aquatic systems consists of a relevant pollution pattern causing drastic impacts on human, animal, and ecosystem health [1]. Among the various chemical contaminants, mercury plays an important role and once released in aquatic systems, mercury can resist to natural degradation processes and persist for a long time in these environments without losing its toxicity [2].

The concern about environmental contamination by this metal is due to its high toxicity, especially to the nervous system, and its bioaccumulation and biomagnification, providing persistence and wide distribution in global aquatic environment. Even regions with no mercury discharging may be affected [2–7].

Mercury toxicity to humans and other organisms is related to the chemical form to which the organisms were exposed, the route and time of exposure, dose, nutritional status, individual susceptibility, and genetic predisposition [3, 4, 6, 8]. Symptoms and contamination sources are rather different in exposure to elemental mercury, inorganic or organic mercury compounds [3, 4]. Human contamination by this metal may occur by different pathways such as vapors inhalation, contaminated food and/or water consumption, and to a lesser extent through skin contact [3, 6]. Mercury exposure triggers a series of effects including neurological, renal, cardiovascular, respiratory, gastrintestinal, hepatic, genotoxic, immunological, dermal, reproductive, and neoplasic disorders. Exposure during pregnancy may lead to

TABLE 1: Origin of *Escherichia coli* strains included in this study.

E. coli strains	Strains (*n*)	Aquatic system[*]	Sampling site	Sampling period
RM 1–RM 30 RM 33–RM 77	75	RWW	Canal do Mangue, Rio Jacaré, Canal do Cunha, Rio Faria, Rio Irajá, Canal do Meriti, Rio Sarapuí, Lagoa Rodrigo de Freitas, Lagoa da Tijuca, Lagoa de Marapendi, Rio São João	December/2009 to August/2010
RM 31-RM 32 RM 78–RM 84	09	IWW	Rio Saracuruna, Rio Imbariê, Rio Iguaçú	October/2010
RM 85–RM 110	26	AWW	Rio Vargem Grande, Córrego das Pedras	October/2010
RM 111–RM 149	29	HWW	Lagoa de Jacarepaguá	January/2011
RM 150–RM 154 RM 156–RM 179	39	RW	Parque Nacional da Praia de Ramos	January/2011

[*]RWW: residential wastewater; IW: industrial wastewater; AWW: agricultural wastewater; HWW: hospital wastewater, and RW: recreational waters.

malformations, mental retard, cerebral palsy, seizures, and death [3, 4, 6, 8].

Mercury resistance is one of the most studied toxic metals resistance mechanisms [7]. It has been reported that some bacteria and fungi isolated from different sources have developed resistance mechanisms that enable them to survive even in environments highly contaminated by mercury [9]. There are several described bacterial mechanisms that confer protection to harmful concentrations of mercury [10]. Among them, we highlight the mercury enzymatic detoxification, promoted by the mercuric reductase protein (MerA), which catalyze the reduction of Hg(II) to volatile Hg(0) [11, 12]. Considering the genetic of MerA expression, the Hg resistance (*mer*) operon presents a fundamental role in regulation, Hg binding, and organomercury degradation. It consists of essential genes as *merR* (responsible for the regulation of the operon), *merT/merP* (transport of mercury into the bacterial cell), *merA* (reduction of ionic mercury), and accessory genes such as *merB*, *merC*, *merD*, *merE*, *merF*, and *merG*, that encode proteins that add other skills to microorganisms [13, 14]. MerR protein can act both as a repressor and activator of transcription. In the absence of Hg^{2+}, MerR acts as repressor by binding to the *mer* operon operator region and preventing the transcription of *merTPCAD*. In presence of Hg^{2+}, it binds to one of two MerR binding sites forming a complex that acts as an activator of *mer* operon transcription [15]. Mer-mediated approaches have had broad applications in the bioremediation of mercury-contaminated environments and industrial waste streams [8, 11, 12, 16, 17].

Mercury resistance in bacteria has been observed in both Gram-positive (*S. aureus*, *Bacillus* sp.) and Gram-negative bacteria (*E. coli*, *P. aeruginosa*, *Serratia marcescens*, and *Thiobacillus ferrooxidans*) [12, 16]. Mercury resistance is encoded on genetic elements such as plasmids and transposons, which contributes to horizontal dissemination among different bacteria and widespread occurrence in different bacterial groups and environments [12].

In Gram-negative bacteria, including *E. coli*, the *mer* operon has already been described [18]. However, epidemiological and genetic studies related to mercury resistance are scarce. Therefore, the investigation of the mercury resistance features has been crucial to improve bioremediation processes in contaminated environments in order to minimize human exposure and consequent adverse health effects.

In the present study, *E. coli* isolates from aquatic systems, in the city of Rio de Janeiro, Brazil, were characterized by phenotypic and genotypic traits related to mercury resistance. Bacteriological tests were carried out in order to determine mercury susceptibility, and molecular approaches based on amplification assays were used to investigate the presence and diversity of mercury resistance gene (*merA*).

2. Materials and Methods

2.1. Water Sampling. Samples were selected and grouped according to potential contamination sources in the city of Rio de Janeiro, Brazil. We studied five aquatic environments: residential, industrial, agricultural, and hospital wastewaters and recreational waters (Table 1).

2.2. Sample Collection. Collection procedure consisted of membrane filtration method with some modifications [19]. An aliquot of 60 mL of water was aspirated from the upper layer of the water column to a depth of approximately 30 centimeters with a syringe holder adapted to sterile filtration. The aspirate was filtered on a $0.22\,\mu m$ cellulose acetate membrane (Milllipore) and transported under refrigeration for immediate laboratory processing.

2.3. Escherichia coli Isolation and Identification. The membrane containing the retained cells was incubated in 20 mL of tryptic soy broth (TSB, Difco) for 18–24 hr at 37°C. After a period of bacterial growth, an aliquot of the broth, diluted (1:10, 1:50, and 1:100) in saline 0.9% NaCl (w/v), was streaked on eosin methylene blue agar (EMB, Difco). After 18–24 h of incubation at 37°C 10–15 bacterial colonies, lactose positive and lactose negative, were selected based on morphological and physiological characteristics suggestive of *E. coli*. For confirmation of genus and species, the selected colonies were inoculated in culture medium for biochemical identification (Probac of Brazil). *E. coli* biochemical pattern includes gas production from glucose (+), glucose utilization (+), hydrogen sulfide production (−), urea hydrolysis (−) and tryptophan deamination (−), motility (variable), indole production (+), decarboxylation of lysine (variable), and citrate (−) [20]. Bacterial cells identified as *E. coli* were stored at −20°C in TSB plus 15% glycerol (v/v) until analysis. This study included a total of 178 *E. coli* isolates (RM 1 to RM 179) (Table 1).

2.4. Mercury Resistance Phenotype. In order to classify *E. coli* as resistant or sensitive to mercury, each strain was tested on nutrient agar (NA, Difco) supplemented with 5 μM of Hg^{2+}. Evidence of bacterial growth after a period of 24–48 h at 37°C allowed to classify *E. coli* strains as Hg resistant (Hg^R). *E. coli* ATCC 35218 (Hg resistant) and *E. coli* ATCC 23724 (Hg susceptible) were used as control strains. When no growth was observed, a strain was considered as sensitive. These tests were done in duplicate.

2.5. Minimal Inhibitory Concentration (MIC). MIC determination was performed following the methodology described by Andrews [21] with some modifications. Overnight cultures of the isolates in nutrient broth (NB, Difco) containing 1 μM Hg were adjusted in saline NaCl 0.9% (w/v) in order to contain 1.5×10^9 bacterial cells/mL (McFarland 0.5). An aliquot of 50 μL was inoculated in nutrient agar plates containing 10 to 40 μM Hg. After 24–48 h at 37°C, the MIC value was determined by observing bacterial growth on agar plates in the presence of the lowest Hg concentration. MIC tests were performed with those *E. coli* strains exhibiting mercury resistance phenotype ≥5 μM Hg. All experiments were performed in duplicate.

2.6. merA Detection. All strains were screened for the presence of *merA* sequence by PCR amplifications as described by Ní Chadhain and colleagues [22], with some modifications. Each reaction was carried out in 25 μL PCR mixture containing 3 μL of bacterial DNA obtained through thermal extraction of 18–24 h bacterial growth in tryptic soy broth (TSB, Difco), 2.5 μL of 10X buffer (Invitrogen), 2 mM $MgCl_2$ (Invitrogen), 0.2 mM dNTP (Invitrogen), 30 μM of each primer, and 1 U of Platinum *Taq* DNA polymerase (Invitrogen). The pair of primers used was A1s-n.F (5'-TCCGCAAGTNGCVACBGTNGG-3') and A5-n.R (5'-ACCATCGTCAGRTARGGRAAVA-3'). PCR reaction was conducted in *Mastercycler Personal* thermocycler (Eppendorf) under the following amplification conditions: initial denaturing step at 94°C for 5 min, followed by 45 cycles at 94°C for 10 sec, 68°C for 40 sec, and 72°C for 1 min with a final extension at 72°C for 7 min. Approximately 10 μL of the resulting amplification products was added to 2 μL of running buffer (*gel loading buffer*, Invitrogen) and separated by electrophoresis on agarose gel at 1.3% concentration (w/v) prepared in Tris-Borate-EDTA 0.5X (5X-0.89 M Tris-HCl (LGC Biotech) 0.89 M boric acid (Merck) and 0.024 M EDTA (LGC Biotech) (pH 8.4)) at a constant voltage of 70 V. Electrophoresis gel was stained with 0.5 μg/mL ethidium bromide solution (Invitrogen) over a period of 15 min and washed in

distilled water for about 30 min. Gel was visually inspected by using an ultraviolet light transilluminator (UVITec, Cambridge, UK) and photographed in digital image capture system (silver UVIPro, Cambridge, UK). To estimate the size of the fragments a 100 bp DNA ladder standard (Invitrogen) was used. *E. coli* strains ATCC 35218 (Hg resistant) and ATCC 23724 (Hg sensitive) were used as controls.

2.7. Random Amplification of Polymorphic DNA (RAPD-PCR). RAPD-PCR analysis was performed according to the methodology described by Pacheco and colleagues [23]. Each reaction was carried out in a 30 μL PCR mixture containing 2 μL of bacterial DNA, 3 μL of 10X buffer (Invitrogen), 250 μM each dNTP (Invitrogen), 3 mM $MgCl_2$ (Invitrogen), 1 U of *Taq* DNA polymerase (Invitrogen), and 30 μM of each primer. Primers used were 1247 (5'-AAGAGCCCGT-3'), 1254 (5'-CCGCAGCCAA-3'), 1290 (5'-GTGGATGCGA-3'), and A04 (5'-AATCGGGCTG-3'). The reaction was conducted in a *Mastercycler Personal* thermocycler (Eppendorf) under the following amplification conditions: an initial denaturing step at 94°C for 1 min, followed by 4 cycles at 94°C for 4 min, 37°C for 4 min, and 72°C for 4 min, 30 cycles at 94°C for 1 min, 37°C for 1 min, and 72°C for 2 min with a final extension at 72°C for 10 min. Reaction products were analyzed by electrophoresis in 1.5% agarose gels and stained with ethidium bromide. RAPD profiles were inspected visually and defined according to the presence or absence and intensity of polymorphic bands. A 1 kb DNA ladder was used as a molecular weight marker (GIBCO, BRL, Gaithersburg, MD, USA). Semiautomated analysis used the UVI Soft Image Acquisition and Analysis Software, program UVIPro bandmap version 11.9 (UVItec, Cambridge, UK). Cluster analysis was done by using the unweighted pair group method with arithmetic averages (UPGMA) of the Image Analysis System. The percentages of similarity were estimated by the Dice coefficient. The reproducibility of the RAPD amplifications was assessed using the selected primers with different DNA samples isolated independently from the same strain and amplified at different times.

2.8. Denaturing Gradient Gel Electrophoresis (DGGE). DGGE analysis was performed according to the methodology described by Muyzer and colleagues with some modifications [24]. AxyPrep DNA Gel Extraction kit (Axygen Biosciences) was used for purificating the PCR-*merA* DNA fragment (285 bp). For PCR-DGGE reaction a final volume of 25 μL in amplification reactions containing 3 μL of purified DNA, 2.5 μL 10X buffer (Invitrogen), 2 mM $MgCl_2$ (Invitrogen), 0.2 mM dNTPs (Invitrogen), 30 μM of each primer, 1% formamide, and 1 U of Platinum *Taq* DNA polymerase (Invitrogen) was used. The pair of primers for amplification was A1s-n.F (5'-TCCGCAAGTNGCVACBGTNGG-3') and A5-n.R (5'-ACCATCGTCAGRTARGGRAAVA-3'). The reaction was conducted in *Mastercycler Personal* thermocycler (Eppendorf) and programmed for an initial denaturation of 94°C for 5 min followed by 45 cycles of 94°C for 10 sec, 68°C for 40 sec, and 72°C for 1 min, with a final extension of 72°C for 7 min. Approximately 25 μL of amplified PCR product was added to 15 μL of DNA electrophoresis dye (0.005 g

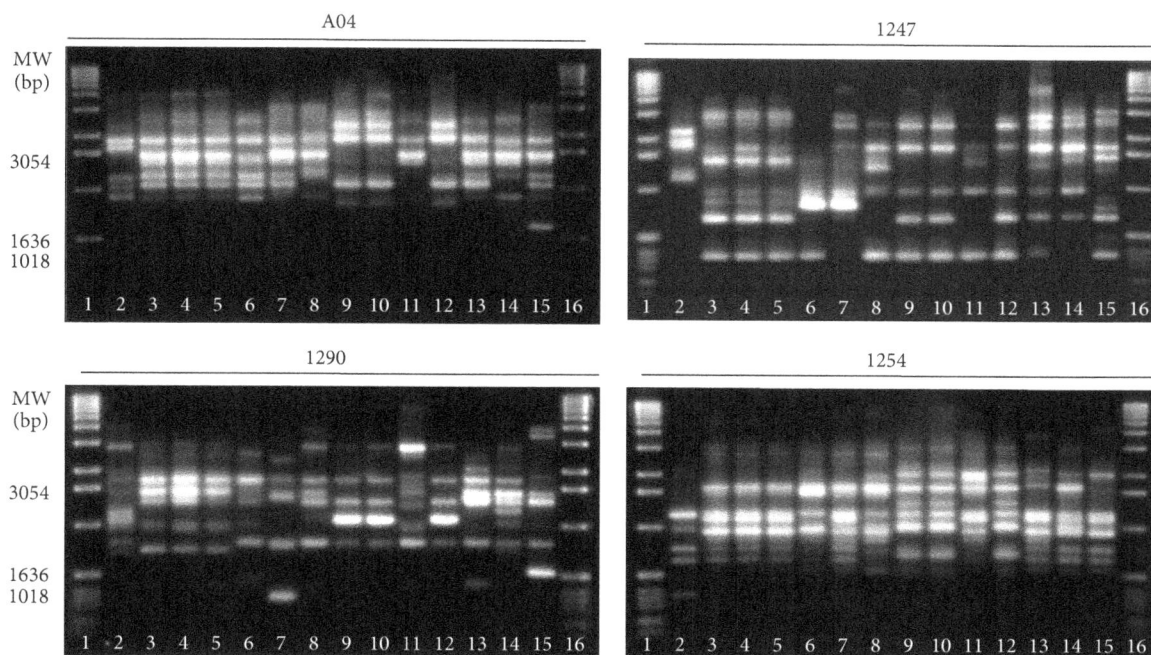

FIGURE 1: RAPD-PCR profiles of representative *E. coli merA*+ strains obtained by using 4 different primers (A04, 1247, 1290, and 1254). Lanes 1, 16: 1 Kb DNA ladder; Lane 2: strain RM 1; Lane 3: strain RM 7; Lane 4: strain RM 8; Lane 5: strain RM 9; Lane 6: strain RM 17; Lane 7: strain RM 20; Lane 8: strain RM 31; Lane 9: strain RM 37; Lane 10: strain RM 44; Lane 11: strain RM 45; Lane 12: strain RM 46; Lane 13: strain RM 61; Lane 14: strain RM 150; Lane 15: strain RM 165.

Bromophenol blue, 0.005 g xylene cyanol, 7 mL glycerol P.A., and 3 mL deionized water) and ran on a polyacrylamide gel (8% w/v of acrylamide/bisacrylamide ratio 37.5 : 1) with a linear denaturant gradient ranging from 55% to 80% (where 100% is a solution of 7 M urea and 40% formamide v/v). Electrophoresis was performed in equipment using the *Dcode Universal Mutation System* (BIO-Rad) and conducted at constant voltage of 100 V at 60°C for 6 h in 0.5X Tris-acetate (10 mM Tris-acetate, 5 mM Sodium Acetate, 25 mM EDTA, and pH 7.4). After electrophoresis the gel was stained with Sybr Green (Molecular Probes, OR, USA) for 30 minutes and visualized under UV transilluminator. The reproducibility of the assay was tested by loading three PCR products for each sample on DGGE gels.

3. Results

3.1. Mercury Resistance Phenotype and Minimal Inhibitory Concentration (MIC). A total of 164 strains were classified as mercury resistant (Hg^R) and represented 92.1% of the *E. coli* isolates (164/178). All Hg^R exhibited the Hg MIC value of 10 μM.

3.2. PCR Amplification of merA Gene. Among *E. coli* strains analyzed in this study, 14 harbored the 285 bp *merA* gene fragment described by Ní Chadhain and colleagues [22]. *E. coli* strains carrying the 285 bp *merA* gene corresponded to 14.7% (11/75) of the isolates obtained from residential wastewaters samples, 11.1% (1/9) from industrial wastewaters samples, and 6.9% (2/29) from hospital wastewaters samples.

3.3. Random Amplification of Polymorphic DNA (RAPD-PCR). The diversity within the *E. coli merA*+ strains was investigated by RAPD-PCR using the primers A04, 1247, 1290, and 1254 (Figure 1). RAPD typing revealed a high degree of diversity among *E. coli* strains. Reactions performed with primers A04, 1247, 1290 (60% GC, each), and 1254 (70% GC) resulted in 11, 10, 10, and 10 different RAPD profiles, respectively. The total number of polymorphic bands was 5–9 bands (A04), 4–11 bands (1247), 5–10 bands (1290), and 8–12 bands (1254) ranging from 600–4100 bp, 200–5600 bp, 450–8000 bp, and 250–9000 bp, respectively. There was no direct correlation between higher G+C content and the ability of the primer to detect polymorphism. The different primers used to investigate the overall chromosomal relatedness among *E. coli* strains were strongly correlated. The cluster analysis revealed a bacterial population arranged into separate branches or small clonal groups, exhibiting Dice similarity index ranging from 6–100%, 18–100%, 6–100%, and 6–100% for primers 1290, 1254, 1247, and A04 (Figure 2), respectively. Close relatedness was specially observed among *merA*+ *E. coli* strains isolated from the same aquatic system (Table 2, Figure 2). Identical RAPD profiles were observed among residential wastewaters isolates: RM 7, RM 8, and RM 9, isolated from Canal do Cunha, and RM 37, RM 44, and RM 46 from Lagoa Rodrigo de Freitas.

3.4. Denaturing Gradient Gel Electrophoresis (DGGE). Electrophoresis technique on denaturing gradient gel enabled the detection of variability within the 285 bp gene fragment associated with mercury resistance (*merA*). Supporting the results obtained from RAPD-PCR, RM 7, RM 8, and RM

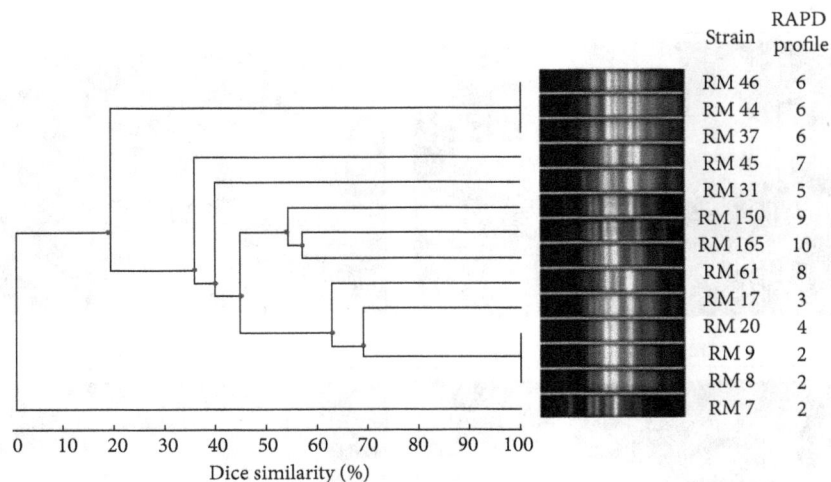

FIGURE 2: Dendrogram generated by the Dice coefficient and clustering by unweighted pair group method with arithmetic mean and respective RAPD profiles of *merA+ Escherichia coli* isolates using 1254 primer.

TABLE 2: Genetic and phenotypic traits of *E. coli* strains carrying the 285 bp *merA* fragment according to aquatic systems and sampling sites.

E. coli strain	Aquatic system*	Sampling site	MIC	RAPD profile			
				A04	1247	1290	1254
RM 1		Canal do Mangue	10 μM	1	1	1	1
RM 7		Canal do Cunha	10 μM	2	2	2	2
RM 8	RWW	Canal do Cunha	10 μM	2	2	2	2
RM 9		Canal do Cunha	10 μM	2	2	2	2
RM 17		Rio Irajá	10 μM	3	3	3	3
RM 20		Rio Irajá	10 μM	4	4	4	4
RM 31	IWW	Rio Iguaçú	10 μM	5	5	5	5
RM 37		Lagoa Rodrigo de Freitas	10 μM	6	6	6	6
RM 44		Lagoa Rodrigo de Freitas	10 μM	6	6	6	6
RM 45	RWW	Lagoa Rodrigo de Freitas	10 μM	7	7	7	7
RM 46		Lagoa Rodrigo de Freitas	10 μM	8	6	6	6
RM 61		Lagoa de Marapendi	10 μM	9	8	8	8
RM 150	HWW	Lagoa de Jacarepaguá	10 μM	10	9	9	9
RM 165		Lagoa de Jacarepaguá	10 μM	11	10	10	10

*RWW: residential wastewater; IW: industrial wastewater; HWW: hospital wastewater.

9 isolates also showed identical DGGE pattern (Figure 3). Despite the diversity observed, no significant differences among the DGGE band patterns were observed.

4. Discussion

4.1. Mercury Resistance Phenotype and Minimal Inhibitory Concentration (MIC). Many studies have been conducted in order to determine the mercury resistance in environmental bacteria by testing the minimum inhibitory concentration [25–28]. There is not a standard protocol for determining the MIC of heavy metals. Liquid and/or solid media with different chemical compositions have been commonly used for these assays, as well as variation of metals concentrations. Methodology itself may offer some obstacles such as precipitation and volatilization of the solution and

complexes between the metal and culture medium components. These variations, if not minimized before its application, may directly influence the result obtained [26]. So, it is very difficult to compare the obtained results with previous studies because of the great diversity of MIC values and the procedures adopted, especially considering the broad spectrum of mercury resistant bacteria that require specific conditions for growing and laboratory processing.

The ubiquity of bacterial mercury resistance has been observed in environments worldwide and is supposed to be the result of external interference by humans and other animals through environmental contamination for several years [5, 12, 26]. There were no reports about mercury contamination in the sampling sites; however, Hg resistance was widely detected. Bacterial resistance to mercury present in the environment is considered as one of many examples of

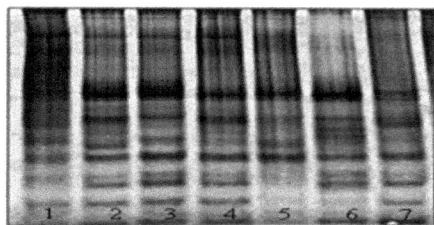

FIGURE 3: DGGE profiles of PCR-amplified *merA*+ gene fragment from *E. coli* strains. Lane 1: strain RM 1; Lane 2: strain RM 7; Lane 3: strain RM 8; Lane 4: strain RM 9; Lane 5: strain RM 17; Lane 6: strain RM 20; Lane 7: strain RM 31.

genetic and physiological adaptation of microbial communities exposed to contaminants. Several factors have been found to contribute to this phenotype in the rural and urban areas including the use of mercury-based fungicide in the paper industry, agriculture, and hospital disinfectants. These factors may encourage selective activities and result in mercury resistance in open environment [29]. Additionally, toxic metal resistance genes are commonly found in environmental bacteria, and these genes may confer coresistance or crossresistance to antimicrobial drugs codified on the same genetic element [26, 30]. So, selection of microbial communities exposed to toxic levels of the metal or submitted to the coselection mechanisms has led to high rates of circulation of these resistant bacteria in aquatic systems [31, 32].

4.2. PCR Amplification of merA Gene.

The genetic system evolved as *mer* operon is the only well-known bacterial mercury resistance system with high yield transformation of its toxic target into volatile nontoxic forms [27, 31, 33–35], particularly in Gram-negative bacteria [14, 22]. The *mer* locus is found to be widely distributed among bacterial lineages, and *mer*-like sequences have been described. Several biochemical mechanisms are identified, and the complexity among the ecological niche of mercury-resistant microbes is still not fully described [10, 35]. *merA* plays a key role on mercury resistance of bacterial community exposed to mercury contamination, but the combinatorial action of genetic determinants seems to confer a broad spectrum mercury detoxification system [10, 35]. So, the involvement of additional genetic determinants not investigated here, acting as effectors or regulators genes, must be considered for the expression of mercury resistance phenotype among *merA* negative *E. coli* strains. *merA* gene was detected *in E. coli* isolated from residential wastewaters (11/75), industrial wastewaters (1/9), and hospital wastewaters (2/29). The higher frequency of *merA*+ *E. coli* strains obtained from residential wastewaters compared to industrial and hospital wastewaters may be related to several factors such as the representative sampling of each area investigated and involvement of additional genetic determinants as well as related to the intrinsic characteristics of the rural and urban locations.

4.3. Random Amplification of Polymorphic DNA (RAPD-PCR).

RAPD-PCR is a recognized powerful tool showing high discriminatory potential, reproducibility, sensibility,

and specificity under well-standardized protocols. Random amplification of polymorphic DNA (RAPD) has been successfully used as a molecular typing system for studies on diversity of *E. coli* population [23, 36].

RAPD typing revealed levels of polymorphism that are consistent with previously reported observations for *E. coli* and has been attributed to the high plasticity of this bacterial species. Several molecular approaches mainly based on genetic techniques have been successfully applied in order to assess the clonal nature and variability within species [23, 36]. The occurrence of distinct patterns of *E. coli* phylogenetic distribution provides evidence of both vertical and horizontal transmission [37–39]. The mechanisms of genetic diversification contribute to *E. coli* evolution and creation of new variants, as this bacterial species is often subjected to DNA rearrangements, excisions, transfers, and acquisitions [37, 40]. There are several highly adapted clones that have acquired specific virulence elements which confer an increased ability to adapt to new niches. Such plasticity may confer ability to acclimate environmental bacteria to new niches allowing these microorganisms to become members of microbial communities in a variety of environments, even facing conditions very different from their primary habitat [36, 38, 39, 41]. RAPD-PCR approach was used to investigate the overall chromosomal relatedness among *merA*+ strains and revealed a high genetic diversity population suggesting that mercury resistance is widely dispersed in *E. coli*. The observed genotypic diversity led us to suppose that, in Rio de Janeiro, *merA*+ *E. coli* isolates consist of nonrelated epidemiological strains and may represent distinct evolutionary lineages. Despite the genetic variability, clustering analysis revealed that the degree of diversity was to a lesser extent among *E. coli* strains obtained from the same aquatic environment evidencing the circulation of closely related strains.

4.4. Denaturing Gradient Gel Electrophoresis (DGGE).

DGGE fingerprinting is a technique widely used in microbial ecology studies and has been focused on studies of genetic diversity and bacterial communities from several environments [17, 42]. Variability within *merA* gene has been described, and diverse MerA protein homologs have been identified in both archaeal as well as bacterial genomes but not in eukaryal genomes [17]. The increased complexity of *mer* operons can be attributed to the gradual addition of functions involved in the regulation of the operon by Hg, Hg transport, and organomercury resistance [17]. The diversity *of merA* gene in Gram-negative and Gram-positive bacteria has been accessed by several approaches including those using restriction fragment assays [27, 31, 33, 34]. In all these studies, a high genetic variability was detected in *merA* determinant carried by bacterial species from different environments. However, RFLP technique is limited since it relies on specific target, requiring prior knowledge of the sequences to be analyzed. In the present study, DGGE was used to investigate the *merA*+ variability among *E. coli* mercury resistant.

DGGE typing revealed diversity within the 285 bp *merA* fragment corroborating previous findings that described the occurrence of genetic exchanges in mercury resistance gene as a result of addition, rearrangements, excisions, and horizontal transfer.

Ní Chadhain and colleagues [22] developed a protocol using degenerated primers and detected high diversity within *merA* sequence from evolutionary distinct Gram-bacteria. In our study, this methodology allowed the detection of variability in the 285 bp *merA* fragment among 14 *E. coli* strains (Figure 3). These results are in agreement with previous findings regarding the widespread occurrence and diversity of mercury resistance markers among distinct microbial populations from several environments, including soils and sediments, aquatic systems, animals, and clinical isolates [13, 14, 27, 28, 31, 32]. The high plasticity found in the bacterial genome contributes to the diversity and dissemination of genetic markers favoring its circulation in geographically dispersed environments, even between distinct evolutionary lineages.

E. coli isolates sharing similar RAPD profiles were found to exhibit the same *merA* DGGE pattern suggesting the circulation of conserved or partially conserved *merA* sequence among closely related strains. The molecular approaches used as fingerprint tools were found to be accurate and useful methods in distinguishing between closely related bacteria. The obtained results are relevant to our understanding on the characteristics of mercury resistant *E. coli* circulating in natural environments in aquatic systems in Rio de Janeiro. Our findings substantially expand our knowledge about *mer* evolution and biodiversity of these microorganisms, and contribute to studies on bioremediation process and environmental management of Hg contamination.

5. Conclusions

The present study detected a wide dissemination of *E. coli* isolates resistant to mercury in distinct aquatic systems in the city of Rio de Janeiro possibly due to selective activities with varying patterns of exposure to Hg. Genetic analysis of *merA+* strains revealed high degree of diversity among the bacterial population indicating that mercury resistance is widely dispersed in *E. coli*. These findings suggest that, in the city of Rio de Janeiro, *merA+ E. coli* may constitute bacterial communities epidemiologically independent and may represent distinct evolutionary lineages. The variability detected within the 285 bp *merA* fragment possibly reflects the occurrence of specific genetic events. *E. coli* strains sharing RAPD profile and DGGE band pattern reinforce the hypotheses of circulation of conserved *merA* sequence among closely related strains. In the light of the pathogenicity attributed to *E. coli* population, more accurate analyses are required for applications in bioremediation processes.

Conflict of Interests

The authors have declared that no conflict of interests exists.

Acknowledgments

This work was supported by a Grant from Fundação Carlos Chagas Filho de Amparo à Pesquisa do Estado do Rio de Janeiro (E-26/110.787/2010) and Coordenação de Aperfeiçoamento de Pessoal de Nível Superior (CAPES). The authors would like to thank Adriana de Lima Bezerra, Marcelo Sampaio, and Thiago Figueiredo for technical assistance in the collection of water samples. The authors also thank the laboratory at Departamento de Saneamento e Saúde Ambiental for conducting the mercury resistance phenotypic assays.

References

[1] A. Bafana, "Mercury resistance in *Sporosarcina* sp. G3," *BioMetals*, vol. 24, no. 2, pp. 301–309, 2011.

[2] F. A. Azevedo, *Toxicologia do Mercúrio*, Editora Rima, São Paulo, Brazil, 2003.

[3] United Nations Environmental Programme, "Global mercury assessment," 2002, http://www.chem.unep.ch/mercury/report/gma-report-toc.htm.

[4] A. L. Oliveira Da Silva, P. R. G. Barrocas, S. Do Couto Jacob, and J. C. Moreira, "Dietary intake and health effects of selected toxic elements," *Brazilian Journal of Plant Physiology*, vol. 17, no. 1, pp. 79–93, 2005.

[5] V. M. Câmara, A. P. Silva, and J. A. Cancio, "Notas para 17 a constituição de um programa de vigilância ambiental dos riscos e efeitos da exposicão do mercúrio metálico em áreas de produção de ouro," *Informe Epidemiológico Do SUS*, vol. 2, pp. 35–44, 1998.

[6] A. T. Jan, I. Murtaza, A. Ali, and Q. M. R. Haq, "Mercury pollution: an emerging problem and potential bacterial remediation strategies," *World Journal of Microbiology and Biotechnology*, vol. 25, no. 9, pp. 1529–1537, 2009.

[7] F. M. M. Morel, A. M. L. Kraepiel, and M. Amyot, "The chemical cycle and bioaccumulation of mercury," *Annual Review of Ecology and Systematics*, vol. 29, pp. 543–566, 1998.

[8] P. B. Tchounwou, W. K. Ayensu, N. Ninashvili, and D. Sutton, "Environmental exposure to mercury and its toxicopathologic implications for public health," *Environmental Toxicology*, vol. 18, no. 3, pp. 149–175, 2003.

[9] M. M. Ball, P. Carrero, D. Castro, and L. A. Yarzábal, "Mercury resistance in bacterial strains isolated from tailing ponds in a gold mining area near El Callao (Bolívar State, Venezuela)," *Current Microbiology*, vol. 54, no. 2, pp. 149–154, 2007.

[10] D. W. Boening, "Ecological effects, transport, and fate of mercury: a general review," *Chemosphere*, vol. 40, no. 12, pp. 1335–1351, 2000.

[11] T. Barkay, S. M. Miller, and A. O. Summers, "Bacterial mercury resistance from atoms to ecosystems," *FEMS Microbiology Reviews*, vol. 27, no. 2-3, pp. 355–384, 2003.

[12] A. M. Osborn, K. D. Bruce, P. Strike, and D. A. Ritchie, "Distribution, diversity and evolution of the bacterial mercury resistance (*mer*) operon," *FEMS Microbiology Reviews*, vol. 19, no. 4, pp. 239–262, 1997.

[13] L. F. Caslake, S. S. Harris, C. Williams, and N. M. Waters, "Mercury-resistant bacteria associated with macrophytes from a polluted lake," *Water, Air, and Soil Pollution*, vol. 174, no. 1–4, pp. 93–105, 2006.

[14] C. A. Liebert, J. Wireman, T. Smith, and A. O. Summers, "Phylogeny of mercury resistance (*mer*) operons of gram-negative bacteria isolated from the fecal flora of primates," *Applied and Environmental Microbiology*, vol. 63, no. 3, pp. 1066–1076, 1997.

[15] M. T. Madigan, J. M. Martinko, P. V. Dunlap, and D. P. Clark, *Microbiologia de Brock*, Artmed, Porto Alegre, Brazil, 2010.

[16] T. K. Misra, "Bacterial resistances to inorganic mercury salts and organomercurials," *Plasmid*, vol. 27, no. 1, pp. 4–16, 1992.

[17] E. S. Boyd and T. Barkay, "The mercury resistance operon: from an origin in a geothermal environment to an efficient detoxification machine," *Frontiers in Microbiology*, vol. 3, pp. 1–13, 2012.

[18] M. Zeyaullah, G. Nabi, R. Malla, and A. Ali, "Molecular studies of *E. coli* mercuric reductase gene (*merA*) and its impact on human health," *Nepal Medical College Journal*, vol. 9, no. 3, pp. 182–185, 2007.

[19] N. Ramaiah and J. De, "Unusual rise in mercury-resistant bacteria in coastal environs," *Microbial Ecology*, vol. 45, no. 4, pp. 444–454, 2003.

[20] E. W. Koneman, S. D. Allen, V. R. Dowell, and H. M. Sommers, *Diagnóstico Microbiológico: Texto e Atlas Colorido*, Medicina Panamericana Editora do Brasil ltda., São Paulo, Brazil, 2008.

[21] J. M. Andrews, "Determination of minimum inhibitory concentrations," *Journal of Antimicrobial Chemotherapy*, vol. 48, pp. S5–S16, 2001.

[22] S. M. Ní Chadhain, J. K. Schaefer, S. Crane, G. J. Zylstra, and T. Barkay, "Analysis of mercuric reductase (*merA*) gene diversity in an anaerobic mercury-contaminated sediment enrichment," *Environmental Microbiology*, vol. 8, no. 10, pp. 1746–1752, 2006.

[23] A. B. F. Pacheco, B. E. C. Guth, K. C. C. Soares, L. Nishimura, D. F. De Almeida, and L. C. S. Ferreira, "Random amplification of polymorphic DNA reveals serotype-specific clonal clusters among enterotoxigenic *Escherichia coli* strains isolated from humans," *Journal of Clinical Microbiology*, vol. 35, no. 6, pp. 1521–1525, 1997.

[24] G. Muyzer, E. C. De Waal, and A. G. Uitterlinden, "Profiling of complex microbial populations by denaturing gradient gel electrophoresis analysis of polymerase chain reaction-amplified genes coding for 16S rRNA," *Applied and Environmental Microbiology*, vol. 59, no. 3, pp. 695–700, 1993.

[25] R. A. I. Abou-Shanab, P. van Berkum, and J. S. Angle, "Heavy metal resistance and genotypic analysis of metal resistance genes in gram-positive and gram-negative bacteria present in Ni-rich serpentine soil and in the rhizosphere of *Alyssum murale*," *Chemosphere*, vol. 68, no. 2, pp. 360–367, 2007.

[26] A. Hassen, N. Saidi, M. Cherif, and A. Boudabous, "Resistance of environmental bacteria to heavy metals," *Bioresource Technology*, vol. 64, no. 1, pp. 7–15, 1998.

[27] M. Narita, K. Chiba, H. Nishizawa et al., "Diversity of mercury resistance determinants among *Bacillus* strains isolated from sediment of Minamata Bay," *FEMS Microbiology Letters*, vol. 223, no. 1, pp. 73–82, 2003.

[28] M. Zeyaullah, B. Islam, and A. Ali, "Isolation, identification and PCR amplification of *merA* gene from highly mercury polluted Yamuna river," *African Journal of Biotechnology*, vol. 9, no. 24, pp. 3510–3514, 2010.

[29] A. O. Summers, "Organization, expression, and evolution of genes for mercury resistance," *Annual Review of Microbiology*, vol. 40, pp. 607–634, 1986.

[30] F. Matyar, T. Akkan, Y. Uçak, and B. Eraslan, "Aeromonas and Pseudomonas: antibiotic and heavy metal resistance species from Iskenderun Bay, Turkey (northeast Mediterranean Sea)," *Environmental Monitoring and Assessment*, vol. 167, no. 1–4, pp. 309–320, 2010.

[31] M. C. Hart, G. N. Elliott, A. M. Osborn, D. A. Ritchie, and P. Strike, "Diversity amongst *Bacillus merA* genes amplified from mercury resistant isolates and directly from mercury polluted soil," *FEMS Microbiology Ecology*, vol. 27, no. 1, pp. 73–84, 1998.

[32] J.-B. Ramond, T. Berthe, R. Duran, and F. Petit, "Comparative effects of mercury contamination and wastewater effluent input on Gram-negative *merA* gene abundance in mudflats of an anthropized estuary (Seine, France): a microcosm approach," *Research in Microbiology*, vol. 160, no. 1, pp. 10–18, 2009.

[33] A. M. Osborn, K. D. Bruce, P. Strike, and D. A. Ritchie, "Polymerase chain reaction-restriction fragment length polymorphism analysis shows divergence among mer determinants from gram-negative soil bacteria indistinguishable by DNA-DNA hybridization," *Applied and Environmental Microbiology*, vol. 59, no. 12, pp. 4024–4030, 1993.

[34] K. D. Bruce, "Analysis of mer gene subclasses within bacterial communities in soils and sediments resolved by fluorescent-PCR-restriction fragment length polymorphism profiling," *Applied and Environmental Microbiology*, vol. 63, no. 12, pp. 4914–4919, 1997.

[35] V. B. Mathema, B. C. Thakuri, and M. Sillanpää, "Bacterial *mer* operon-mediated detoxification of mercurial compounds: a short review," *Archives of Microbiology*, vol. 193, no. 12, pp. 837–844, 2011.

[36] A. H. Regua-Mangia, T. A. T. Gomes, M. A. M. Vieira, K. Irino, and L. M. Teixeira, "Molecular typing and virulence of enteroaggregative *Escherichia coli* strains isolated from children with and without diarrhoea in Rio de Janeiro city, Brazil," *Journal of Medical Microbiology*, vol. 58, no. 4, pp. 414–422, 2009.

[37] U. Dobrindt, "(Patho-)genomics of *Escherichia coli*," *International Journal of Medical Microbiology*, vol. 295, no. 6-7, pp. 357–371, 2005.

[38] A. H. Regua-Mangia, B. C. Guth, J. R. Da Costa Andrade et al., "Genotypic and phenotypic characterization of enterotoxigenic *Escherichia coli* (ETEC) strains isolated in Rio de Janeiro city, Brazil," *FEMS Immunology and Medical Microbiology*, vol. 40, no. 2, pp. 155–162, 2004.

[39] A. H. Regua-Mangia, T. A. Tardelli Gomes, J. R. Costa Andrade et al., "Genetic analysis of *Escherichia coli* strains carrying enteropathogenic *Escherichia coli* (EPEC) markers, isolated from children in Rio de Janeiro City, Brazil," *Brazilian Journal of Microbiology*, vol. 34, no. 1, pp. 38–41, 2003.

[40] M. A. Schmidt, "LEEways: tales of EPEC, ATEC and EHEC," *Cellular Microbiology*, vol. 12, no. 11, pp. 1544–1552, 2010.

[41] S. Ishii and M. J. Sadowsky, "*Escherichia coli* in the environment: implications for water quality and human health," *Microbes and Environments*, vol. 23, no. 2, pp. 101–108, 2008.

[42] J. L. Sanz and T. Köchling, "Molecular biology techniques used in wastewater treatment: an overview," *Process Biochemistry*, vol. 42, no. 2, pp. 119–133, 2007.

5

Fish Diversity and Abundance of Lake Tanganyika: Comparison between Protected Area (Mahale Mountains National Park) and Unprotected Areas

Emmanuel A. Sweke,[1,2] **Julius M. Assam,**[1] **Takashi Matsuishi,**[3] **and Abdillahi I. Chande**[1,4]

[1] *Tanzania Fisheries Research Institute, P.O. Box 90, Kigoma, Tanzania*
[2] *Graduate School of Fisheries Sciences, Hokkaido University, 3-1-1 Minato-cho, Hakodate 041-8611, Japan*
[3] *Faculty of Fisheries Sciences, Hokkaido University, 3-1-1 Minato-cho, Hakodate 041-8611, Japan*
[4] *Marine Parks and Reserves, P.O. Box 7565, Dar es Salaam, Tanzania*

Correspondence should be addressed to Emmanuel A. Sweke; esweke@yahoo.com

Academic Editor: Marco Milazzo

High biodiversity is the most remarkable characteristic of Lake Tanganyika including vertebrates, invertebrates, and plants. A few protected areas have been created along the lake to protect its biodiversity. However, limited studies have been carried out to ascertain their effectiveness. The current study aimed at assessing and comparing fish diversity and abundance of Lake Tanganyika in a protected area (Mahale Mountains National Park (MMNP)) and unprotected areas surrounding it. The data were collected in the near shore zone at 5 m and 10 m depths using stationary visual census (SVC) technique. The protected area recorded higher fish richness and abundance than unprotected areas ($P < 0.05$). It was concluded that the protected area is effective in conserving the fish diversity and abundance of the lake. However, more studies should be carried out regularly to explore the efficacy of the protected area in conservation of aquatic biodiversity and abundance.

1. Introduction

High biodiversity is the most remarkable characteristic of Lake Tanganyika including vertebrates, invertebrates, and plants [1–3]. It contains more than 1300 species of fish, invertebrates, and plants among which 500 species do not exist anywhere else on earth (endemic species) making it an important contributor to global biodiversity [4, 5]. The lake has received relatively less human impact than many other African lakes [6] and could serve as an example of managing lakes in other developing countries. The complex ecosystem of the lake in terms of number of species as well as their complex interactions is without any doubt unique in the world [5].

The fishery of Lake Tanganyika is of great importance to the surrounding region where protein is scarce [1, 7]. Fishing is the greatest simple economic activity depended upon by the communities surrounding the lake and it is perceived to be one of the greatest threats to biodiversity. The removal of large quantities of fish (app. 130,000 to 170,000 tons of fish^{-year}) might be expected to have a direct impact on the biodiversity of the fish (and other aquatic organisms) in the lake [8]. The productivity of an ecosystem promotes its quality whereby living organisms are manufactured through interactions of community and environment. Standing crop, rate of removal of resources, and rate of production are the measures of the quality of an ecosystem, [9, 10] recommended strengthening of prohibition of fishing activities within the park to enhance biodiversity and biomass within park boundaries.

The reputation for high biodiversity in Lake Tanganyika is best demonstrated by fish of the family cichlidae and the mollusks. Both of these groups have a high number of species with a substantial proportion of endemic species [1] and a considerable genetic variability within species.

The lake, however, is vulnerable to pollution [11–13] and there are currently few efforts being made to conserve

Fish Diversity and Abundance of Lake Tanganyika: Comparison between Protected Area
(Mahale Mountains National Park) and Unprotected Areas

35

its biodiversity. The most immediate threats to the lake's unique environment and biota are pollution from various sources [14] and intensive fishing with illegal methods. These problems and their effects are increasing and immediate attention is required to assess and control these problems and conserve the biodiversity. One of the methods of conserving biodiversity of the lake is to control human activities by establishing protected areas [10, 15, 16]. Mahale Mountains National Park (MMNP) is one the of no-take zones in the lake. Establishment of protected areas is a proactive measure of mitigation, ensuring that some essentially unmodified sites exist within the lake for buffering against uncertainty [17, 18].

Paley et al. [19] reported that 53% of all the species known to inhabit Lake Tanganyika are found in MMNP. However, there are limited studies on fish abundance and diversity carried out in the region [19, 20]. Such studies need to be carried out on a spatial and temporal scale to determine the change in diversity and abundance. In addition, they can assist to ascertain the degree of anthropogenic influence and pollution in such critical ecosystems. The aim of this study was to assess and compare spatial pattern of fish abundance and biodiversity within and outside MMNP in order to verify effectiveness of the protected area and provide reference information as a baseline for such a critical habitat.

2. Materials and Methods

The study was conducted in May to June 2008 and the sampling sites were located within Mahale Mountains National Park (Figure 1) and areas outside the park. The MMNP is located at the southern edge of Kigoma (North North West to South South East), with an elevation ranging from 2,000 to 2,400 m [21]. The MMNP was put in place in 1985, and it covers an area of 1,613 km^2 of which 96 km^2 is aquatic covering a strip of water along the shore of Lake Tanganyika and extends 1.6 km into the lake [22] with many bays and few small rocky outcrops [21]. Of all the areas along Lake Tanganyika, Mahale is one of the richest in the topographical variation [21]. The park is one of the four national parks or natural reserves bordering the lake; the others are Rusizi River Nature Reserve (Burundi), Gombe River National Park (Tanzania), and Nsumbu National Park (Zambia) [15].

The sampling design of the study was categorized into three factors: (1) status of the area with two levels (protected and unprotected); (2) habitat with two fixed and orthogonal levels (rocky and sandy), and (3) sampling site. A total of twelve sampling sites (8 sites within the protected area, that is, MMNP and 4 sites in unprotected areas) were selected. Of these twelve sites six (4 within MMNP and 2 in unprotected areas) were sandy and the other six (4 within MMNP and 2 in unprotected areas) were rocky habitats. The four outermost sites (i.e., the first two and last two) were located in unprotected areas (Buhingu and Sibwesa villages) and the rest were within MMNP. The sampling sites were selected at least 5 km apart. In the survey, we noted that rocky and sandy habitats were not equally interspersed within and outside the protected areas. The geographical positioning system (GPS) locations of the sites were recorded (Table 1).

Environmental parameters, namely, dissolved oxygen (DO), temperature, and transparency, were also recorded at 5 m and 10 m depths (Table 1). Water transparency was measured using a Secchi disc while DO and temperature were recorded using a multiparameter analyser WTW 340i [23].

Fish abundance and diversity data were collected in the near shore zone using stationary visual census (SVC) technique which involved SCUBA diving. A pair of divers conducted censuses of fish population within a quadrant of 8.5 m by 5 m which was laid on a lakebed. The divers first sampled at the deeper point (10 m) and then moved towards the shore to 5 m water depth. At each depth, two points located 10 m apart were sampled. The species present in the column of water were identified and individuals of each species were counted and recorded on slates. It took about 45–50 minutes for divers to count and record fish in the quadrant. Fish were identified to the species level as per Konings [24]. The data obtained were used to calculate fish diversity by the Shannon-Wiener diversity index.

Data were recorded in Microsoft excel packages and further analyzed in STATISTICA (version 8, Inc., 2010). The mean values of water transparency, dissolved oxygen concentration, and temperature were tested independently with one-way ANOVA for each parameter at a significant level ($P = 0.05$). In addition, species diversity and richness were derived by using Shannon-Weiner diversity index (H') formula as shown below:

$$H' = -\sum_{i=1}^{S} P_i \ln P_i, \qquad (1)$$

where S is number of species in the sample, and P_i is relative importance values obtained as the squared ration of the importance values of S individual value for all species to N the total importance.

3. Results

There was no significant variation ($P > 0.05$) in water parameters tests between sampling sites in protected area and unprotected areas; water transparency (ANOVA, $F_{1,22} = 0.12$, $P = 0.73$), DO (ANOVA, $F_{1,22} = 3.67$, $P = 0.07$), and temperature (ANOVA, $F_{1,22} = 0.16$, $P = 0.69$) (Table 1). In MMNP, five sampling sites (63%) had relatively flat beds, two (25%) had inclined beds, and one site (12%) showed a mixture of flat and inclined bed. However, as one moves from the shore towards the open waters, there was a sharp increase in depth. Purely flat rocky bottom was not encountered in the survey. Riverine was another type of habitat found in the park although survey could not be conducted due to crocodiles' menace.

There was a large degree of overlap between fish species found in protected area (inside MMNP) and unprotected areas (outside MMNP) (Table 2). A total of 70 and 55 fish species were recorded inside MMNP and outside MMNP, respectively.

Generally, dominant fish species (with their percentage of individuals on the total individuals counted in the survey in parentheses) included *Lepidiolamprologus*

TABLE 1: Environmental parameters mean (±SD) of the sampled sites in Lake Tanganyika during May to June 2008 survey. Sites S1, S2, S11, and S12 are within unprotected areas and sites S3–S10 are within the protected area (Mahale Mountains National Park).

Site ID	Habitat	GPS location	Water transparency (m)	Water DO (mg/l)		Water temperature (°C)	
				5 m depth	10 m depth	5 m depth	10 m depth
S1	Sandy	S 05°58.408′ E 029°50.391′	9.00 ± 0.15	7.35 ± 1.90	7.75 ± 1.10	27.00 ± 0.00	26.80 ± 0.20
S2	Rocky	S 06°00.537′ E 029°45.509′	9.00 ± 0.24	6.30 ± 1.80	6.50 ± 1.80	26.85 ± 0.10	26.95 ± 0.10
S3	Sandy	S 06°02.393′ E 029°43.889′	10.00 ± 0.30	7.25 ± 1.70	6.70 ± 2.40	26.85 ± 0.30	26.90 ± 0.40
S4	Rocky	S 06°05.262′ E 029°43.051′	10.00 ± 0.21	7.80 ± 0.40	6.95 ± 0.90	27.35 ± 0.30	27.10 ± 0.80
S5	Sandy	S 06°08.052′ E 029°42.839′	10.00 ± 0.16	7.25 ± 0.70	6.70 ± 0.80	26.90 ± 0.40	26.85 ± 0.30
S6	Rocky	S 06°10.461′ E 029°43.699′	10.00 ± 0.13	6.50 ± 2.00	6.45 ± 1.90	27.15 ± 0.70	26.80 ± 0.40
S7	Sandy	S 06°13.950′ E 029°42.915′	10.00 ± 0.11	7.35 ± 0.10	6.95 ± 0.30	27.00 ± 0.40	26.95 ± 0.50
S8	Rocky	S 06°17.157′ E 029°45.323′	10.00 ± 0.12	7.45 ± 0.70	7.00 ± 1.20	25.55 ± 0.50	25.40 ± 0.40
S9	Sandy	S 06°19.733′ E 029°46.983′	10.00 ± 0.14	8.05 ± 0.10	7.50 ± 1.40	26.60 ± 0.00	26.60 ± 0.20
S10	Rocky	S 06°23.340′ E 029°49.702′	10.00 ± 0.20	6.60 ± 1.00	6.60 ± 1.20	26.40 ± 0.20	26.30 ± 0.00
S11	Rocky	S 06°25.183′ E 029°51.359′	10.00 ± 0.18	7.25 ± 0.30	6.70 ± 0.80	26.00 ± 0.00	26.00 ± 0.00
S12	Sandy	S 06°27.393′ E 029°53.934′	10.00 ± 0.30	7.40 ± 0.60	6.35 ± 1.90	26.25 ± 0.10	26.25 ± 0.10

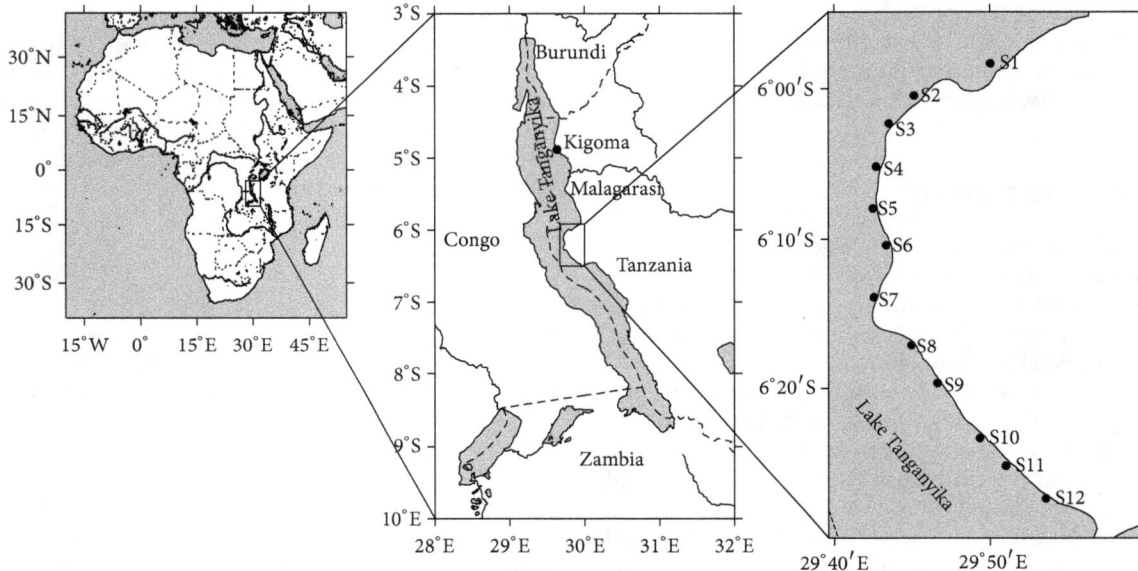

FIGURE 1: Map showing study sites (labeled S1–S12) along the shore of Lake Tanganyika during May to June 2008 survey. S1–S2 and S11–S12 represent study sites in unprotected areas and S3–S10 are study sites inside the protected area (Mahale Mountains National Park).

attenuates (5.47%), Neolamprologus brichardi (4.78%), Ophthalmotilapia nasuta (4.22%), Lepidiolamprologus elongatus (3.94%), Neolamprologus splendens (3.88%), Cyathopharynx foai (3.32%), Neolamprologus savoryi (3.18%), Plecodus paradoxus (2.77%), Xenotilapia ochrogenys (2.70%), and Ophthalmotilapia ventralis (2.63%). Species with lowest species composition (0.07%) included Telmatochromis brichardi, Lamprologus leleupi, Mastacembelus platysoma, Bathybates ferox, Limnotilapia dardennii, Variabilichromis moorii, Neolamprologus bifasciatus, Tanganicodus irsacae, Simochromis marginatus, Simochromis pleurospilus, and Neolamprologus modestus. Inside MMNP, Lepidiolamprologus attenuates and Neolamprologus brichardi were the most dominant species each contributing 6.48%, whereas Perissodus

microlepis, Simochromis diagramma, Eretmodus cyanostictus, Limnotilapia dardennii, Variabilichromis moorii, Neolamprologus bifasciatus, and Tanganicodus irsacae were infrequent species and each contributed 0.09% in composition (Table 2). While Ophthalmotilapia ventralis (10%) was the most dominant species in the areas outside MMNP, Plecodus paradoxus, Haplotaxodon microlepis, Petrochromis texas, Julidochromis regani, Plecodus microlepis, Simochromis diagramma, Eretmodus cyanostictus, Lamprologus leleupi, Mastacembelus platysoma, Bathybates ferox, Simochromis marginatus, Simochromis pleurospilus, and Neolamprologus modestus recorded lowest species composition of 0.29% each.

Species diversity was high in rocky habitats as compared to sandy habitats ($F = 16.71$, df = 537, $P = 0.001$) (Figure 2).

Fish Diversity and Abundance of Lake Tanganyika: Comparison between Protected Area
(Mahale Mountains National Park) and Unprotected Areas

37

TABLE 2: Species composition (%) and mean abundance (±SE) of fish species in rocky and sandy habitats within protected area and unprotected areas in Lake Tanganyika during May to June 2008 survey.

Fish species	Species composition (%)				Mean abundance ± SE			
	Protected area		Unprotected areas		Protected area		Unprotected areas	
	Rocky	Sandy	Rocky	Sandy	Rocky	Sandy	Rocky	Sandy
Neolamprologus brichardi	8.83	0	0	0	17.25 ± 12.75	0	0	0
Neolamprologus splendens	7.17	0	0	0	14.00 ± 14.00	0	0	0
Neolamprologus savoryi	5.25	0	1.75	0	10.25 ± 3.68	0	2.50 ± 0.50	0
Lepidiolamprologus elongatus	4.99	2.46	3.86	0	9.75 ± 1.32	1.75 ± 0.63	5.50 ± 2.50	0
Lepidiolamprologus attenuatus	4.87	10.92	3.51	0	9.50 ± 1.85	7.75 ± 5.45	5.00 ± 1.00	0
Ophthalmotilapia nasuta	4.87	0	8.07	0	9.50 ± 4.91	0	11.50 ± 7.50	0
Paracyprichromis nigripinnis	3.84	0	0	0	7.50 ± 4.41	0	0	0
Telmatochromis temporalis	3.46	0	1.75	0	6.75 ± 0.85	0	2.50 ± 2.50	0
Plecodus paradoxus	3.33	4.58	0.35	0	6.50 ± 1.26	3.25 ± 1.80	0.50 ± 0.50	0
Cyathopharynx foai	3.33	0	7.72	0	6.50 ± 1.76	0	11.0 ± 7.00	0
Petrochromis Moshi	3.33	0	2.81	0	6.50 ± 1.85	0	4.00 ± 1.00	0
Lamprologus callipterus	2.82	1.76	2.81	0	5.50 ± 4.01	1.25 ± 0.75	4.00 ± 2.00	0
Lamprologus lemairii	2.82	0	0	0	5.50 ± 3.18	0	0	0
Neolamprologus caudopunctatus	2.82	0	0	0	5.50 ± 2.47	0	0	0
Petrochromis orthognathus	2.43	0	5.26	0	4.75 ± 0.25	0	7.50 ± 4.50	0
Telmatochromis vittatus	2.30	0	1.75	13.68	4.50 ± 1.85	0	2.50 ± 0.50	6.50 ± 6.50
Cyprichromis leptosoma	2.30	0	0	0	4.50 ± 4.50	0	0	0
Neolamprologus mustax	2.18	2.46	2.11	0	4.25 ± 2.66	1.75 ± 1.44	3.00 ± 3.00	0
Lobochilotes labiatus	2.18	0	1.40	0	4.25 ± 1.70	0	2.00 ± 1.00	0
Xenotilapia spiloptera	2.18	0	2.11	0	4.25 ± 1.44	0	3.00 ± 3.00	0
Tropheus annectens	1.92	0	1.05	0	3.75 ± 2.25	0	1.50 ± 1.50	0
Xenotilapia spilopterus	1.79	0	4.56	0	3.50 ± 3.50	0	6.50 ± 6.50	0
Petrochromis famula	1.79	0	0	0	3.50 ± 1.32	0	0	0
Microdontochromis tenuidentatus	1.66	0	0	0	3.25 ± 3.25	0	0	0
Xenotilapia sima	1.41	1.41	0	0	2.75 ± 1.89	0	0	0
Tropheus brichardi	1.28	0	1.75	0	2.50 ± 0.87	0	2.50 ± 2.50	0
Barbus sp.	1.15	0	0	0	2.25 ± 0.95	0	0	0
Neolamprologus toae	0.90	0	1.75	0	1.75 ± 1.03	0	2.50 ± 2.50	0
Asprotilapia leptura	1.02	0	1.40	0	2.00 ± 1.08	0	2.00 ± 2.00	0
Altolamprologus compressiceps	0.90	0	0.70	1.05	1.75 ± 0.63	0	0	0.50 ± 0.50
Neolamprologus fasciatus	1.02	0	0	0	2.00 ± 0.82	0	0	0
Neolamprologus gracilis	0.90	0	0	0	2.00 ± 1.75	0	0	0
Plecodus straeleni	0.77	0.35	0.70	0	1.50 ± 0.87	0.25 ± 0.25	1.00 ± 0.00	0
Tropheus moorii	0.77	0	0	0	1.50 ± 1.19	0	0	0
Petrochromis texas	0.64	0	0.35	0	1.25 ± 0.75	0	0.50 ± 0.50	0
Neolamprologus tretocephalus	0.64	0	0	0	1.25 ± 0.48	0	0	0
Lamprologus toae	0.64	0	0	0	1.25 ± 1.25	0	0	0

TABLE 2: Continued.

Fish species	Species composition (%)				Mean abundance ± SE			
	Protected area		Unprotected areas		Protected area		Unprotected areas	
	Rocky	Sandy	Rocky	Sandy	Rocky	Sandy	Rocky	Sandy
Chalinochromis ndobhoi	0.51	0	1.05	0	1.00 ± 1.00	0	1.50 ± 0.50	0
Telmatochromis brichardi	0.51	0	0	0	1.00 ± 1.00	0	0	0
Neolamprologus nigriventris	0.51	0	0	0	1.00 ± 0.41	0	0	0
Tropheus duboisi	0.51	0	0	0	1.00 ± 1.00	0	0	0
Lates angustifrons	0.51	0	0	0	1.00 ± 1.00	0	0	0
Juridochromis regani	0.38	0	0.35	0	0.75 ± 0.48	0	0.50 ± 0.50	0
Limnotilapia dardennii	0.38	0	0	0	0.75 ± 0.48	0	0	0
Chalinochromis brichardi	0.38	0	0	0	0.75 ± 0.48	0	0	0
Haplotaxodon microlepis	0.26	1.76	0.35	0	0.50 ± 0.50	1.25 ± 1.25	0.50 ± 0.50	0
Gnathochromis pfefferi	0.26	0	1.05	0	0.50 ± 0.50	0	1.50 ± 0.50	0
Neolamprologus foai	0.26	0	2.81	0	0.50 ± 0.50	0	4.00 ± 4.00	0
Plecodus microlepis	0.26	0	0.35	0	0.50 ± 0.50	0	0.50 ± 0.50	0
Perissodus microlepis	0.13	0	2.81	0	0.25 ± 0.25	0	4.00 ± 4.00	0
Simochromis diagramma	0.13	0	0.35	0	0.25 ± 0.25	0	0.50 ± 0.50	0
Eretmodus cyanostictus	0.13	0	0.35	0	0.25 ± 0.25	0	0.50 ± 0.50	0
Variabilichromis moorii	0.13	0	0	0	0.25 ± 0.25	0	0	0
Neolamprologus bifasciatus	0.13	0	0	0	0.25 ± 0.25	0	0	0
Tanganicodus irsacae	0.13	0	0	0	0.25 ± 0.25	0	0	0
Xenotilapia ochrogenys	0	13.73	0	0	0	9.75 ± 7.88	0	0
Enantiopus melanogenys	0	11.97	0	0	0	8.50 ± 6.93	0	0
Grammatoria lemairii	0	10.21	0	0	0	7.25 ± 3.97	0	0
Xenotilapia bathyphilus	0	9.86	0	0	0	7.00 ± 7.00	0	0
Lepidiolamprologus cunningtoni	0	6.34	0	10.53	0	4.50 ± 2.25	0	5.00 ± 3.00
Neolamprologus tetracanthus	0	5.28	0	17.89	0	3.75 ± 2.59	0	8.50 ± 8.50
Lepidiolamprologus boulengeri	0	3.52	0	24.21	0	2.50 ± 2.50	0	11.5 ± 11.5
Neolamprologus callipterus	0	3.52	0	0	0	2.50 ± 2.50	0	0
Ectodus descampsii	0	3.17	7.72	8.42	0	2.25 ± 2.25	11.00 ± 11.00	0
Neolamprologus kungweensis	0	2.11	0	0	0	1.50 ± 1.19	0	4.00 ± 4.00
Lamprologus microlepis	0	1.76	0	0	0	1.25 ± 1.25	0	0
Boulengerochromis microlepis	0	1.06	0	0	0	0.75 ± 0.48	0	0
Telmatochromis dhonti	0	0.70	0	0	0	0.50 ± 0.29	0	0
Plecodus mustax	0	0.70	0	0	0	0.50 ± 0.50	0	0
Limnotilapia dardennii	0	0.35	0	0	0	0.25 ± 0.25	0	0
Ophthalmotilapia ventralis	0	0	13.33	0	0	0	19.00 ± 19.00	0
Lamprichthys tanganicanus	0	0	4.56	0	0	0	6.50 ± 6.50	0
Lepidiolamprologus lemairii	0	0	1.05	0	0	0	1.50 ± 1.50	0
Aulonocranus dewindti	0	0	1.05	0	0	0	1.50 ± 1.50	0
Neolamprologus mondabu	0	0	0.70	0	0	0	1.00 ± 1.00	0

Fish Diversity and Abundance of Lake Tanganyika: Comparison between Protected Area (Mahale Mountains National Park) and Unprotected Areas

39

TABLE 2: Continued.

Fish species	Species composition (%)				Mean abundance ± SE			
	Protected area		Unprotected areas		Protected area		Unprotected areas	
	Rocky	Sandy	Rocky	Sandy	Rocky	Sandy	Rocky	Sandy
Tropheus moorii	0	0	0.70	0	0	0	1.00 ± 1.00	0
Neolamprologus falcicula	0	0	0.70	0	0	0	1.00 ± 1.00	0
Neolamprologus furcifer	0	0	0.70	0	0	0	1.00 ± 0.00	0
Telmatochromis brichardi	0	0	0.35	0	0	0	0.50 ± 0.50	0
Lamprologus leleupi	0	0	0.35	0	0	0	0.50 ± 0.50	0
Mastacembelus platisoma	0	0	0.35	0	0	0	0.50 ± 0.50	0
Bathybates ferox	0	0	0.35	0	0	0	0.50 ± 0.50	0
Simochromis marginatus	0	0	0.35	0	0	0	0.50 ± 0.50	0
Simochromis pleurospilus	0	0	0.35	0	0	0	0.50 ± 0.50	0
Neolamprologus modestus	0	0	0.35	0	0	0	0.50 ± 0.50	0
Neolamprologus brevis	0	0	0	10.53	0	0	0	5.00 ± 5.00
Neolamprologus boulengeri	0	0	0	9.47	0	0	0	4.50 ± 4.50
Neolamprologus similis	0	0	0	2.11	0	0	0	1.00 ± 1.00
Telmatochromis bifrenatus	0	0	0	2.11	0	0	0	1.00 ± 1.00

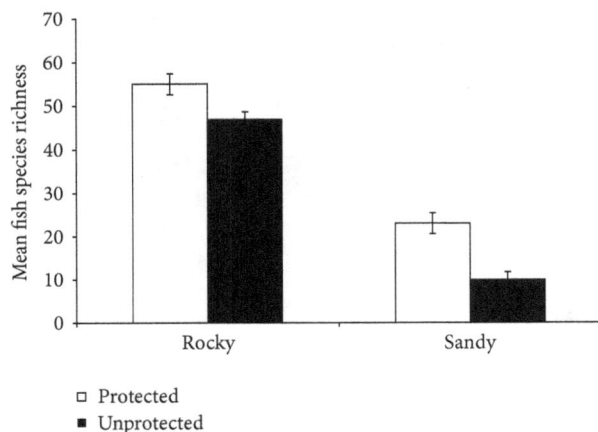

FIGURE 2: Mean fish species richness (±SE) for rocky and sandy habitats in protected area and unprotected areas in Lake Tanganyika during May to June 2008 survey.

Cichlids dominated in both areas. A small number of species ($n = 4$) were recorded in the outermost sampling site of the unprotected area in Sibwesa. Of the sites inside the park, site 5 recorded few species diversity ($n = 7$). Generally, percentage of individuals of fish species on the total individuals counted in a sampling site was higher in sites with low species richness than sites with high species richness. For instance in sampling site S12, *X. ochrogenys* and *N. tetracanthus* dominated by 50% and 47%, respectively.

The area inside the park had higher fish biodiversity than areas in the unprotected areas (outside the park), ($P < 0.05$) (Table 3). Comparison of rocky and sandy habitats showed that the former had high mean fish abundance per stationary visual census and $1\,m^2$ in both protected and unprotected areas (Table 3).

Shannon-Wiener diversity index was used to describe the fish species diversity in rocky and sandy habitats within the protected area and unprotected areas (Table 3).

Although the two divers recorded many fish species in the protected area as compared to unprotected areas, there was no significant difference ($P > 0.05$) in fish diversity between the areas. Nonetheless, pooling all data from both protected and unprotected areas showed that fish species diversity was significantly difference ($P < 0.01$) (Table 4). In addition, a significant variation ($P < 0.001$) in species diversity between rocky and sandy habitats was evident in protected area, and vice versa for unprotected areas (Table 4).

4. Discussion

Environmental parameters, namely, dissolved oxygen, water temperature, and transparency, seemed to have no significant difference between areas inside and outside the park hence their influence on diversity and abundance of both areas was negligible. This could ascertain the fact that the variation in the fish abundance and diversity did not depend on the water quality parameters, but probably due to mainly other factors

such as the management mechanism and the location of the sites.

Stationary visual census (SVC) employed in this study was noted to be size nondependent, unlike gillnet which relies on the mesh sizes and thereby affects the size of fish sampled; this method depends only on the visual capabilities of divers. In addition, the estimated sample of fish does not revolve around the captured and the noncaptured samples, but rather based upon the random throw of transects which appear to be representative of the study location. It was pointed out by Watson and Harvey [25] that SCUBA diver is likely to obtain accurate measures of species richness. However, the major setbacks of this method involved the look-alike of cichlids, mobility of fish [25, 26] due to disturbance, and the schooling phenomenon of fish which was minimized by throwing the transects randomly at least 5 m apart by the thoroughly trained experts on cichlids taxonomy. The avoidance of such sampling errors seemed to make the method robust.

Generally, rocky habitats recorded high fish species diversity than the sandy habitats (Figure 2 and Table 3). For example the Shannon-Weiner diversity index for sampling sites outside the park indicated that fish diversity in rocky habitats was twice of that of sandy habitats. This could have been attributed to the favorable environment on the rocks for the growth of algae [27] which is the food of many herbivorous fish species. The same finding was reported by Paley et al. [19]. This implied that rocky areas had good biodiversity complexity hence attracted and harbored various aquatic organisms including fish [8]. Another reason could probably be the suitability of the rocks to provide hiding sites for cichlids to avoid predators. The hiding sites are mainly crevices which form microhabitats that are used as nests for breeding purposes. The fish species that use rocky habitats for breeding include many *Neolamprologus* species [24] which were among the dominant species in the park. Therefore, variations in fish diversity among the habitats could be a result of differences in food availability and habitat preference of fish. On the other hand, sandy habitat defined as a predominantly sandy bottom with less than one tenth of the area covered with rocks is known to support relatively few cichlids [24]. Only those species capable of forming schools venture out over bare sandy bottoms. This could probably be the reason for school foraging species were infrequently recorded in the survey.

Protected area indicated high mean species richness than unprotected areas. Surprisingly, the protected area failed to show significant difference in fish species diversity under the Shannon-Weiner. This might be due to some reasons including relatively few species recorded in both protected and unprotected areas during the survey; the fact that protected areas do not present equilibrium points [28] hence other factors apart from human predation such as physico-chemical properties of water can affect the biodiversity. In addition, nonvulnerable species could be negatively affected by the protection effect through their ecological relationship with other fishes, such as competition or predation [29]. Species richness and individuals within a species declined as one moves out of the park (Figure 2). This might have been contributed by intensive fishing [30] and environmental

Fish Diversity and Abundance of Lake Tanganyika: Comparison between Protected Area
(Mahale Mountains National Park) and Unprotected Areas

41

TABLE 3: Fish mean abundance (individual m^{-2}± SE) and Shannon-Wiener diversity index (H$'$) for protected area and unprotected areas in Lake Tanganyika during May to June 2008 survey.

	Protected area			Unprotected areas		
	Rocky	Sandy	Total	Rocky	Sandy	Total
Mean abundance (ind. m^{-2})	4.59 ± 2.44	1.67 ± 1.28	3.13 + 2.01	3.35 ± 2.60	1.13 ± 1.08	2.24 ± 1.98
H$'$	2.99	1.96	2.47	2.85	1.31	2.08

TABLE 4: One-way ANOVA tests for species diversity in protected area and unprotected areas in Lake Tanganyika during May to June 2008 survey.

Test value	Protected area		Unprotected areas		Protected and unprotected		Site S1–S12
	Rocky	Sandy	Rocky	Sandy	Rocky	Sandy	
F	1.4362	1.4362	1.0775	0.9817	1.8534	1.4848	3.1271
P	0.0002***	0.0151*	0.3631	0.5342	2.83E^{-5}***	0.0056**	0.0004***
df	355	355	177	177	533	533	1067

*$P < 0.05$, **$P < 0.01$, ***$P < 0.001$.

degradation [31] such as poor agricultural practice [14] in the unprotected areas which might have led to siltation consequently destruction of the breeding grounds of fish. Tierney et al. [32] pointed out that decline in primary production is likely to further impact particularly the clupeid fishery of the lake. During the survey, beach seine nets were witnessed being illegally operated along the shore beaches in Sibwesa (sampling sites S11 and S12). Furthermore, poor fishery management attributed to insufficient fisheries managers and limited patrols by the surveillance and control unit situated some distance from these areas could be the reasons for these illicit practices. High abundance of delicious and expensive fish species like *B. microlepis* (kuhe) sometimes referred to as "the lake's chicken" and *L. dardennii* (kungura) within the park waters was believed to be "a calling factor" for poaching practices in the park by the villagers surrounding the park. Many of such cases had been reported (pers. comm.).

Fish abundance was unevenly distributed between species in habitats, inside MMNP, and outside MMNP. All the dominant fish species (Table 2) reported to be present in the park by Paley et al. [19] were recorded in the current study except *X. flavipinnis* which was recorded neither inside MMNP nor outside MMNP. Almost all the dominant species are endemic to Lake Tanganyika [33]. Four of the current most dominant species, namely, *L. attenuates*, *N. brichardi*, *L. elongates*, and *E. melanogenys*, were recorded in the top ten dominant species of the report in the same sampling technique, that is, SVC whereas another two species, *P. paradoxus* and *X. ochrogenys* were recorded in gillnets. That is to say, three species of the current top 10 dominant species are reported for the first time by this study. Furthermore, 5 of the 10 top dominant species inside MMNP were not recorded outside MMNP, the rest were recorded in relatively low numbers. In contrast, only 2 of the top 10 abundant species outside MMNP were not recorded in areas within MMNP. Therefore, fish species dominated inside MMNP were not dominant outside MMNP and the vice-versa except for *O. nasuta* which was the second dominant species in areas outside MMNP and ranked eighth

inside MMNP. Interestingly, the most dominant fish species *O. ventralis* outside MMNP was not recorded inside MMNP. It was recorded in high numbers probably because it was in aggregation [34] or low abundance of its predator, *Plecodus straeleni* (0.53%) [35]. Furthermore, species from some families such as Mastacembelidae and Bagridae which are highly endemic [19] were recorded in the park only justifying that the park serves as a "safety valve" in fisheries resource conservation.

High diversity of fish species indicated by the Shannon-Wiener diversity index and abundance in the park especially in rocky habitats means that the park continues to nurse and serve as a "gene pool" for many fish species of Lake Tanganyika as compared to its neighborhood unprotected areas.

5. Conclusion

Fish abundance and diversity in the protected area were higher than in unprotected areas. Restriction of any fishing activities within the protected area seemed to contribute much to availability of such good stock of fish. However, there were some variations in biodiversity within the park and habitats. There were higher species abundance and diversity in rocky habitats than in sandy habitats. The aquatic environment in unprotected areas seemed to be threatened by many factors such as siltation due to uncontrolled agricultural activities and illegal fishing methods that could have resulted into reduced fish species diversity. There is a need to create awareness on the importance of the MMNP aquatic component and its conservation to the riparian communities to avoid conflict of interest between the stakeholders. Because of limited studies on fish biodiversity and abundance in the region, the findings of the current study serve as baseline information for future references as an attempt to revive the fisheries resources in the area and the lake at large. Further regular and comprehensive studies are required to verify the efficacy of the protected area in conservation of aquatic biodiversity and propose sound management practices.

Conflict of Interests

The authors declare that there is no relation with the mentioned software enterprises (Microsoft and STATISTICA) which might lead to conflict of interests with the companies therein. In addition, they are trained researchers with the information herein meant merely for scientific rationale and therefore they account their output based on research and information dissemination.

Acknowledgments

This work was part of the Mahale Ecosystem Management Project (MEMP). The authors are greatly indebted to the Tanzania National Parks (TANAPA), European Union (UE), and Frankfurt Zoological Society (FZS) for funding the project. The authors owe much thanks to the Mahale Mountains National Park for the coherent assistance they offered during the whole exercise of data collection especially the logistics. They are grateful to the field team members, Athanasio Mbonde, Robert Wakafumbe, George Kazumbe, Abel Mtui, and Magnus Mosha, for their hard work and dedication in data collection which made this paper available. They extend their thanks to Christopher Mulanda Aura for advice and encouragement. Moreover, they thank the editor and two anonymous reviewers for their thorough views, numerous comments, and suggestions that improved the paper.

References

[1] H. Mölsä, J. E. Reynolds, E. J. Coenen, and O. V. Lindqvist, "Fisheries research towards resource management on Lake Tanganyika," *Hydrobiologia*, vol. 407, pp. 1–24, 1999.

[2] J. Snoeks, L. Ruber, and E. Verheyen, "The Tanganyika problem: comments on the taxonomy and distribution patterns of its cichlid fauna," *Archiv Für Hydrobiologie-Beiheft Ergebnisse Der Limnologie*, vol. 44, pp. 355–372, 1994.

[3] J. Snoeks, "How well known is the ichthyodiversity of the large East African lakes?" *Advances in Ecological Research*, vol. 31, pp. 17–38, 2000.

[4] L. C. Beadle, *The Inland Waters of Tropical Africa: An Introduction to Tropical Limnology*, Longman, London, UK, 1981.

[5] G. Hanek, E. J. Coenen, and P. Kotilainen, "Aerial frame survey of Lake Tanganyika fisheries," *FAO/ FINNIDA research for the management of the fisheries on Lake Tanganyika. GCP/RAF/ 271/FIN-TD/09 (En)*, 1993.

[6] S. E. Jorgensen, G. Ntakimazi, and S. Kayombo, Experience and lessons learned brief report, 2006, http://www.ilec.or.jp/eg/lbmi/pdf/22_Lake_Tanganyika_27February2006.pdf.

[7] E. J. Coenen, "Historical data report on the fisheries, fisheries statistics, fishing gears and water quality of Lake Tanganyika (Tanzania)," FAO/FINNIDA research for the management of the fisheries on Lake Tanganyika. GCP/RAF/271/FIN-TD/15 (En & Fr), 1994.

[8] G. Patterson and J. Makin, *The State of Biodiversity in Lake Tanganyika: A Literature Review*, Natural Resources Institute, Chatham, UK, 1998.

[9] K. R. George, *Ecology of Inland Waters and Estuaries*, Pittsburgh, Pennsylvania, Pa, USA, 1961.

[10] S. E. Lester, B. S. Halpern, K. Grorud-Colvert et al., "Biological effects within no-take marine reserves: a global synthesis," *Marine Ecology Progress Series*, vol. 384, pp. 33–46, 2009.

[11] A. S. Cohen, R. Bills, C. Z. Cocquyt, and A. G. Caljon, "The impact of sediment pollution on biodiversity in Lake Tanganyika," *Conservation Biology*, vol. 7, no. 3, pp. 667–677, 1993.

[12] E. B. Worthington and R. Lowe-McConnell, "African lakes reviewed: creation and destruction of biodiversity," *Environmental Conservation*, vol. 21, no. 3, pp. 199–213, 1994.

[13] H. H. Nkotagu, "Lake Tanganyika ecosystem management strategies," *Aquatic Ecosystem Health and Management*, vol. 11, no. 1, pp. 36–41, 2008.

[14] I. Donohue, R. W. Duck, and K. Irvine, "Land use, sediment loads and dispersal pathways from two catchments at the southern end of Lake Tanganyika, Africa: implications for lake management," *Environmental Geology*, vol. 44, no. 4, pp. 448–455, 2003.

[15] D. Pauly, V. Christensen, S. Guénette et al., "Towards sustainability in world fisheries," *Nature*, vol. 418, no. 6898, pp. 689–695, 2002.

[16] F. Vandeperre, R. M. Higgins, J. Sánchez-Meca et al., "Effects of no-take area size and age of marine protected areas on fisheries yields: a meta-analytical approach," *Fish and Fisheries*, vol. 12, pp. 412–426, 2011.

[17] G. W. Allison, S. D. Gaines, J. Lubchenco, and H. P. Possingham, "Ensuring persistence of marine reserves: catastrophes require adopting an insurance factor," *Ecological Applications*, vol. 13, no. 1, pp. S8–S24, 2003.

[18] F. R. Gell and C. M. Roberts, "Benefits beyond boundaries: the fishery effects of marine reserves," *Trends in Ecology and Evolution*, vol. 18, no. 9, pp. 448–455, 2003.

[19] R. Paley, G. Ntakimazi, N. Muderhwa et al., "Biodiversity special study (BIOSS) final report: Mahale National Park aquatic March/April 1999 survey," Pollution Control and other Measures to Protect Biodiversity in Lake Tanganyika (RAF/92/G32), 2000.

[20] P. H. Skelton and E. R. Swartz, "Walking the tightrope: trends in African freshwater systematic ichthyology," *Journal of Fish Biology*, vol. 79, pp. 1413–1435, 2011.

[21] Anonymous, *Study For the Proposed Mahale Mountains National Park*, Japan International Cooperation Agency, 1980.

[22] T. Nishida, *The Chimpanzees of the Mahale Mountains: Sexual and Life History Strategies*, University of Tokyo Press, Tokyo, Japan, 1990.

[23] R. G. Wetzel and G. E. Likens, *Limnological Analyses*, Spinger, New York, NY, USA, 2000.

[24] A. Konings, *Tanganyika Cichlids in Their Natural Habitat*, Cichlid Press, 1998.

[25] D. L. Watson and E. S. Harvey, "Behaviour of temperate and subtropical reef fishes towards a stationary SCUBA diver," *Marine and Freshwater Behaviour and Physiology*, vol. 40, no. 2, pp. 85–103, 2007.

[26] C. Ward-Paige, J. M. Flemming, and H. K. Lotze, "Overestimating fish counts by non-instantaneous visual censuses: consequences for population and community descriptions," *PLoS ONE*, vol. 5, no. 7, Article ID e11722, 9 pages, 2010.

[27] G. W. Coulter, *Lake Tanganyika and Its Life*, Oxford University Press, London, UK, 1991.

[28] E. Sala, M. Ribes, B. Hereu et al., "Temporal variability in abundance of the sea urchins Paracentrotus lividus and Arbacia lixula in the northwestern Mediterranean: comparison between

Fish Diversity and Abundance of Lake Tanganyika: Comparison between Protected Area
(Mahale Mountains National Park) and Unprotected Areas

43

a marine reserve and an unprotected area," *Marine Ecology Progress Series*, vol. 168, pp. 135–145, 1998.

[29] V. Dufour, J. Y. Jouvenel, and R. Galzin, "Study of a Mediterranean reef fish assemblage. Comparison of population distributions between depths in protected and unprotected areas over one decade," *Aquatic Living Resources*, vol. 8, no. 1, pp. 17–25, 1995.

[30] I. A. Kimirei, Y. D. Mgaya, and A. I. Chande, "Changes in species composition and abundance of commercially important pelagic fish species in Kigoma area, Lake Tanganyika, Tanzania," *Aquatic Ecosystem Health and Management*, vol. 11, no. 1, pp. 29–35, 2008.

[31] R. L. Welcomme, "An overview of global catch statistics for inland fish," *ICES Journal of Marine Science*, vol. 68, pp. 1751–1756, 2011.

[32] J. E. Tierney, M. T. Mayes, N. Meyer et al., "Late-twentieth-century warming in Lake Tanganyika unprecedented since AD 500," *Nature Geoscience*, vol. 3, no. 6, pp. 422–425, 2010.

[33] R. Froese and D. Pauly Editors, "FishBase," World Wide Web electronic publication, http://www.fishbase.org/, version, 2011.

[34] T. Kuwamura, "Parental care and mating systems of cichlid fishes in Lake Tanganyika: a preliminary field survey," *Journal of Ethology*, vol. 4, no. 2, pp. 129–146, 1986.

[35] M. Nshombo, "Foraging behavior of the scale-eater Plecodus straeleni (Cichlidae, Teleostei) in Lake Tanganyika, Africa," *Environmental Biology of Fishes*, vol. 39, no. 1, pp. 59–72, 1994.

6

Activity Budgets of Impala (*Aepyceros melampus*) in Closed Environments: The Mukuvisi Woodland Experience, Zimbabwe

Muposhi Victor Kurauwone,[1] **Muvengwi Justice,**[2] **Utete Beven,**[1] **Kupika Olga,**[1]
Chiutsi Simon,[3] **and Tarakini Tawanda**[1]

[1] *Department of Wildlife & Safari Management, Chinhoyi University of Technology, P. Bag 7724, Chinhoyi, Zimbabwe*
[2] *Department of Environmental Science, Bindura University of Science Education, P. Bag 1020, Bindura, Zimbabwe*
[3] *Department of Travel & Recreation Management, Chinhoyi University of Technology, P. Bag 7724, Chinhoyi, Zimbabwe*

Correspondence should be addressed to Muposhi Victor Kurauwone; vmuposhi@cut.ac.zw

Academic Editor: Masashi Sekino

Activity pattern plasticity in ungulates serves as an evolutionary adaptation to optimize fitness in inconsistent environments. Given that time is a limited and valuable resource for foraging wildlife species, provisioning and attraction may affect the activity pattern plasticity and reduce complexities of time partitioning for different activities by impala in closed environments. We assessed activity budgets of free-ranging impala social groups in a closed environment. Social group type had an influence on the activity budgets of impala except for foraging and moving activity states. Both the harem and bachelor groups spent more than 30% of their daily time foraging. Bachelor groups spent more time exhibiting vigilance tendencies than the harem groups. Season influenced the activity budgets of social groups other than vigilance and foraging activity states. Foraging time was highly correlated with vigilance, resting, and grooming. We concluded that provisioning and attraction may have reduced the influence of seasonality on the proportion of time spent on different activity states by impala social groups. There is a need to establish long-term socioecological, physiological, and reproductive consequences of provisioning and habituation on impala under closed environments.

1. Introduction

Impalas (*Aepyceros melampus melampus*, Lichtenstein, 1812) are regarded as the most common, widely distributed, and abundant medium-sized antelope species throughout southern and east Africa [1, 2]. Classified as intermediate feeders, impalas are adapted to browsing and grazing, thus making them successful inhabitants of the savanna ecosystems [3, 4]. Favoured for game farming as well as hunting, the subspecies has been widely introduced to privately owned land and game reserves in Zimbabwe, South Africa, and Namibia [1]. For that reason, impalas are extremely important to the game ranching and conservation sector of southern Africa.

In natural ecosystems, time is a valuable limited resource for all animals, and its partitioning might be influenced by sociality and as such may constrain sociality of free-ranging individuals [5]. Nakayama et al. [6] assert that the allocation of time for multiple activities has significant effects on the survival of wildlife species. Consequently, individuals adapt to environmental changes, such as food availability and temperature, by adjusting the amount of time spent in different behavioural activities [7]. The seasonality of activity budgets might be highly flexible in response to seasonal fluctuations in food supply and corresponding temperature [8]. However, the influence of seasonality on food quality and availability in some environments seems to be affected by the current trends of attraction, provisioning, and habituation of some species [9–12]. Consequently, we expect that provisioned individuals would ultimately spend less time searching for food and foraging during the dry season compared to those occurring in non-provisioned environments.

Attraction is the process of luring wild animals with food handouts to a strategic site, "feeding spot," to increase the likelihood of viewing the animals [13]. Closely related to attraction is the concept of provisioning which is an interaction where humans exploit the animal's appetites and desire for food to offset or neutralise their aversion to humans [11, 14]. Habituation is the waning of response to repeated, neutral stimuli such as human presence that ultimately render hitherto elusive animals susceptible to regular, proximate and protracted human viewing [11]. We argue that the level of provisioning and attraction for wildlife species in some systems may reduce seasonal variations in activity budgets of impala social groups. Observations made by Pays et al. [15] indicate that improving forage patch quality modifies the trade-offs between vigilance and foraging in favour of feeding. Animals invest time in the acquisition of information about forage resources within their environmental setting thus affecting the proportion of time allocated to other activities [16]. It is essential to know how impala social groups interact with their environment and invest energy as well as time for survival and reproduction by exploring their activity budgets.

Pollard and Blumstein [5] assert that time budgets can be divided into four mutually exclusive and exhaustive behavioural categories, namely, (1) subsistence (foraging or feeding), (2) locomotion (moving or traversing), (3) rest (inactivity), and (4) "other" which includes active social and nonsocial behaviours. However, other researchers [17–19] have used specific behavioural states (e.g., foraging, vigilance, resting, grooming, ruminating, moving, flight, excretion, mating, and social interaction among others) to infer the contribution of a set of certain treatments on wildlife species. We conducted an ethological study of free-ranging impala social groups in a closed environment, Mukuvisi Woodlands, an environmental education centre and ecotourism facility where attraction and provisioning are practised. We hypothesised that impala social groups at Mukuvisi exhibit different activity budgets according to seasons and that the activity budgets of bachelor and harem groups are different.

2. Materials and Methods

2.1. Study Area. The study was conducted in Mukuvisi Woodlands Wildlife & Environmental Centre (17° 50′ 10.39″ S and 31° 05′ 18.41″ E), located southeast of the city of Harare in Zimbabwe. The Centre is a 263-hectare woodland preserve home to a variety of Zimbabwe's indigenous flora and fauna including impala, zebra (*Equus burchellii*), giraffe (*Giraffa camelopardalis*), eland (*Taurotragus oryx*), and common duiker (*Sylvicapra grimmia*). Average rainfall ranges between 650–850 mm/annum, and mean annual temperatures is 9°C for winter and 40°C for summer [20]. The woodland are a typical Miombo and open savanna grassland. Due to the size of the preserve and the number of resident species, management interventions such as provision of dietary supplements were introduced.

2.2. Behaviour Definition. The activities of impala social groups were classified into nine categories based on other studies [17–19, 21] and personal observations. In this study, social interaction, mating behaviour, and nursing were combined (Table 1).

2.3. Behavioural Observations. Observations on the activity budgets of four impala social groups (2 harem herds and 2

TABLE 1: Ethogram for *Aepyceros melampus* activity states used in the study.

Activity	Operational definition
Foraging	Actively ingesting food or drink, or processing (chewing) food items during a grazing bout and or food searching with head below the vertebral column
Vigilant	Individuals scanning their surroundings and exhibiting agonistic displays
Resting	Standing or sleeping in the sun or shade, neither ruminating nor scanning its environment
Grooming	Scratching, stroking, massaging self or others
Ruminating	Chewing cud while standing, lying, or in locomotion
Moving	Locomotion between foraging source or within study area
Excretion	Defecating or urinating
Flight	Animal running away galloping out of observer sight
Others	Social interaction (necking), nursing, and mating

bachelor herds) were done during the wet season (7 January–27 March 2012) and dry season (4 July–24 September 2012). The group sizes for the harem herds were 24 and 16 whilst; those of the bachelor herds were 5 and 7 individuals. We combined the focal animal sampling and instantaneous scan sampling techniques [22, 23] to collect data on the activity budgets of impala social groups. Using two observation teams, we monitored each group type two times a week simultaneously for the wet and dry season. Since individuals were free ranging and not marked, we arbitrarily selected an active animal from a group as suggested by other researchers [24]. Focal individuals were rapidly scanned instantaneously for 30 minutes at thirty seconds intervals as described by Martin and Bateson [25]. We systematically shifted our focus, with a time lag of two minutes, to different animals in a group to avoid resampling of the same individual, whilst different groups were observed on different days of the week to avoid pseudoreplication [24].

The behavioural states of each focal animal were observed with the aid of Nikon 10 × 50 binoculars and reported to an assistant for recording to reduce errors. We spread observations across the daylight hours, (0700 Hours to 1700 Hours), to avoid over estimating or underestimating behavioural activities associated with time budgets of ungulates [26]. Observations were carried out either from a platform or on foot from a hidden position to reduce observer interferences on the behaviour of the group under observation. Accordingly, care was taken not to disturb the animals prior to or during the observations. If the animals were disturbed, behavioural recording was delayed until they appeared to ignore the observers. A total of 1344 hours of focal animal observations were recorded across all groups during the study.

2.4. Data Analysis. We calculated time of activity by determining proportion of time, expressed as a percentage, that each focal group or individual spent on an activity state. To

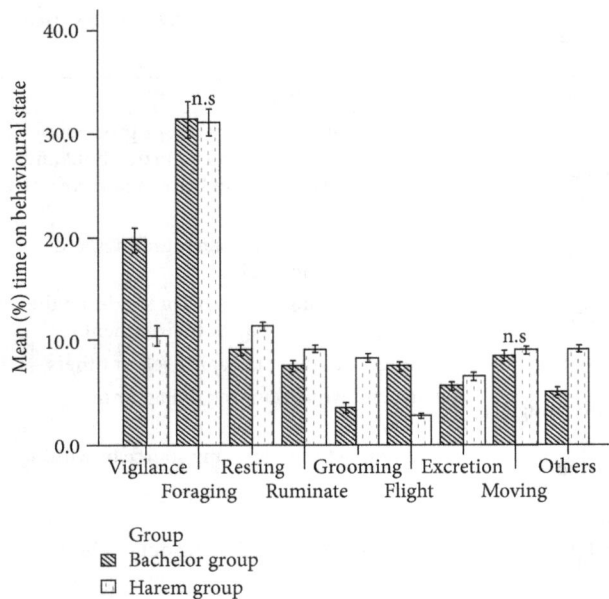

FIGURE 1: Proportion of time spent on behavioural states by the social groups, (n.s = not significant, $P > 0.05$).

TABLE 2: The mean ($\pm$ SE) proportion of time (%) spent on different activities by impala social groups for the dry and wet season.

Activity	Dry season	Wet season	$F_{1,262}$	P
Vigilant	15.35 ± 0.53	15.013 ± 0.54	0.197	0.658
Foraging	30.88 ± 0.75	31.723 ± 0.77	0.613	0.434
Resting	9.51 ± 0.23	10.978 ± 0.24	19.340	0.0001**
Ruminating	8.34 ± 0.22	8.356 ± 0.23	0.003	0.953
Grooming	5.59 ± 0.23	6.323 ± 0.24	4.830	0.029*
Flight	5.69 ± 0.15	4.473 ± 0.15	33.944	0.0001**
Excretion	6.96 ± 0.17	5.140 ± 0.18	54.029	0.0001**
Moving	9.33 ± 0.22	8.155 ± 0.22	14.035	0.0001**
Others	6.89 ± 0.19	7.422 ± 0.19	3.975	0.047*

Statistical significance (P value), $^*P < 0.05$, $^{**}P < 0.001$.

derive activity budgets for the wet and dry season, data on behavioural state occurrences from the three months of each season under each category (i.e., harem and bachelor group) were pooled to produce two data sets: wet season (January–March) and dry season (July–September). Data were tested for normality using one sample Kolmogorov-Smirnov test and satisfied the normality assumptions. We computed a general linear model to test the effect of group type and season on the activity budgets. Pearson correlation was done to test the relationship between activities. All statistical analyses were performed using SPSS release 16.0 (SPSS Inc., 2007).

3. Results

The proportion of time spent on different behavioural states, (e.g., vigilance, resting, ruminating, grooming, flight, excretion, and others) by the harem and bachelor groups were significantly different (Post Hoc test, $P < 0.05$; Figure 1). However, the proportion of time spent foraging by the harem groups (31.157 ± 0.757) and bachelor groups (31.450 ± 0.757) was not significantly different ($F_{(1, 262)} = 0.075$, $P = 0.784$). Similarly, no significant differences ($F_{(1, 262)} = 3.189$, $P = 0.075$) were noted on the proportion of time spent moving by harem (9.023 ± 0.222) and bachelor groups (8.463 ± 0.222).

The proportion of time apportioned for different behavioural states by impala varied with season except for vigilance, foraging, and ruminating as shown in Table 2.

Although group type had no effect on the proportion of time spent moving, we noted that the season had an influence on the proportion of time spent by the impala groups moving. Comparable to group type, season had a significant effect ($P < 0.005$) on the activities like resting, grooming, flight, excretion, and other behavioural states. Females generally

spent more time resting, grooming, and other social activities than their male counterparts did for the two seasons (Figure 2).

The proportion of time spent by impala social groups being vigilant was negatively correlated with foraging, resting, ruminating, grooming, excretion, and others (see Table 3). However, there was no correlation between the time spent foraging and flight behavioural states.

4. Discussion

The activity budgets of harem and bachelor groups in Mukuvisi Woodlands were significantly different except for the proportion of time spent foraging and moving. Generally, our findings are similar to observations made elsewhere [27, 28]. It is acknowledged that males tend to spend more of their time being vigilant compared to their female counterparts [21, 29]. Similar observations have been witnessed in impala (e.g., [27, 28]), springbok (*Antidorcas marsupialis*, e.g., [30]), gazelles (*Procapra picticaudata* e.g., [31, 32]), ring-tailed coati (*Nasua nasua*, e.g., [33]), among others. However, contrary to our findings, Burger and Gochfeld [34] reported no significant differences in the levels of vigilance in male and female springbok. We argue that the proportion of time spent on vigilance and consequently other behavioural states (e.g., foraging, resting, locomotion, and grooming, among others), tends to vary with the degree of disturbance stimuli [35, 36] and perceived predation risk [37–39] within an environments and that it varies with time and space. However, observations elsewhere indicate that nursing female ungulates exhibit elevated vigilance tendencies compared to their non-nursing counterparts [28, 40]. This phenomenon is as an adaptation mechanism to protect and defend the calves.

Although the time spent on vigilance by impala males in this study was significantly different from the females, the proportion spent on foraging did not differ. Other researchers have noted that vigilance comes as a cost to individuals by conflicting with other activities such as feeding, resting, and grooming [16, 41–44]. Although Frid and Dill [36] consider ecotourism as a form of predation risk that reduces time spent on other important activities, arguing that impalas in

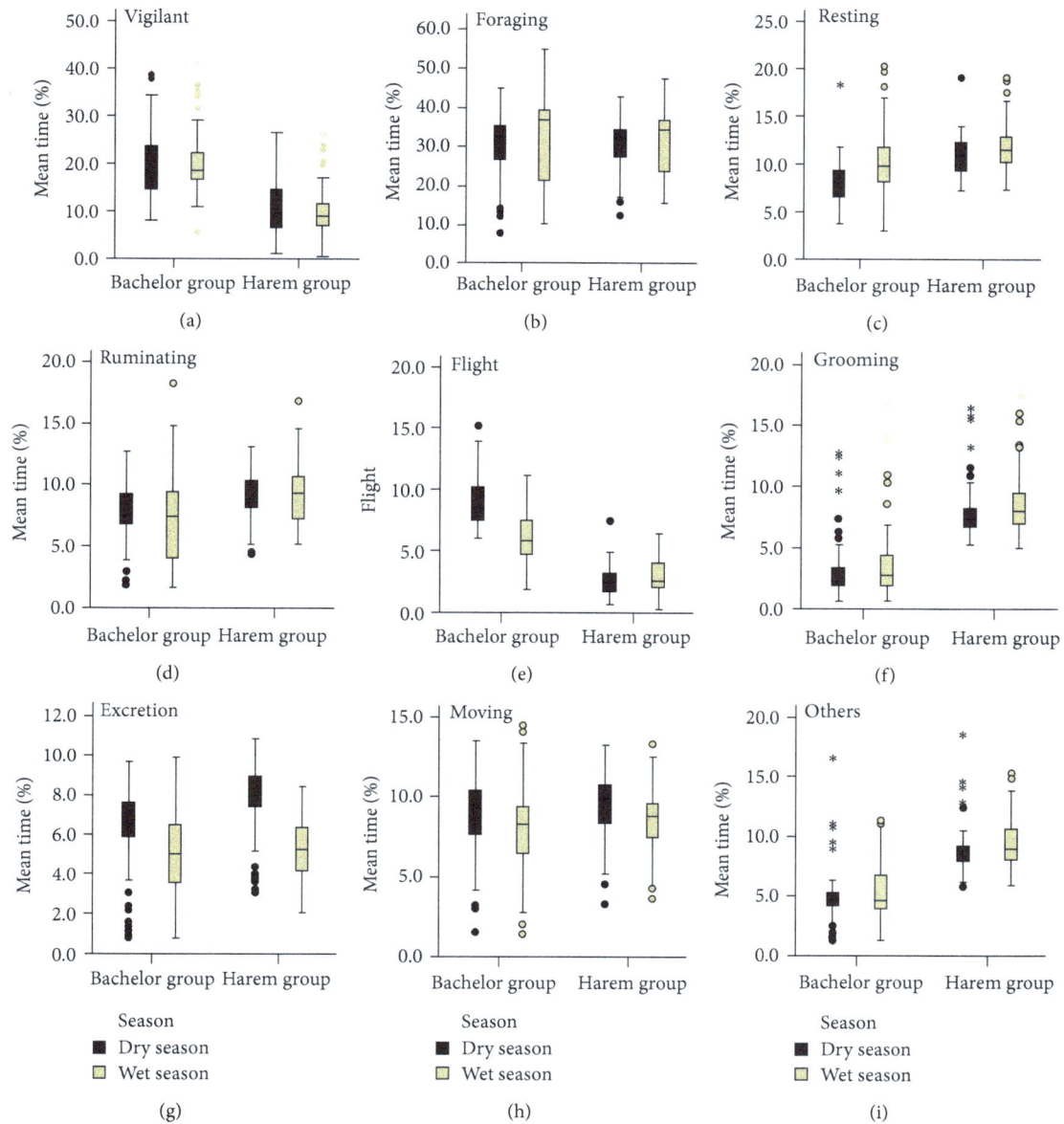

FIGURE 2: Influence of season and group type on the proportion of time spent on activity states by impala social groups.

TABLE 3: Pearson correlation of proportion of time spent on different activity states by impala social groups.

	Vigilant	Foraging	Resting	Ruminating	Grooming	Flight	Excreting	Moving	Others
Vigilant		−.488**	−.189**	−.125*	−.361**	.417**	−.423**	−.080	−.446**
Foraging	—		−.287**	−.409**	−.366**	.019	−.028	−.308**	−.251**
Resting	—	—		.296**	.390**	−.439**	−.092	−.164**	.222**
Ruminating	—	—	—		.306**	−.296**	.019	.017	.224**
Grooming	—	—	—	—		−.556**	.103	.044	.536**
Flight	—	—	—	—	—		−.186**	−.071	−.539**
Excreting	—	—	—	—	—	—		.362**	.244**
Moving	—	—	—	—	—	—	—		.128*
Others	—	—	—	—	—	—	—	—	—

** Correlation is significant at the 0.01 level (2-tailed). * Correlation is significant at the 0.05 level (2-tailed).

Mukuvisi Woodlands seem to have habituated themselves to neutral disturbance stimuli (i.e., the presence of humans). According to Whittaker and Knight [45], habituation occurs when individuals are constantly exposed to repeated neutral stimuli over time. This therefore implies that disruptions on resting, foraging, or other activities may not be altered by the level of vigilance given that the impala social groups would not consider presence of humans as a threat. Behavioural habituation has also been observed in Serengeti National Park where the flight initiation distance (FID) for nonprovisioned impala, topi, Thomson's gazelle, zebra, and wildebeest in the Central Serengeti was less than fifty metres compared to the Western corridor with FID of above 150 m [46].

We consider Mukuvisi Woodlands as a "predation-free" environment where anthropogenic disturbances are the sole disturbance stimuli source from elevated levels of ecotourism and related activities. We argue that the combination of attraction [46] and provisioning [47] and ultimately human habituation through supplementary feeding have altered the activity budgets of impala at the centre. Although these interventions increase visitor satisfaction [47, 48], the long-term socioecological implications and unintended consequences remain uncertain. Knight [11] asserts that habituated or provisioned animals are not brought only within viewing range but also within nuisance range. This challenge occurs when human invitation to animals to come closer ends up as an animal intrusion into human space where they tend to exhibit "begging" behaviours towards tourists [49]. In some cases, long-term provisioning of wild animals may lead to aggressive violent behaviours towards people [50]. The time it would take for these and many other unintended consequences of provisioning and habituation to be expressed by the impala in Mukuvisi Woodland is uncertain. It is important therefore to have monitoring and control mechanisms to deal and reduce the chances of such inadvertent consequences.

Season had no significant effect on the proportion of time spent on foraging, vigilance, and ruminating. These findings are different from observations made in Hwange National Park, Zimbabwe for impala where the group size and season influenced the frequency of vigilance [51]. Likewise, Wronski [52] revealed that impala in Mburo National Park, Uganda spent less time browsing during the wet season than in the dry season and increased the foraging time during the dry season. These observations buttress the notion that in natural systems feeding time by ungulates tends to increase during the dry season, a period when feed quality and quantity will be limiting [33]. Animals therefore spend relatively more time searching for food to fulfil their daily energy requirements. The effect of seasonality in forage quality [53] on the forging and vigilance activity by impala was not visible for impala in Mukuvisi Woodland. We therefore argue that the forage quality hypothesis as described by Blanchard et al. [54] may not apply in closed environments where attraction and provisioning is practised. However, our findings are similar to observations noted in goitered gazelle (*Gazella subgutturosa*) by Xia et al. [29] where seasonal factors had no considerable effect on the level of vigilance but affected other behavioural states. Our findings are contrary to those of Dunham [55] who argue that during the dry season individuals are supposed to spend more time moving and foraging due to insufficient food supply compared to the wet season. We attribute this deviation to the level of provisioning in Mukuvisi, which may neutralise the effects of seasonal variation in feed quality for the impala. However, the variations in the proportion of time spent moving may be due to the location of water and the supplementary points in relation to the respite areas or shaded areas during the dry season. Although bachelor groups spent more time on flight in the dry season compared to the wet season, we speculate that this could be related to the rutting season when males are generally aggressive to each other [56].

Impala social groups spent relatively more time resting during the wet season, than during the dry season and these findings are similar with observations made elsewhere [57, 58]. We noted that bachelor groups spent less time resting compared to harem groups. This corresponds to the proportion of time spent on other activities such as flight and vigilance compared to females. Nevertheless, a large component of resting serves no physiological or ecological function other than energy conservation [59]. Although in natural settings, food searching time is high during the dry season [58] due to reduction in quality, we attribute high mobility of males during the dry season due to the mate searching behaviour during rutting. This is essential in males because females are regarded as a seasonally available, fitness-limiting resource [17, 60, 61]. Our findings indicate that males tend to rest more during the wet season, a time when they are compensating for the condition loss during rutting season through foraging, resting, ruminating among other activities.

Grooming is a useful measure of social relationships in impala as it ushers two main functions, (1) removal of ectoparasites that an animal is not able to reach by itself and, (2) maintenance or establishment of social relationships through increase in psychological and physiological wellbeing and the rewarding effect [62, 63]. In this study, the levels of grooming were more pronounced in females than in males. Similar observations where females spend more time grooming compared to males have been reported elsewhere [64, 65]. Given that the bachelor groups were small compared to the harem groups, we argue that the variation exhibited in the time spent grooming was also because of the group size effect. Grooming rate may vary as a function of group size or interindividual spacing, thus lager groups might stimulate grooming through a social facilitation effect compared to smaller groups [66, 67]. Our findings were related to the assertions of Lehmann et al. [66] that grooming seldom exceeds 15% of daytime activity of most social species. Although Bridges et al. [61] argue that activity pattern plasticity of social species varies with seasons, the level of attraction and provisioning practiced at Mukuvisi Woodlands seems to have diluted this effect. The activity budgets of impala in Mukuvisi are slightly different from those in natural settings, where seasonality plays a crucial function in determining how species allocate their time and energy towards different activities for survival. Serious attempts by reserve managers should be made to reduce the effects of provisioning on the socio-ecology of impala in Mukuvisi Woodland.

5. Conclusion

We conclude that impala social groups at Mukuvisi Woodland spend more time foraging in the dry and wet seasons than any other activity. Season had no effect on the time apportioned during foraging and vigilance behaviour by impala social groups. The activity budgets of impala in closed environments under provisioning and attraction seem to be predictable and less dynamic than those in other natural settings. Although management intervention of attraction and provisioning may promote conservation, scientific tourism, and educational initiatives of the centre, the long-term socioecological, physiological, and reproductive behaviour of impala under provisioning and attraction should not be overlooked. We recommend continuous behavioural monitoring of the impala social groups under similar conditions to provide long-term information for use in adaptive management initiatives of closed environments.

Acknowledgments

Chinhoyi University of Technology through the Senate Research Grant financed this research. Thanks to Dr Nduku (Wildlife Environment Zimbabwe) and Mr Chimanikire (Mukuvisi Woodland Association) for their support and valuable insights in which they shared with the research team.

References

[1] R. East, *Antelope Specialist Group Report*, IUCN, Gland, Switzerland, 1998.

[2] E. D. Lorenzen, P. Arctander, and H. R. Siegismund, "Regional genetic structuring and evolutionary history of the impala *Aepyceros melampus*," *Journal of Heredity*, vol. 97, no. 2, pp. 119–132, 2006.

[3] J. D. Skinner and R. H. N. Smithers, *Mammals of the Southern African Sub-Region*, University of Pretoria, Pretoria, South Africa, 1990.

[4] J. Kingdom, *Field Guide To African Mammals*, Academic Press, London, UK, 1997.

[5] K. A. Pollard and D. T. Blumstein, "Time allocation and the evolution of group size," *Animal Behaviour*, vol. 76, no. 5, pp. 1683–1699, 2008.

[6] Y. Nakayama, S. Matsuoka, and Y. Watanuki, "Feeding rates and energy deficits of juvenile and adult Japanese monkeys in a cool temperate area with snow coverage," *Ecological Research*, vol. 14, no. 3, pp. 291–301, 1999.

[7] M. F. Jaman and M. A. Huffman, "Enclosure environment affects the activity budgets of captive Japanese macaques (*Macaca fuscata*)," *American Journal of Primatology*, vol. 70, no. 12, pp. 1133–1144, 2008.

[8] N. Vasey, "Activity budgets and activity rhythms in red ruffed lemurs (Varecia rubra) on the Masoala Peninsula, Madagascar: seasonality and reproductive energetics," *American Journal of Primatology*, vol. 66, no. 1, pp. 23–44, 2005.

[9] L. G. Rapaport, "Provisioning in wild golden lion tamarins (*Leontopithecus rosalia*): benefits to omnivorous young," *Behavioral Ecology*, vol. 17, no. 2, pp. 212–221, 2006.

[10] J. W. K. Parr and J. W. Duckworth, "Notes on diet, habituation & sociality of yellow-throated Marten (*Martes flavigula*)," *Small Carnivore Conservation*, vol. 36, pp. 27–29, 2007.

[11] J. Knight, "Making wildlife viewable: habituation & attraction," *Society and Animals*, vol. 17, no. 2, pp. 167–184, 2009.

[12] V. Geist, "Wildlife habituation: advances in understanding and management application," *Human-Wildlife Interactions*, vol. 5, pp. 9–12, 2011.

[13] M. B. Orams, "Feeding wildlife as a tourism attraction: a review of issues and impacts," *Tourism Management*, vol. 23, no. 3, pp. 281–293, 2002.

[14] R. B. Gill, "Build an experience and they will come: managing the biology of wildlife viewing for benefits to people and wildlife," in *Wildlife Viewing: A Management Handbook*, M. J. Manfredo, Ed., pp. 218–253, Oregon State University Press, Corvallis, Ore, USA, 2002.

[15] O. Pays, P. Blanchard, M. Valeix et al., "Detecting predators and locating competitors while foraging: an experimental study of a medium-sized herbivore in an African savanna," *Oecologia*, vol. 169, pp. 419–430, 2011.

[16] D. Fortin, M. S. Boyce, E. H. Merrill, and J. M. Fryxell, "Foraging costs of vigilance in large mammalian herbivores," *Oikos*, vol. 107, no. 1, pp. 172–180, 2004.

[17] J. L. Koprowski and M. C. Corse, "Time budgets, activity periods, and behavior of Mexican fox squirrels," *Journal of Mammalogy*, vol. 86, no. 5, pp. 947–952, 2005.

[18] E. Donadio and S. W. Buskirk, "Flight behavior in guanacos and vicuñas in areas with and without poaching in western Argentina," *Biological Conservation*, vol. 127, no. 2, pp. 139–145, 2006.

[19] D. W. S. Challender, N. V. Thai, M. Jones, and L. May, "Time-budgets and activity patterns of captive Sunda pangolins (*Manis javanica*)," *Zoo Biology*, vol. 31, no. 2, pp. 206–218, 2012.

[20] R. Bulton, *The Makabusi Historical Background*, Mukuvisi Woodlands Association, Quick Print Publishers, Harare, Zimbabwe, 1995.

[21] M. G. Dyck and R. K. Baydack, "Vigilance behaviour of polar bears (*Ursus maritimus*) in the context of wildlife-viewing activities at Churchill, Manitoba, Canada," *Biological Conservation*, vol. 116, no. 3, pp. 343–350, 2004.

[22] J. Altmann, "Observational study of behavior: sampling methods," *Behaviour*, vol. 49, no. 3-4, pp. 227–267, 1974.

[23] H. F. Xu and E. D. Zhang, *Wildlife Conservation and Management Principles and Techniques*, East China Normal University Press, Shanghai, China, 1998.

[24] T. Namgail, J. L. Fox, and Y. V. Bhatnagar, "Habitat shift and time budget of the Tibetan argali: the influence of livestock grazing," *Ecological Research*, vol. 22, no. 1, pp. 25–31, 2007.

[25] P. Martin and P. Bateson, *Measuring Behaviour: An Introductory Guide*, Cambridge University Press, 2nd edition, 1993.

[26] K. E. Ruckstuhl, "Foraging behaviour and sexual segregation in bighorn sheep," *Animal Behaviour*, vol. 56, no. 1, pp. 99–106, 1998.

[27] B. Shorrocks and A. Cokayne, "Vigilance and group size in impala (*Aepyceros melampus* Lichtenstein): a study in Nairobi National Park, Kenya," *African Journal of Ecology*, vol. 43, no. 2, pp. 91–96, 2005.

[28] Z. Li, Z. Jiang, and G. Beauchamp, "Vigilance in Przewalski's gazelle: effects of sex, predation risk and group size," *Journal of Zoology*, vol. 277, no. 4, pp. 302–308, 2009.

[29] C. Xia, W. Xu, W. Yang, D. Blank, J. Qiao, and W. Liu, "Seasonal and sexual variation in vigilance behavior of goitered gazelle (*Gazella subgutturosa*) in western China," *Journal of Ethology*, vol. 29, no. 3, pp. 443–451, 2011.

[30] J. Burger, C. Sallna, and M. Gochfeld, "Factors affecting vigilance in springbok: importance of vegetative cover, location in herd, and herd size," *Acta Ethologica*, vol. 2, no. 2, pp. 97–104, 2000.

[31] Z. Li and Z. Jiang, "Group size effect on vigilance: evidence from Tibetan gazelle in Upper Buha River, Qinghai-Tibet Plateau," *Behavioural Processes*, vol. 78, no. 1, pp. 25–28, 2008.

[32] J. Shi, D. Li, and W. Xiao, "Influences of sex, group size, and spatial position on vigilance behavior of Przewalski's gazelles," *Acta Theriologica*, vol. 56, pp. 73–79, 2010.

[33] Y. Di Blanco and B. T. Hirsch, "Determinants of vigilance behavior in the ring-tailed coati (Nasua nasua): the importance of within-group spatial position," *Behavioral Ecology and Sociobiology*, vol. 61, no. 2, pp. 173–182, 2006.

[34] J. Burger and M. Gochfeld, "Vigilance in African mammals: differences among mothers, other females, and males," *Behaviour*, vol. 131, no. 3-4, pp. 153–169, 1994.

[35] R. J. Steidl and R. G. Anthony, "Experimental effects of human activity on breeding bald eagles," *Ecological Applications*, vol. 10, no. 1, pp. 258–268, 2000.

[36] A. Frid and L. M. Dill, "Human-caused disturbance stimuli as a form of predation risk," *Conservation Ecology*, vol. 6, no. 1, p. 11, 2002.

[37] S. L. Lima and L. M. Dill, "Behavioral decisions made under the risk of predation: a review and prospectus," *Canadian Journal of Zoology*, vol. 68, no. 4, pp. 619–640, 1990.

[38] S. L. Lima, "Stress and decision making under the risk of predation: recent developments from behavioural, reproductive and ecological perspectives," *Advances in the Study of Behavior*, vol. 27, pp. 215–290, 1998.

[39] C. M. Papouchis, F. J. Singer, and W. B. Sloan, "Responses of desert bighorn sheep to increased human recreation," *Journal of Wildlife Management*, vol. 65, no. 3, pp. 573–582, 2001.

[40] X. Lian, T. Zhang, Y. Cao, J. Su, and S. Thirgood, "Group size effects on foraging and vigilance in migratory Tibetan antelope," *Behavioural Processes*, vol. 76, no. 3, pp. 192–197, 2007.

[41] M. S. Mooring and B. L. Hart, "Costs of allogrooming in impala: distraction from vigilance," *Animal Behaviour*, vol. 49, no. 5, pp. 1414–1416, 1995.

[42] A. G. McAdam and D. L. Kramer, "Vigilance as a benefit of intermittent locomotion in small mammals," *Animal Behaviour*, vol. 55, no. 1, pp. 109–117, 1998.

[43] A. Treves, "Theory and method in studies of vigilance and aggregation," *Animal Behaviour*, vol. 60, no. 6, pp. 711–722, 2000.

[44] C. M. Bealle, "The behavioural ecology of disturbance responses," *International Journal of Comparative Psychology*, vol. 20, pp. 111–120, 2007.

[45] D. Whittaker and R. L. Knight, "Understanding wildlife responses to humans," *Wildlife Society Bulletin*, vol. 26, no. 2, pp. 312–317, 1998.

[46] J. W. Nyahongo, "Flight initiation distances of five herbivores to approaches by vehicles in the Serengeti National Park, Tanzania," *African Journal of Ecology*, vol. 46, no. 2, pp. 227–229, 2008.

[47] D. Newsome and K. Rodger, "To feed or not to feed: a contentious issue in wildlife tourism," *Australian Zoologist*, vol. 34, pp. 255–270, 2008.

[48] W. McGrew, *The Cultured Chimpanzee: Reflections on Cultural Primatology*, Cambridge University Press, Cambridge, UK, 2004.

[49] J. Knight, "Feeding Mr. Monkey: cross-species food "exchange" in Japanese monkey parks," in *Animals in Person: Cultural Perspectives on Human-Animal Intimacies*, J. Knight, Ed., pp. 231–253, Berg, Oxford, UK, 2005.

[50] Q. K. Zhao and Z. Y. Deng, "Dramatic consequences of food handouts to Macaca thibethana at Mount Emei, China," *Folia Primatologica*, vol. 58, pp. 24–31, 1992.

[51] S. Periquet, L. Todd-Jones, M. Valeix et al., "Influence of immediate predation risk by lions on the vigilance of prey of different body size," *Behavioural Ecology*, vol. 23, pp. 970–976, 2012.

[52] T. Wronski, "Feeding ecology and foraging behaviour of impala *Aepyceros melampus* in Lake Mburo National Park, Uganda," *African Journal of Ecology*, vol. 40, no. 3, pp. 205–211, 2002.

[53] P. Blanchard and H. Fritz, "Seasonal variation in rumination parameters of free-ranging impalas *Aepyceros melampus*," *Wildlife Biology*, vol. 14, no. 3, pp. 372–378, 2008.

[54] P. Blanchard, R. Sabatier, and H. Fritz, "Within-group spatial position and vigilance: a role also for competition? The case of impalas (*Aepyceros melampus*) with a controlled food supply," *Behavioral Ecology and Sociobiology*, vol. 62, no. 12, pp. 1863–1868, 2008.

[55] K. M. Dunham, "The foraging behaviour of impala *Aepyceros melampus*," *South African Journal of Wildlife Research*, vol. 12, pp. 36–40, 1982.

[56] H. Robbel and G. Child, "Notes on the 1969 Rut in the Moremi," *Botswana Notes & Records*, vol. 2, pp. 95–97, 1970.

[57] R. A. Norberg, "An ecologigical theory on foraging time and energetic a choice of optional food searching method," *Journal of Animal Ecology*, vol. 46, pp. 511–529, 1977.

[58] P. E. Komers, F. Messier, and C. C. Gates, "Search or relax: the case of bachelor wood bison," *Behavioral Ecology and Sociobiology*, vol. 31, no. 3, pp. 192–203, 1992.

[59] J. T. Du Toit and C. A. Yetman, "Effects of body size on the diurnal activity budgets of African browsing ruminants," *Oecologia*, vol. 143, no. 2, pp. 317–325, 2005.

[60] C. Vanpé, N. Morellet, P. Kjellander, M. Goulard, O. Liberg, and A. J. M. Hewison, "Access to mates in a territorial ungulate is determined by the size of a male's territory, but not by its habitat quality," *Journal of Animal Ecology*, vol. 78, no. 1, pp. 42–51, 2009.

[61] A. S. Bridges, M. R. Vaughan, and S. Klenzendorf, "Seasonal variation in American black bear Ursus americanus activity patterns: quantification via remote photography," *Wildlife Biology*, vol. 10, no. 4, pp. 277–284, 2004.

[62] K. Taira and E. T. Rolls, "Receiving grooming as a reinforcer for the monkey," *Physiology and Behavior*, vol. 59, no. 6, pp. 1189–1192, 1996.

[63] C. Lazaro-Perea, M. F. De Arruda, and C. T. Snowdon, "Grooming as a reward? Social function of grooming between females in cooperatively breeding marmosets," *Animal Behaviour*, vol. 67, no. 4, pp. 627–636, 2004.

[64] B. M. Md-Zain, N. A. Sha'ari, M. Mohd-Zaki et al., "A comprehensive population survey and daily activity budget on long-tailed macaques of Universiti Kebangsaan Malaysia," *Journal of Biological Sciences*, vol. 10, no. 7, pp. 608–615, 2010.

[65] M. Y. Akinyi, J. Tung, M. Jeneby, N. B. Patel, J. Altmann, and S. C. Alberts, "Role of grooming in reducing tick load in wild baboons (*Papio cynocephalus*)," *Animal Behaviour*, vol. 85, pp. 559–568, 2013.

[66] J. Lehmann, A. H. Korstjens, and R. I. M. Dunbar, "Group size, grooming and social cohesion in primates," *Animal Behaviour*, vol. 74, no. 6, pp. 1617–1629, 2007.

[67] G. Schino, F. Di Giuseppe, and E. Visalberghi, "The time frame of partner choice in the grooming reciprocation of *Cebus apella*," *Ethology*, vol. 115, no. 1, pp. 70–76, 2009.

7

Dynamics and Conservation Management of a Wooded Landscape under High Herbivore Pressure

Adrian C. Newton, Elena Cantarello, Natalia Tejedor, and Gillian Myers

Centre for Conservation Ecology and Environmental Science, School of Applied Sciences, Bournemouth University, Talbot Campus, Poole, Dorset BH12 5BB, UK

Correspondence should be addressed to Adrian C. Newton; anewton@bournemouth.ac.uk

Academic Editor: James T. Anderson

We present the use of a spatially explicit model of woodland dynamics (LANDIS-II) to examine the impacts of herbivory in the New Forest National Park, UK, in relation to its management for biodiversity conservation. The model was parameterized using spatial data and the results of two field surveys and then was tested with results from a third survey. Field survey results indicated that regeneration by tree species was found to be widespread but to occur at low density, despite heavy browsing pressure. The model was found to accurately predict the abundance and richness of tree species. Over the duration of the simulations (300 yr), woodland area increased in all scenarios, with or without herbivory. While the increase in woodland area was most pronounced under a scenario of no herbivory, values increased by more than 70% even in the presence of heavy browsing pressure. Model projections provided little evidence for the conversion of woodland areas to either grassland or heathland; changes in woodland structure and composition were consistent with traditional successional theory. These results highlight the need for multiple types of intervention when managing successional landscape mosaics and demonstrate the value of landscape-scale modelling for evaluating the role of herbivory in conservation management.

1. Introduction

Identification of an appropriate approach for managing disturbance regimes represents one of the most significant challenges to conservation management. Disturbance can be considered as a cause of biodiversity loss, and in such cases management responses might be developed which reduce the frequency or intensity of disturbance events. However, many species are dependent on disturbance to complete one or more parts of their life cycle, and the persistence of such species within a given area will depend on maintenance of an appropriate disturbance regime [1, 2]. This is particularly the case in successional habitats, such as heathland or grassland, which account for a significant proportion of areas of high conservation value in countries such as the UK [3]. Many management interventions, such as the cutting, burning, or grazing of vegetation, are designed to prevent the succession of such habitats to woodland. In order for effective management approaches to be developed, an understanding of the potential impacts of different disturbance regimes is required, including the interactions between different types of disturbance [4, 5].

To ensure that appropriate disturbance regimes are identified, tools are required that enable the ecological impacts of disturbance to be forecast. Such tools would inform the development of effective conservation management plans, by enabling the relative impacts of different management interventions to be explored. A range of approaches can potentially be used to predict the impacts of different forms of environmental change on biodiversity, including extrapolation, experiments, population models, expert opinion, and scenarios [6]. Given the fact that disturbance regimes are spatially heterogeneous and many of the processes influencing biodiversity have a spatial dimension [7], such tools should ideally be spatially explicit and support the identification of impacts at the landscape scale. In this context, we employ here a spatially explicit model of vegetation dynamics in order to examine the impacts of disturbance on a temperate wooded landscape of high biodiversity value. Although a number of process-based models of forest dynamics have been developed [8, 9], relatively few of these are spatially explicit, and few have been applied specifically to support conservation management [10]. Here we used the model LANDIS-II, which simulates the dynamics of wooded landscapes through the incorporation of a number of ecological processes including succession, disturbance and seed dispersal [11, 12]. LANDIS-II is based on an object-oriented modelling approach operating on raster maps, with each cell containing species, environment, disturbance, and harvesting information. LANDIS-II models have been widely applied in different parts of the world [12], increasingly in a conservation context [13]. For example, Newton et al. [4] explored the application of LANDIS-II to support systematic conservation planning in a dryland environment in Chile. Here we build on this previous research, by focusing explicitly on analyzing the impacts of herbivory as a form of disturbance.

In recent years, there has been a widespread increase in the use of large herbivores as a tool for conservation management, particularly in successional habitats. For example, grazing animals have recently been reintroduced to many lowland heathlands in the UK, despite the fact that evidence is lacking regarding their effectiveness in improving habitat quality [14]. Large herbivores are also central to the concept of "rewilding," which involves the reintroduction of populations of animals such as deer, cattle, and ponies to provide "naturalistic grazing" [15]. This approach has been advocated as the "optimal conservation strategy for the maintenance and restoration of biodiversity in Europe" [15, 16]. Examples of large-scale naturalistic grazing initiatives include the Oostvaardersplassen [17] and Veluwezoom National Park [18] in the Netherlands and Knepp Estate, Wicken Fen, and Ennerdale in the UK [19]. These have parallels in the concept of "Pleistocene rewilding" currently being explored in both North and South America [20, 21].

The deployment of large herbivores as a conservation management approach has been supported by the theory developed by Vera [16], who hypothesized that in prehistory, the intense browsing pressure exerted by populations of large herbivores could have maintained extensive areas largely free of tree cover. Vera [16] also proposed that vegetation dynamics are cyclical under high herbivore pressure, owing to the prevention of tree regeneration under a forest canopy by herbivory. Instead, tree species would establish outside woodland through protection of seedlings from herbivory by spiny shrubs. Groves of trees would, therefore, become established within shrub vegetation, providing an example of tree regeneration by facilitation. Such groves would mature over time and eventually collapse, being replaced by grassland, which would subsequently be colonised by shrubs, reinitiating the vegetation cycle. The result would be a dynamic park-like mosaic of woodland and grassland, a pattern that would be created and maintained by populations of large wild herbivores [16].

While Vera's theory has had a major influence on conservation policy and management, it has also stimulated a great deal of debate among both researchers and management practitioners [18, 22, 23]. Much of this has focused on palaeoecological evidence regarding whether the early postglacial vegetation of northern Europe was densely forested, as traditionally believed, or more open in character. Relatively little research has been conducted on contemporary vegetation with the explicit objective of testing Vera's theory [24, 25]. Much of this evidence is based on inference from existing vegetation structure and composition from which dynamics are inferred, as a form of space-for-time substitution. As noted by Fukami and Wardle [26], such approaches have a number of limitations, which can potentially be addressed by integrating survey data with modelling approaches, as explored here.

In this investigation, we employed LANDIS II supported by the collection and analysis of field survey data to examine the potential impact of populations of large herbivores on the spatial dynamics and composition of vegetation at the landscape scale. The overall aim of the research was to inform conservation management plans, both in relation to grazing and to other forms of disturbance undertaken as part of management, including the cutting and burning of vegetation. The research was conducted in the New Forest National Park, UK, an extensive area of seminatural vegetation of high biodiversity value that has been subjected to high herbivore pressure for many centuries [27, 28], and was cited by Vera [16] as evidence in support of his theory. Field surveys and modelling approaches were, therefore, used to test the following hypotheses, which were identified from Vera's theory [16, 24]: (i) within woodlands there is little or no regeneration of trees, even in canopy gaps, because of high herbivore pressure; (ii) regeneration occurs on the periphery of woodlands in blackthorn or holly scrub, which results in a concentric expansion of forest; (iii) woodland groves will tend to break up with maturity and be converted to either grassland or heathland; (iv) cyclic dynamics of vegetation at the site scale produces shifting mosaics of vegetation structural types at the landscape scale, dependent on the intensity, extent, and frequency of herbivory.

2. Methods

2.1. Study Area. The New Forest is situated on the south coast of England in the counties of Hampshire and Wiltshire (longitude from $1°17'59''$ to $1°48'8''$ W, latitude from $50°42'19''$ to $51°0'17''$ N). Its present character is strongly dependent on its history as a medieval hunting forest and the survival of a medieval commoning system. As a result, this landscape has developed under the influence of large, free-ranging herbivores, including deer as well as livestock, over a prolonged period [27]. The current research focused on the New Forest Special Area of Conservation (SAC), a Natura 2000 site, which forms the core of the National Park and is approximately 29,000 ha in area. The vegetation is a mosaic of pasture woodland, heathland, grassland, scrub, and mire communities. Woodlands occur both within inclosed areas that are legally designated for silviculture ("silvicultural inclosures") and in noninclosed areas ("open forest"). The difference between inclosed and noninclosed woodlands primarily reflects contrasting management histories, with timber extraction and plantation establishment largely restricted to the inclosed areas. While many inclosures have been fenced at some time in their past, recent management approaches have tended towards fence removal. In recent years, some 6000–7400 livestock, principally ponies and cattle, have been depastured in the New Forest [27, 28]. Five species of deer occur in the New Forest, with total numbers regulated at around 2000 animals through culling [28], although the actual number of deer present within the Forest at any one time is not known with precision.

2.2. Field Survey. A series of three field surveys were conducted over the period 2005–2010, as detailed below.

Survey 1 (woodland). Woodland structure and species composition were assessed throughout the study area. The sampling approach adopted the woodland units (WUs) used as a basis for monitoring habitat condition and included ancient pasture woodlands, exotic and native tree plantations, within both the open forest and the silvicultural inclosures. A total of 173 WUs were sampled, representing all areas classified as woodland. In each unit, a 50 × 50 m plot was established and surveyed for woodland structure and composition. The plots were located randomly within the WUs using ArcGIS (v8.2 1999–2002, ESRI, California) and were located in the field using a Global Position System (GPS) device (Garmin III Plus, Garmin Europe Ltd., Southampton, UK). Within each plot, the number of individuals of each tree species was counted, and the diameter at breast height (dbh) of each tree >10 cm dbh was measured using a diameter tape. The total number of saplings in the plot was measured, with saplings defined as trees <10 cm dbh but >1.3 m height, and also the total number of tree seedlings (i.e., individuals <1.3 m height) of each species was recorded in a 10 m × 10 m subplot. Canopy closure (*sensu* Newton [10]) was estimated in the centre of each plot, using a spherical densiometer (Forestry Suppliers Inc., Missouri, USA). In each plot, a series of ten variables were assessed by visual observation to provide a measure of browsing pressure, following Reimoser et al. [29]. The categorization of browsing pressure, based on these assessments, was derived from the same source [29] (see Appendix 1 (in Supplementary Material available online at http://dx.doi.org/10.1155/2013/273948/)). Trees of each species were also cored for age determination in the laboratory.

Survey 2 (heathland and grassland). In order to parameterise LANDIS-II for the entire study area, information was also required on the structure and composition of vegetation in grassland and heathland areas. In total, 50 grassland and heathland units (GHUs) were surveyed, including 99% units >50 ha in size and 92% units >10 ha in size (those representing road verges being excluded). In each unit, a 50 × 50 m plot was located randomly and surveyed as described in Survey 1. The relationship between age and stem diameter in *Ulex europea* (gorse) and in the ericaceous shrubs *Calluna vulgaris* (common heather), *Erica cinerea* (bell heather), and *Erica tetralix* (cross-leaved heath) were determined by measuring the stem diameter of 20 individuals of each species at the base

using a digital calliper. Increment cores (or stem sections for the ericaceous shrubs) were taken at the stem base for age determination in the laboratory.

Survey 3 (woodland expansion). A third field survey ("Survey 3") was undertaken to provide an independent data set with which to test the LANDIS-II model. For this purpose, field data were required for sites where woodland expansion is occurring outside the boundary of the WUs and were, therefore, not included in model parameterization. Thirty locations were identified based on aerial photographs and field observations. Sites for survey were identified by selecting areas of woodland expansion greater than 50 × 50 m in area, lying outside but adjacent to WU boundaries. At each location, a 50 m × 50 m plot was established and surveyed as described for Survey 1. In addition, a record was made of whether individual tree seedlings were associated with protective shrub cover or not.

2.3. Model Parameterization and Calibration. In LANDIS II, tree species are simulated as the presence or absence of species age cohorts in each cell, at a time step specified by the user. A detailed description of the LANDIS-II model is provided elsewhere ([12], http://www.landis-ii.org/). The inputs required by the model include a land type or ecoregion map, which describes the ecological conditions influencing tree establishment, and an initial communities map, which describes the distribution and age of cohorts of each species at year 0 of the simulations. The initial communities map was produced using the WUs defined by Natural England and the field data collected in Surveys 1 and 2.

The ecoregion map was produced by defining 30 ecoregions on the basis of elevation and soil type. This was achieved by mapping combinations of six elevation classes (0–29, 30–59, 60–89, 90–119, and ≥120 m a.s.l.) and eight soil types that occur in the study area, at 50 m resolution. Ecoregions were mapped by identifying combinations of elevation classes and soil type. Elevation data were derived from OS Land-Form PROFILE DTM 1 : 10000 using DIGIMAP (http://edina.ac.uk/digimap/) and converted to raster format. Soil data were obtained from (NATMAP National Soil Map; National Soil Resources Institute (NSRI), Silsoe, Bedfordshire, UK). A land cover map obtained from the Hampshire Biodiversity Information Centre (HBIC), together with a map of water bodies obtained from DIGIMAP, was used to identify areas of water, sea shore habitats, urban development, quarries, and arable or horticultural land, which were excluded from all model simulations. All maps were produced and manipulated using ArcGIS 9.2 (1999–2006 ESRI Inc., Redlands, California) and Idrisi Andes (Clark Labs, Clark University, Worcester, MA, USA) projected using British National Grid. The maps were 800 × 800 pixels in area.

In LANDIS-II, forest succession is a competitive process governed by the probability of establishment in different ecoregions and the life history characteristics of each species. The life history characteristics of the 37 woody species encountered in the field survey (Table 1) were obtained from the scientific literature, supplemented by field observations. In LANDIS-II, forest succession interacts with several

TABLE 1: Ecological characteristics for the tree and shrub species encountered in the New Forest study area, which were included in model simulations.

Name	Long	Mat	ShT	FiT	EffSD	MaxSD	VRP	Min VRP	Max VRP	P-FiR	DB1	DB2	DB3	PB	DB1
Abies grandis	300	20	4	1	60	120	0	0	0	None	Y				Y
Acer campestre	200	10	3	1	80	120	1	10	120	None		Y		Y	
Acer pseudoplatanus	150	12	4	1	120	400	1	10	100	None	Y			Y	Y
Alnus glutinosa	250	12	3	1	120	200	1	10	200	None		Y		Y	
Betula pendula	160	18	2	1	200	1600	1	10	120	None		Y		Y	
Calluna vulgaris	30	1	1	1	100	250	1	0	30	Resprout	Y			Y	Y
Carpinus betulus	250	20	4	1	90	130	1	10	150	None		Y		Y	
Castanea sativa	300	35	3	1	300	700	1	10	250	None		Y		Y	
Chamaecyparis lawsoniana	200	20	3	1	80	120	0	0	0	None			Y		
Corylus avellana	80	10	4	1	300	700	1	10	80	None	Y			Y	Y
Crataegus monogyna	150	4	2	1	300	700	1	10	100	None					
Fagus sylvatica	500	55	5	1	300	700	1	10	300	None		Y		Y	
Frangula alnus	80	3	2	2	300	700	1	10	30	None	Y			Y	Y
Fraxinus excelsior	200	17	3	1	90	120	1	10	200	None	Y			Y	Y
Ilex aquifolium	300	10	3	1	300	700	1	10	300	None			Y	Y	
Larix decidua	200	20	2	1	120	400	0	0	0	None		Y			
Malus sylvestris	130	8	2	1	300	700	1	10	100	None		Y		Y	
Picea abies	300	40	2	1	100	120	0	0	0	None			Y		
Picea sitchensis	300	22	2	1	100	120	0	0	0	None			Y		
Pinus nigra	350	22	2	1	100	150	0	0	0	None			Y		
Pinus ponderosa	300	7	2	1	100	150	0	0	0	None			Y		
Pinus sylvestris	300	12	2	1	100	1000	0	0	0	None		Y			
Populus alba	250	7	2	1	500	1600	1	10	250	None	Y			Y	Y
Prunus spinosa	60	4	2	1	300	700	1	10	60	None					
Pseudotsuga menziesii	400	12	2	3	120	380	0	0	0	None		Y			
Quercus robur	500	60	2	1	300	700	1	10	400	None	Y			Y	Y
Quercus rubra	200	22	4	3	300	700	0	0	0	None	Y			Y	Y
Salix cinerea	90	35	2	1	1000	1600	1	10	70	None	Y			Y	Y
Sorbus aria	150	6	2	1	300	700	0	0	0	None	Y			Y	Y
Sorbus aucuparia	100	15	2	1	300	700	1	10	100	None	Y			Y	Y
Sorbus torminalis	100	13	4	1	300	700	1	10	100	None	Y			Y	Y
Taxus baccata	3000	20	4	1	300	700	0	0	0	None			Y		
Tsuga heterophylla	400	15	2	1	120	160	0	0	0	None			Y		
Ulex europea	40	2	1	1	1	700	1	0	40	Resprout					
Viburnum opulus	50	5	2	1	300	700	1	10	40	None	Y			Y	Y

Long: longevity (years); Mat: age of sexual maturity (years); ShT: shade tolerance (1–5); FiT: fire tolerance (1–5); EffSD: effective seed dispersal distance (m); MaxSD: maximum seed dispersal distance (m); VRP: vegetative reproduction probability (0-1); MinVRP: minimum age of vegetative reproduction (years); MaxVRP: maximum age of vegetative reproduction (years); P-FiR: postfire regeneration form (none, resprouting, or serotiny). Principal literature sources included [39, 40, 51, 53–58]. BP refers to browsing preference; in these simulations, those species indicated with a "Y" were considered to be of either high (DB1), moderate (DB2), or low (DB3) preference by deer. PB indicates whether the species were browsed by ponies (Y) or not in the scenarios where browsing was included. These groups were identified by reference to the available literature [14, 32, 59–65].

spatial components (i.e., seed dispersal, fire, and harvesting disturbances). The establishment probability of each tree species in each ecoregion was derived from the Ecological Site Classification (ESC) decision-support system developed for British forests (v1.7; [30]; see also Appendix 1). Model calibration was performed by examining the spatial dynamics of each species in repeated simulations over a 1000 year interval, with browsing and fire disturbances approximating current values, and comparing results with field observations.

The LANDIS-II Base Fire v2.1 extension was used to explore fire dynamics. Calibration involves the systematic adjustment of fire model parameters over multiple simulations, until the average area burned per time interval is within a small percentage of the target value [31]. A fire ecoregion map was created by dividing the study area into 11 units, coinciding with management units subjected to rotational heathland burning as part of current management practices [27]. On average, around 400 ha of heathland are burned annually. Maximum fire size was set to 10 ha to be consistent with current management practice (Mr. David Morris, Forestry Commission, personal communication). Fire ignition probabilities were varied proportionally to the area of each management unit, so that the proportion of each management unit that burned in a given year was roughly equal. Fire "spread age" (which represents the expected fire rotation period) was set to 15 years, to coincide with the rotation period used in current management practice. Fire ignition probabilities were adjusted through multiple simulations to achieve an expected average annual burned area close to 400 ha. Calibration was completed when a value within 1% of this total was obtained.

The harvest module of LANDIS-II (Base Harvest extension v1.2) was used to simulate the impacts of browsing by livestock and deer. Browsing was modelled by the removal of the youngest cohort (1–10 years old) of plants. Browsing was implemented using the "Patch Cutting" option in Base Harvest, such that randomly selected groups of sites within a stand were harvested. Browsing pressure was distributed evenly among different vegetation types. The percentage of cells with a stand to be harvested as a result of deer browsing was set at 90% for the most palatable species, 80% for intermediate species, and 70% for the least palatable species (Table 1). For browsing by ponies, the percentage value was 50%. The size of the patches that were harvested varied according to the typical home range size of the animals concerned and was set at 80 ha for deer and 150 ha for ponies [32]. Base Harvest was calibrated through a sensitivity analysis, which was performed by systematically varying both the percentage of cells within a forest stand to be harvested and the patch size. In order to be consistent with field observations, the harvesting parameters were adjusted such that the model projected 95% mortality of birch in the open forest (i.e., in areas outside woodland) as a result of browsing. These same parameter values produced a 66% reduction in frequency of beech within woodland areas.

2.4. Scenarios. Once model calibration was completed, final values of the model parameters were used in a series of modelled scenarios. Simulations were conducted for 300 years. Five replicated simulations (with varying random number seed) were performed for each scenario. Two forms of disturbances were explored in the scenarios: the mortality of young trees caused by the activities of large mammals ("browsing") and the effects of burning ("fire"). The time steps were set at 10 years for tree succession, 10 years for fire disturbance, and 1 year for browsing. The scenarios were defined as follows: scenario 1, no disturbance (neither fire nor browsing); scenario 2, browsing only; scenario 3, fire only; scenario 4, fire plus browsing; scenario 5, browsing, fire, and protection from herbivory by presence of spiny shrubs, where mortality of trees as a result of browsing was set to zero if either of the species *Crataegus monogyna* (hawthorn), *Prunus spinosa* (blackthorn), or *Ulex europea* (gorse) was present of

ages 5–100 years old within at least 50% of the cells in the stand.

2.5. Data Analysis. In order to test the factors influencing regeneration, the density of seedlings and saplings of each native woody species recorded in each plot in Survey 1 was correlated with (i) the extent of canopy closure and (ii) the index of browsing pressure. Spearman's rank correlation analyses were performed using SPSS (SPSS Inc., Chicago, USA), as the data were not normally distributed, as indicated by the Shapiro-Wilk tests. The data obtained in Survey 1 were divided into two groups, "inclosed" and "noninclosed," based on whether or not the survey plots were located within silvicultural inclosures, as these two groups have been subjected to different management histories. Stand structure was analysed by calculating size-frequency distributions for each tree species based on the number of individuals within each stem diameter category, pooling survey plots within each group. Statistical differences in stem densities and canopy closure between inclosed and noninclosed plots were examined using Mann-Whitney U tests performed using SPSS. The alpha level of statistical tests was set at 0.05.

Field data from Survey 3, which were independent of those used in model parameterisation, were used to test model predictions. The model was tested by analysing the relationship between the number of survey plots in which each tree species was predicted to occur using the model and the number of plots in which each species was recorded in the field survey. This relationship was tested by linear regression using SPSS v.16 (SPSS Inc., Chicago, USA).

The Age Cohort Statistics v1.0 extension of LANDIS-II was used to produce outputs of (i) minimum and maximum age across all species in each pixel, and (ii) the presence of selected species in each pixel under each of the scenarios. The following species were selected for analysis on the basis of their ecological importance within the study area: *Betula pendula* (birch), *Fagus sylvatica* (beech), *Ilex aquifolium* (holly), *Pinus sylvestris* (Scots pine), and *Quercus robur* (oak). The LANDIS-II outputs consist of raster maps produced for each time step of the simulation (10 years). The total extent of woodland was calculated as the total number of pixels within which one or more of these five species was present, as an adult tree ≥10 years old.

3. Results

3.1. Survey 1. In total, 70 plots were established in noninclosed woodland units and 103 in inclosed units. The plots were predominantly located in woodland stands with high canopy closure, with 79% plots in both inclosed and noninclosed units, demonstrating canopy closure values >80% (Figure 1). Median values did not differ significantly between the two unit types ($P = 0.193$, Mann-Whitney U test), with values recorded of 93.5% and 91.9% for inclosed and noninclosed units respectively. In terms of stem densities of mature trees (>10 cm dbh), the most abundant species in the noninclosed stands were oak, beech, birch, holly, and Scots pine, each with mean values of >11 ind·ha^{-1} (Appendix 1).

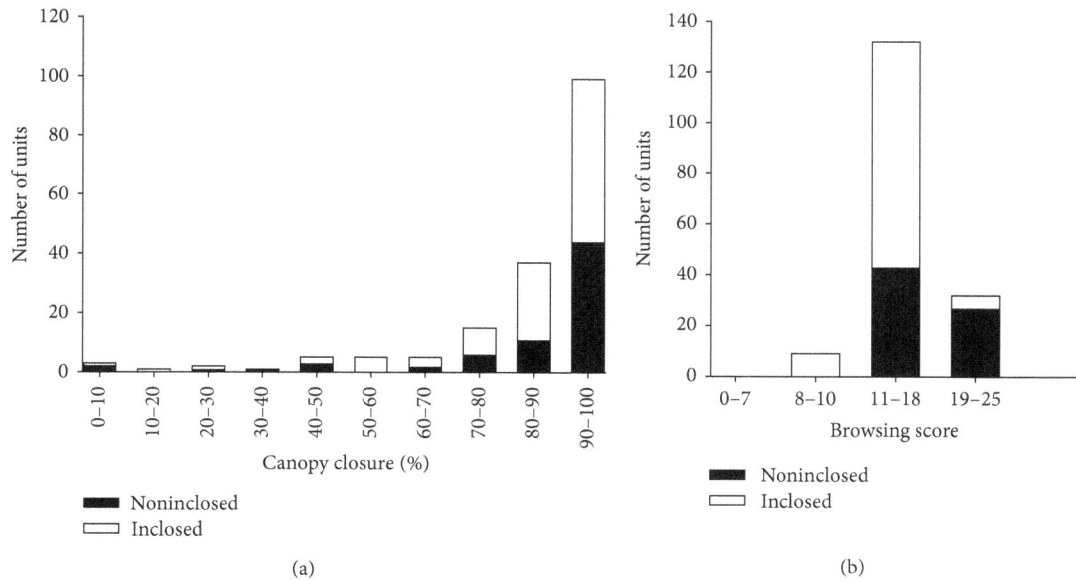

FIGURE 1: Measurements of (a) canopy closure and (b) browsing pressure in the woodlands of the New Forest, as assessed in Survey 1 (see text). Inclosed areas are those designated for silvicultural intervention. Total browsing scores of <7 were associated with light, 8–10 moderate, 11–18 heavy, and 19–25 very heavy browsing pressure, respectively (following [29]).

In the inclosures, a number of nonnative coniferous species were also abundant, including *Picea abies* (Norway spruce), *Pinus nigra* (European black pine), and *Pseudotsuga menziesii* (Douglas fir), each with mean densities of >16 ind·ha^{-1} (Appendix 1).

Individual species displayed a variety of different structural patterns (Figure 2). Stand structure of beech was characterised by an "inverse-J" shape, with relatively large numbers of trees in smaller-diameter size classes and with decreasing frequency as the size class increased. This pattern was evident in both noninclosed and inclosed stands and implied continual recruitment in both woodland types. Inverse-J size-frequency distributions were also encountered in holly, where stem densities were substantially higher in noninclosed than in inclosed stands and in birch, particularly in inclosed stands (Figure 2). Other species characterised by inverse-J stand structures included hawthorn, *Fraxinus excelsior* (ash), and *Salix cinerea* (grey willow), which like holly, occurred at much higher densities in noninclosed than in inclosed stands.

In contrast, a number of species were characterised by unimodal distributions, with relatively low numbers of individuals in the smaller-diameter size classes. For example, highest stem densities of oak were recorded in the 40–45 cm size class in noninclosed sites and in the 20–25 cm size class in inclosed sites (Figure 2). Mean sapling densities of this species (<10 cm dbh) were 4.4 and 1.7 ind·ha^{-1} in noninclosed and inclosed units, respectively. Scots pine also displayed a unimodal size distribution, with highest values recorded in the 20–25 cm size class in inclosed forest; substantially lower values were found in the noninclosed sites. A similar stand structure was displayed by *Alnus glutinosa* (alder).

Of the 38 woody species recorded in the survey with stems >10 cm dbh, 34 (89%) were also recorded as a sapling and 31 (82%) as a seedling. For most species, seedling densities did not differ significantly between inclosed and noninclosed units (Figure 3). The regeneration of most woody species was not related to the extent of canopy closure, when analysed by correlation. In the case of sapling density, a statistically significant relationship with canopy closure was only recorded in one species, namely, alder ($r = -0.15$, $P = 0.045$). In the case of seedlings, statistically significant relationships with canopy closure were recorded in only two species, beech ($r = 0.15$, $P = 0.048$) and holly ($r = 0.016$, $P = 0.032$).

Results of the browsing survey (Figure 1(b)) indicated that all of the inclosed plots were characterised by at least moderate browsing pressure (score 8–10), whereas all of the noninclosed plots were associated with at least heavy browsing pressure (score ≥ 11). Most plots (61% noninclosed and 86% inclosed) were classified as heavy browsing (score 11–18), with very heavy browsing (score 19–25) proportionally being more evident in the noninclosed than in the inclosed plots (39% and 5% of plots resp.). Median browsing score was significantly higher in the noninclosed than in the inclosed plots (values of 18 and 14, resp.; $P < 0.001$, Mann-Whitney U test). Browsing impact score was not significantly related to canopy closure ($r = -0.016$, $P = 0.837$). When sapling density was correlated with browsing impact score, significant relationships were found only in holly ($r = 0.31$, $P < 0.001$), birch ($r = -0.25$, $P = 0.001$), *Frangula alnus* ($r = 0.18$, $P = 0.019$), *P. sylvestris* ($r = -0.21$, $P = 0.005$), and *Prunus spinosa* (alder buckthorn) ($r = 0.21$, $P = 0.004$). In the case of seedling density, significant relationships were

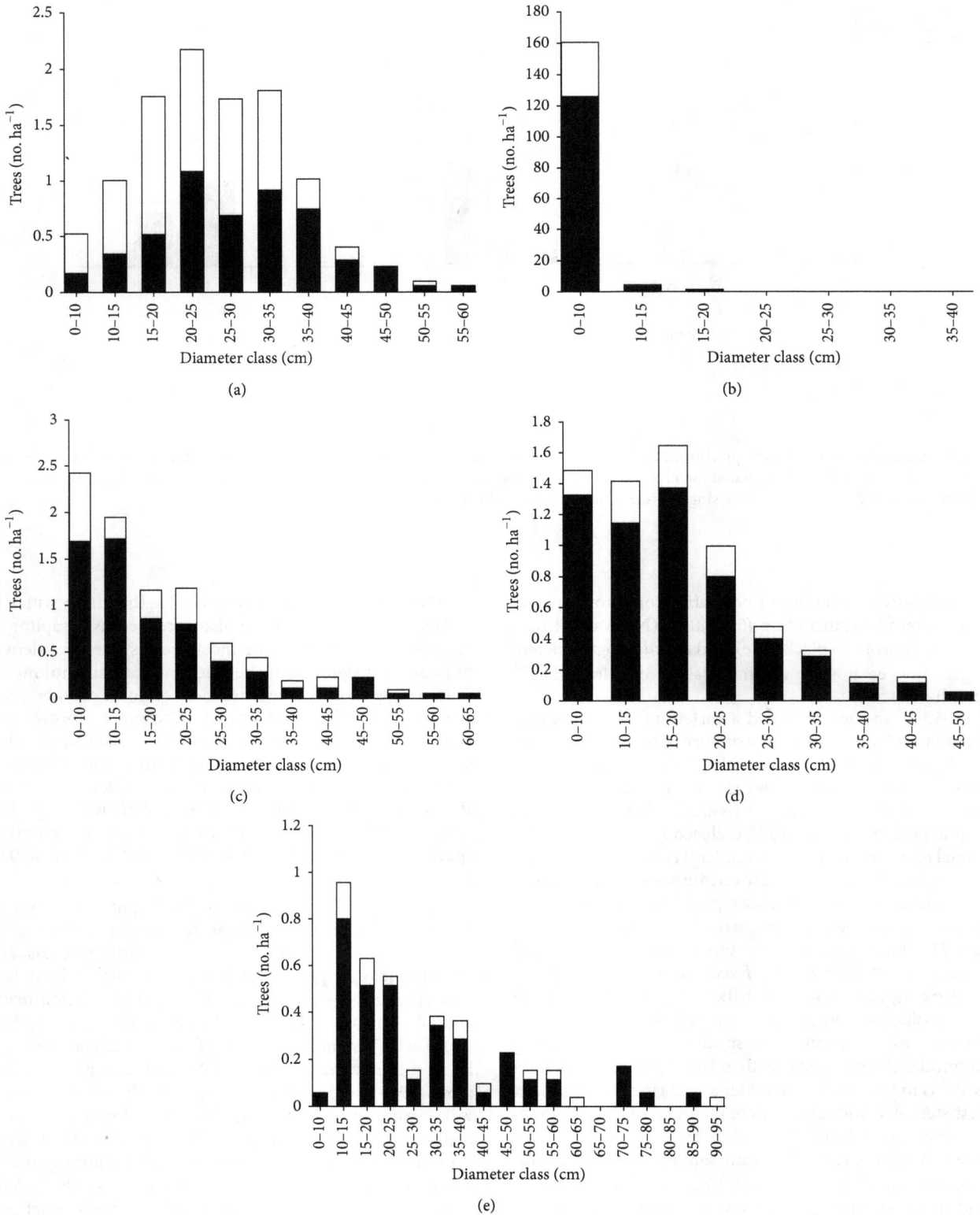

FIGURE 2: Stand structure of the New Forest woodlands, illustrated as the size frequency distributions of selected tree species, assessed in Survey 1 (see text). (a) *Betula pendula*; (b) *Fagus sylvatica*; (c) *Ilex aquifolium*; (d) *Pinus sylvestris*; (e) *Quercus robur*. Filled bars, noninclosed units; open bars, inclosed units.

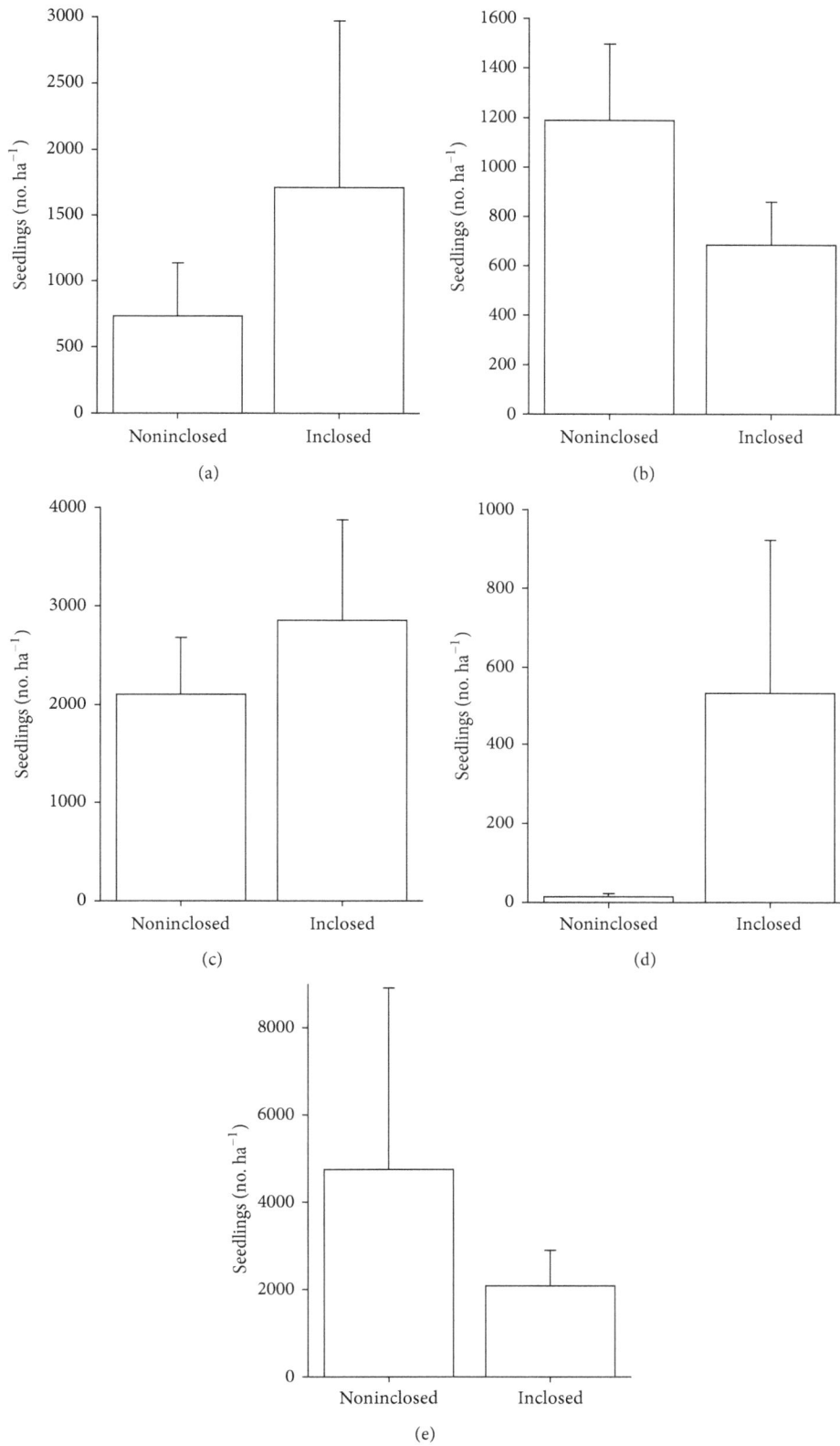

FIGURE 3: Seedling densities of selected tree species in the New Forest woodlands, assessed in Survey 1 (see text). Values presented are means (±SE). (a) *Betula pendula*; (b) *Fagus sylvatica*; (c) *Ilex aquifolium*; (d) *Pinus sylvestris*; (e) *Quercus robur*.

TABLE 2: List of species encountered in the field survey of grassland and heathland sites (Survey 2) as seedlings and/or saplings. Species that are not considered to be native to the study area are asterisked. Note that "seedlings" and "saplings" of *Ulex europea* and *Calluna vulgaris* were not recorded separately in this survey. *Ulex europea* was present in 52.6% of grassland plots and 25.8% of heathland sites, while corresponding values for *Calluna vulgaris* were 26.3% and 61.3%, respectively.

Latin name	English name	Grassland sites		Heathland sites	
		Percentage of plots with seedlings	Percentage of plots with saplings	Percentage of plots with seedlings	Percentage of plots with saplings
Acer campestre	Field maple	5.3	—	—	—
Betula pendula	Silver birch	36.8	52.6	45.2	22.6
Corylus avellana	Hazel	—	—	3.2	—
Crataegus monogyna	Hawthorn	5.3	31.6	—	—
Fagus sylvatica	Beech		5.3	—	—
Frangula alnus	Alder buckthorn	—	10.5	6.5	3.2
Ilex aquifolium	Holly	5.3	15.8	6.5	9.7
*Pinus sylvestris	Scots pine	5.3	5.3	19.4	3.2
Prunus spinosa	Blackthorn	10.5	10.5	6.5	3.2
Quercus robur	Oak	15.8	10.5	22.6	—
Salix cinerea	Grey willow	—	—	—	3.2
Sorbus aucuparia	Rowan	10.5	10.5	6.5	3.2
Viburnum opulus	Guelder rose	—	5.3	—	—

found only in holly ($r = -0.40$, $P < 0.001$), birch ($r = -0.15$, $P = 0.045$), Scots pine ($r = -0.21$, $P = 0.006$), and *Taxus baccata* (yew) ($r = 0.23$, $P = 0.002$).

3.2. Survey 2. In the survey of grassland and heathland areas, seedlings of a total of 13 woody species were encountered in addition to the shrubs gorse and heather, which were both widespread (Table 2). Of these, birch was consistently the most frequent, being present as a sapling on 23% of heathland sites and 53% of grassland sites. Oak was also widespread as a seedling, being present on 23% of heathland sites and 16% of grassland sites. Spiny shrubs such as hawthorn and blackthorn were more frequent on grassland sites than heathland sites, but both species were less frequent than birch on both types of site.

3.3. Survey 3. A total of ten woody species were observed as saplings and nine as seedlings (Table 3). Again, birch was found to be the most frequent species, occurring on 73% of plots either as a seedling and/or as a sapling. Hawthorn and blackthorn were present on 27% and 13% of plots, respectively, indicating the presence of spiny shrubs that could potentially provide protection from herbivory. However, such putative protection was only observed in a minority of plots and for only two species, namely, 10% of plots in the case of birch and 3% in the case of beech.

3.4. Model Testing. Regression analysis indicated a positive relationship between the number of survey plots in which each tree species was predicted to occur by the model and the number of plots in which each species was recorded in Survey 3 ($r^2 = 0.55$, $P = 0.001$). Predicted species richness

for the field plots closely approximated that recorded in the field survey (mean ± SE: 2.7 ± 0.3 and 2.2 ± 0.3, resp.) and median values were identical (2.5; Mann-Whitney U test, $P = 0.34$). The model correctly predicted presence of eight species in the field plots and correctly predicted the absence of a further 17 species. However, the model also predicted the presence of five species that were not recorded in the field survey (*Castanea sativa* (sweet chestnut), European black pine, Douglas fir, grey willow, and *Sorbus aucuparia* (rowan)), although none of these species was predicted to occur in more than 20% of field plots. The model also failed to predict the presence of three species (beech, ash, and blackthorn) that were encountered in this survey, although none of these occurred in more than 13% of field plots.

3.5. Scenarios. Over the duration of the simulations (300 yr), woodland area increased in all scenarios, with or without disturbance. The increase in woodland area was most pronounced in Scenario 1 (no disturbance), with an increase of 175% over the initial value (Figure 4). However, woodland area increased by more than 70% even in Scenario 4, where disturbance was most intense. The inclusion of protection from herbivory by presence of spiny shrubs (Scenario 5) resulted in larger woodland areas than in the other scenarios where disturbance was present, with an increase of 30% over the value recorded in Scenario 4. The substantially lower values recorded in Scenario 4 than either Scenario 2 or 3 suggest a positive interaction between the effects of fire and browsing.

Different tree species displayed contrasting patterns of abundance over time (Figure 5). Scots pine, holly, and birch were characterised by an initial increase in extent followed by a subsequent decline, with peak values recorded after

TABLE 3: (a) List of species encountered in the field survey of woodland expansion (Survey 3) as seedlings and/or saplings. Species that are not considered to be native to the study area are asterisked. (b) Occurrence of protection of seedlings by spiny shrubs, encountered in the field survey of woodland expansion (Survey 3). Species that are not considered to be native to the study area are asterisked.

(a)

Latin name	English name	Percentage of plots with saplings	Percentage of plots with seedlings
Betula pendula	Silver birch	43.3	53.3
Crataegus monogyna	Hawthorn	20.0	16.7
Fagus sylvatica	Beech	3.3	6.7
Frangula alnus	Alder buckthorn	3.3	3.3
Fraxinus excelsior	Ash	3.3	
Ilex aquifolium	Holly	23.3	30.0
*Pinus sylvestris	Scots pine	10.0	10.0
Prunus spinosa	Blackthorn	6.7	6.7
Quercus robur	Oak		33.3
Taxus baccata	Yew	6.7	3.3
Ulex europea	Common gorse	3.3	

(b)

Latin name	English name	Percentage of plots with seedlings present, protected by shrubs	Percentage of plots with seedlings present, not protected by shrubs
Betula pendula	Silver birch	10.0	43.3
Crataegus monogyna	Hawthorn		16.7
Fagus sylvatica	Beech	3.3	3.3
Frangula alnus	Alder buckthorn		3.3
Ilex aquifolium	Holly		30.0
*Pinus sylvestris	Scots pine		10.0
Prunus spinosa	Blackthorn		6.7
Quercus robur	Oak		33.3
Taxus baccata	Yew		3.3

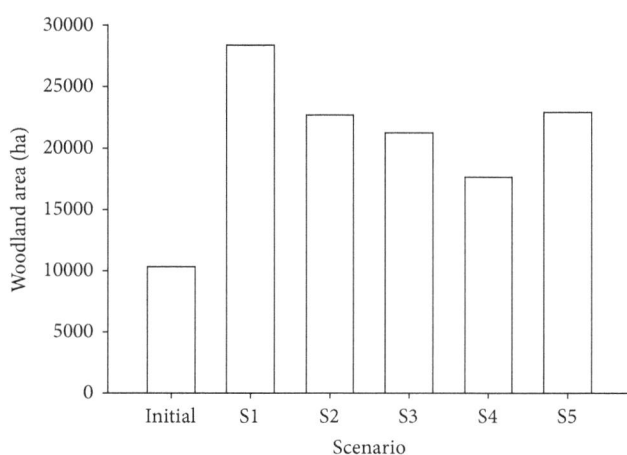

FIGURE 4: Projected woodland extent under different disturbance regimes. Values presented are the areas (ha) occupied by one or more of the five principal tree species (*Betula pendula, Fagus sylvatica, Ilex aquifolium, Pinus sylvestris*, and *Quercus robur*), as individuals ≥10 years old. "Initial" values are those at the onset of model scenarios; the values given under each scenario (S1–5) are those projected to occur after 300 years following the simulations described in the text.

60–90 years in the case of birch, 180–210 years in the case of holly, and 90–180 years in the case of Scots pine, precise values differing between scenarios (Figure 5). In contrast, beech demonstrated a continual increase in extent in all scenarios, which was most pronounced in Scenarios 1 and 5. Oak increased during the initial 30 year interval and thereafter remained relatively constant in extent, with a slight decline recorded after 90–180 years, depending on the scenario. After 300 years, values of oak and beech were higher than initial values in all scenarios, whereas the extent of birch was lower than the initial value in all scenarios. Although the highest final values of beech, oak, and holly were recorded in Scenario 1, the highest final value of birch was recorded in Scenario 5 and that of pine in Scenario 2.

Hypotheses (ii)–(iv) were tested by examining the output maps generated by LANDIS II. These indicated that the pattern of tree colonisation, as hypothesized, tended to occur particularly on the periphery of woodlands, leading to a concentric pattern of woodland expansion. However, this pattern was observed even in Scenario 4, without protection from herbivory by spiny shrubs. This indicates that woodland expansion is projected to occur even in the absence of

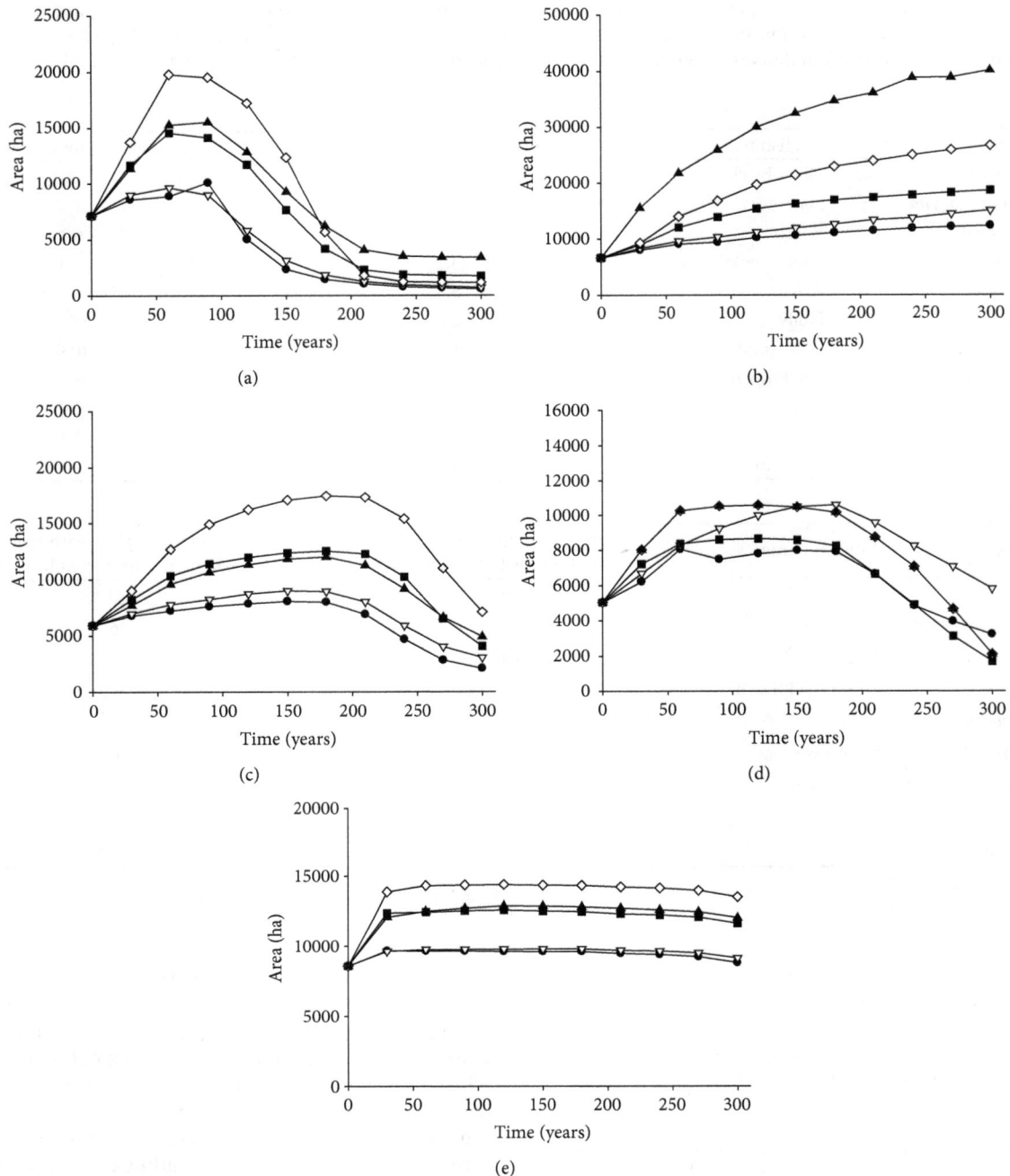

FIGURE 5: Projected extent of occurrence ("area") of selected tree species in the New Forest, using the LANDIS II model to explore different disturbance regimes. (a) *Betula pendula*; (b) *Fagus sylvatica*; (c) *Ilex aquifolium*; (d) *Pinus sylvestris*; (e) *Quercus robur*. Calculation of the extent of occurrence was based on the occupation of a 50 × 50 m site (pixel) by an individual tree of age ≥ 10 years old. Scenario 1, empty diamond; scenario 2, empty triangle; scenario 3, filled square; scenario 4, filled circle; scenario 5, filled triangle (see text for details).

facilitation, although to a lesser degree than when facilitation is present. Overall, woodland expansion was observed in all scenarios, with little evidence for the breakup of groves or their conversion to either grassland or heathland. Stands initially dominated by early successional species such as birch and Scots pine became increasingly dominated by beech over time. Evidence for the breakup or conversion of stands was examined by overlaying output maps for Scenario 4 after 300 years on that for time zero. A total of 2,169 50 × 50 m sites (pixels) were found to be woodland at time zero but were not projected to be wooded after 300 years; this contrasts with 39,133 sites that were wooded at both times and 31,550 that were wooded after 300 years but not at the outset. In other words, 5% of the woodland area present at the outset was not associated with woodland at the end of the simulation, in this most disturbed scenario. The corresponding value for Scenario 1 (no disturbance) was 0.2% (106 sites).

4. Discussion

Although large herbivores are now widely employed as a tool for biodiversity conservation management, their use remains controversial [33], particularly as evidence regarding their effectiveness is often lacking [14]. As noted by Gordon et al. [34], the use of large herbivores to achieve conservation management objectives requires an understanding of the relationships between grazing and biodiversity; specifically, managers require information on the potential impacts of different herbivore densities under different conditions. Such information is often difficult to obtain in a systematic manner because of the high cost of appropriate experimental designs and the need for measurement of multiple driving factors, before widely applicable generalizations can be made [34].

As illustrated here, the impacts of herbivory on the spatial dynamics of vegetation can potentially be explored using spatially explicit models. While progress has been made in understanding the influence of landscape pattern on the dynamics and movement of large herbivores, using a variety of different modelling approaches, relatively few models have been developed that can directly inform conservation management [35]. Key challenges to the development of such models include the need to incorporate interactions among multiple animal and plant species, together with landscape-scale processes such as plant dispersal, cultural features, and land use change and interactions between herbivory and other forms of disturbance [35]. As demonstrated here, LANDIS II can potentially overcome such challenges, enabling the impacts of herbivory on the spatial dynamics of vegetation to be explored in conjunction with other types of disturbance. LANDIS II also connects strongly with real-world data and can incorporate management activities as driving variables, two other features that are useful for addressing questions relating to conservation management [35]. Other landscape-scale models that can be used to explore the role of large herbivores in a management context include SAVANNA [36] and FORSPACE [37, 38]. SAVANNA differs from LANDIS II in that dispersal of individual plant species is not explicitly simulated, whereas FORSPACE differs in that age structure is treated much more simply; all trees are grouped into a single cohort [35]. However, both models explicitly incorporate energetics to simulate the removal of plant biomass by herbivores, rather than purely plant species and age as employed here.

As with any attempt at ecological modelling, the research described here is subject to a number of uncertainties [10], and these should be borne in mind when interpreting the results. While every attempt was made to ensure accurate parameterization of the model, a degree of uncertainty is inevitable given the lack of appropriate information. For example, it is widely recognized that the dispersal characteristics of many plant species remain poorly defined, particularly at larger scales [39, 40]. Greater accuracy might

be achieved if dispersal measurements for each species were conducted within the study landscape, an approach that was beyond the scope of the current investigation. Other potential disturbance factors that were not considered here include wind disturbance, climate change, and damage by *Sciurus caroliniensis* (grey squirrel), pests, and pathogens, all of which could be significant causes of tree mortality. Another key assumption was that browsing was assumed to be evenly distributed across the study area, which is likely not to be the case in reality, although accurate information on the spatial distribution of browsing mammals is currently lacking. Information on the browsing preferences of different mammal species within the study area, and the tolerance of different plant species to browsing damage, is also limited. Uncertainties relating to herbivore behaviour and browsing preferences are common to other approaches involving modelling of herbivore impacts on vegetation [35, 41]. The use of LANDIS II to explore herbivore impacts could potentially be strengthened in future, for example, through integration with individual based models of foraging behaviour [42].

In the New Forest, as in many other conservation areas where large herbivores are present, the identification of appropriate herbivore densities has been highly controversial. The persistence of some species of high conservation value is strongly dependent on maintenance of high herbivore pressure, as in the case of plant species associated with short grassland swards [27]. It is for this reason that recent management plans [43] have placed the maintenance of herbivore populations as a principal objective. However, marked declines in other groups of species, such as Lepidoptera, have been attributed to grazing being excessive [27], a conclusion supported by research into the ecology of large herbivores themselves [32]. Particular concern has focused on the perceived lack of woodland regeneration, which has repeatedly been used as justification for forest management interventions, some of which have resulted in negative environmental impacts [28, 44]. A number of authors have suggested that regeneration of tree species is severely limited or entirely absent in these woodlands as a result of high herbivore pressure [45–48]. However, these previous observations were based on a limited number of sites. The results of the present study, in what represents the first systematic survey of tree regeneration in all New Forest woodlands, indicate that tree regeneration occurs widely but at low density; for example, oak and beech saplings were recorded on 16% and 26% of sites, respectively. This is despite the fact that browsing pressure is uniformly high. While it is widely recognized that large herbivores can substantially reduce tree seedling densities and thereby reduce or eliminate cohorts of young trees [49], in the current study saplings were entirely absent in only 10% of woodland plots (Survey 1), and seedlings were entirely absent in only a single survey plot.

Modelling results projected an increase in forest area under all disturbance regimes, even under high browsing pressure and with management of heathland by rotational burning. The current colonization of grassland and heathland areas by tree species is confirmed by results of Surveys 2 and 3. These results are supported by analyses of historic maps, which indicate that native woodland area increased substantially in the New Forest during much of the 20th century [50]. It is for this reason that scrub clearance by cutting also continues to form part of the management of the New Forest, in addition to grazing and burning [43]. The impact of herbivory on the spatial dynamics of vegetation reflects the balance achieved between rates of woodland expansion and collapse, both of which may be highly variable in both space and time [49]. Vera's theory [16], relating to the cyclic dynamics of vegetation, depends critically on the breakup of woodland groves with maturity and their conversion to either grassland or heathland. Modelling results provided very little support for this phenomenon. Rather, the results indicated the progressive dominance of beech and an associated decline of relatively shade-intolerant species such as birch and Scots pine, through a process of competitive exclusion. This same general trend was observed in all disturbance regimes, although differing in rate between them. Increasing dominance of the more shade-tolerant beech is consistent with traditional successional theory [51] and is also supported by historical observations in the New Forest [50]. In addition, palaeoecological analyses have shown marked increases in the amount of beech pollen in the area over the past 500 years [52]. At the landscape scale, model projections indicated progressive expansion and coalescence of woodland areas even under high herbivore pressure, in contrast to the shifting mosaic of woodland and nonwoodland vegetation hypothesized by Vera [16]. Model exploration during calibration indicated that browsing pressures high enough to largely prevent regeneration within woodlands also prevented woodland expansion, which was inconsistent with field observations.

A further key element of Vera's theory [16] is the role of facilitation, involving the protection of tree seedlings from herbivory by spiny shrubs. Results from Surveys 2 and 3 provided some support for this hypothesis, indicated by the association of young trees with spiny shrubs, but also provided evidence of widespread colonization of both heathland and grassland sites by a range of tree species (especially birch) without such facilitation. This suggests that tree regeneration is neither restricted to the periphery of woodlands nor sites where spiny shrubs are present. These results can be compared with those of Bakker et al. [25], who reported a positive association between spiny shrubs and oak in the New Forest; however, this association was significant only for mature trees (>2 m height and mostly >10 cm dbh), and observations were restricted to a single site. The model simulations indicated that facilitation could potentially have a substantial effect on the relative abundance of individual tree species and on total woodland extent, the latter increasing by 30% over a 300 year interval as a result (Scenarios 4 versus 5, Figure 4). However, facilitation had relatively little impact on the pattern of woodland development; expansion from woodland margins was projected even in the absence of facilitation.

These results question whether it is appropriate to base conservation management plans on Vera's theory, which is currently the case in the New Forest [44], as in many other locations. Further research is required on the role of herbivory in woodland collapse and the quantitative importance of facilitation, in order to define with greater precision the situations under which Vera's theory is likely to apply. Such conditions might prevail under higher herbivore densities than those currently being experienced in the New Forest (approximately 1.9 animals ha^{-1} [25]), although current densities are historically at an exceptionally high value [27, 28].

5. Conclusions

As illustrated here, spatially explicit modelling approaches can potentially be of value in exploring the potential impacts of different conservation management options, including manipulation of herbivore pressure, and for identifying the appropriate location of interventions, such as the cutting or burning of vegetation. Modelling results also provide some insight into the potential long-term implications of high herbivore pressure on the structure and composition of vegetation, as well as its spatial pattern. One of the characteristics of the New Forest, examined here, is that a range of different habitats of high conservation value occur together in an intimate successional mosaic. In such locations, the contrasting requirements of different habitats may create potential management conflicts. For example, while high herbivore pressure is widely viewed as essential for maintenance of successional habitats such as heathland and grassland, in the New Forest this has created concerns about the impacts of herbivory on adjacent woodlands. The current results suggest that such conflicts can potentially be reconciled, as woodland regeneration was projected to occur even under high herbivore pressure. However, successional habitats may require active vegetation management such as cutting and burning, in addition to grazing, for succession to woodland to be prevented.

Conflict of Interests

This is to confirm that none of the authors have a direct financial relation with any of the commercial identities mentioned in this paper that might lead to a conflict of interests.

Acknowledgments

Thanks to Andrew Brown, Richard Reeves (New Forest Centre), Ed Mountford (JNCC), Jonathan Spencer, Berry Stone, and Simon Weymouth (Forestry Commission) for assistance with the research.

References

[1] D. Lindenmayer, R. J. Hobbs, R. Montague-Drake et al., "A checklist for ecological management of landscapes for conservation," *Ecology Letters*, vol. 11, no. 1, pp. 78–91, 2008.

[2] S. T. A. Pickett and P. S. White, Eds., *The Ecology of Natural Disturbance and Patch Dynamics*, Academic Press, New York, NY, USA, 1985.

[3] W. J. Sutherland, *The Conservation Handbook, Research, Management and Policy*, Blackwell Science, Oxford, UK, 2000.

[4] A. C. Newton, C. Echeverría, E. Cantarello, and G. Bolados, "Projecting impacts of human disturbances to inform conservation planning and management in a dryland forest landscape," *Biological Conservation*, vol. 144, no. 7, pp. 1949–1960, 2011.

[5] M. Uriarte, C. D. Canham, J. Thompson et al., "Natural disturbance and human land use as determinants of tropical forest dynamics: results from a forest simulator," *Ecological Monographs*, vol. 79, no. 3, pp. 423–443, 2009.

[6] W. J. Sutherland, "Predicting the ecological consequences of environmental change: a review of the methods," *Journal of Applied Ecology*, vol. 43, no. 4, pp. 599–616, 2006.

[7] D. B. Lindenmayer and J. F. Franklin, *Conserving Forest Biodiversity, A Comprehensive Multiscaled Approach*, Island Press, Washington, DC, USA, 2002.

[8] H. Bugmann, "A review of forest gap models," *Climatic Change*, vol. 51, no. 3-4, pp. 259–305, 2001.

[9] A. Porté and H. H. Bartelink, "Modelling mixed forest growth: a review of models for forest management," *Ecological Modelling*, vol. 150, no. 1-2, pp. 141–188, 2002.

[10] A. C. Newton, *Forest Ecology and Conservation. A Handbook of Techniques*, Oxford University Press, Oxford, UK, 2007.

[11] D. J. Mladenoff, "LANDIS and forest landscape models," *Ecological Modelling*, vol. 180, no. 1, pp. 7–19, 2004.

[12] R. M. Scheller, J. B. Domingo, B. R. Sturtevant et al., "Design, development, and application of LANDIS-II, a spatial landscape simulation model with flexible temporal and spatial resolution," *Ecological Modelling*, vol. 201, no. 3-4, pp. 409–419, 2007.

[13] W. Spencer, H. Rustigian-Romsos, J. Strittholt, R. Scheller, W. Zielinski, and R. Truex, "Using occupancy and population models to assess habitat conservation opportunities for an isolated carnivore population," *Biological Conservation*, vol. 144, no. 2, pp. 788–803, 2011.

[14] A. C. Newton, G. B. Stewart, G. Myers et al., "Impacts of grazing on lowland heathland in north-west Europe," *Biological Conservation*, vol. 142, no. 5, pp. 935–947, 2009.

[15] K. H. Hodder and J. M. Bullock, "Really wild? Naturalistic grazing in modern landscapes," *British Wildlife*, vol. 20, no. 5, pp. 37–43, 2009.

[16] F. W. M. Vera, *Grazing Ecology and Forest History*, CABI Publishing, Wallingford, UK, 2000.

[17] F. W. M. Vera, "Large-scale nature development—the Oostvaardersplassen," *British Wildlife*, pp. 28–36, 2009.

[18] K. H. Hodder, J. M. Bullock, P. C. Buckland, and K. J. Kirby, "Large herbivores in the wildwood and modern naturalistic grazing systems," English Nature Research Reports 648, English Nature, Peterborough, UK, 2005.

[19] P. Taylor, "Re-wilding the grazers: obstacles to the "wild" in wildlife management," *British Wildlife*, pp. 50–55, 2009.

[20] M. Galetti, "Parks of the Pleistocene: recreating the cerrado and the Pantanal with megafauna," *Natureza e Conservação*, vol. 2, no. 1, pp. 93–100, 2004.

[21] D. R. Rubenstein, D. I. Rubenstein, P. W. Sherman, and T. A. Gavin, "Pleistocene Park: does re-wilding North America represent sound conservation for the 21st century?" *Biological Conservation*, vol. 132, no. 2, pp. 232–238, 2006.

[22] K. J. Kirby, "A model of a natural wooded landscape in Britain as influenced by large herbivore activity," *Forestry*, vol. 77, no. 5, pp. 405–420, 2004.

[23] F. J. G. Mitchell, "How open were European primeval forests? Hypothesis testing using palaeoecological data," *Journal of Ecology*, vol. 93, no. 1, pp. 168–177, 2005.

[24] H. Olff, F. W. M. Vera, J. Bokdam et al., "Shifting mosaics in grazed woodlands driven by the alternation of plant facilitation and competition," *Plant Biology*, vol. 1, no. 2, pp. 127–137, 1999.

[25] E. S. Bakker, H. Olff, C. Vandenberghe et al., "Ecological anachronisms in the recruitment of temperate light-demanding tree species in wooded pastures," *Journal of Applied Ecology*, vol. 41, no. 3, pp. 571–582, 2004.

[26] T. Fukami and D. A. Wardle, "Long-term ecological dynamics: reciprocal insights from natural and anthropogenic gradients," *Proceedings of the Royal Society B*, vol. 272, no. 1577, pp. 2105–2115, 2005.

[27] A. C. Newton, Ed., *Biodiversity in the New Forest*, Pisces, Newbury Park, Calif, USA, 2010.

[28] A. C. Newton, "Social-ecological resilience and biodiversity conservation in a 900-year-old protected area," *Ecology and Society*, vol. 16, no. 4, article 13, 2011.

[29] F. Reimoser, H. Armstrong, and R. Suchant, "Measuring forest damage of ungulates: what should be considered," *Forest Ecology and Management*, vol. 120, no. 1–3, pp. 47–58, 1999.

[30] D. Ray, *Ecological Site Classification, a PC-Based Decision Support System for British Forests, Users Guide*, Forestry Commission, Edinburgh, UK, 2001.

[31] A. D. Syphard, J. Yang, J. Franklin, H. S. He, and J. E. Keeley, "Calibrating a forest landscape model to simulate frequent fire in Mediterranean-type shrublands," *Environmental Modelling and Software*, vol. 22, no. 11, pp. 1641–1653, 2007.

[32] R. J. Putman, *Grazing in Temperate Ecosystems, Large Herbivores and their Effects on the Ecology of the New Forest*, Croom Helm/Chapman and Hall, London, UK, 1986.

[33] M. F. Wallis De Vries, J. P. Bakker, and S. E. Van Wieren, Eds., *Grazing and Conservation Management*, Kluwer Academic, Dordrecht, The Netherlands, 1998.

[34] I. J. Gordon, A. J. Hester, and M. Festa-Bianchet, "The management of wild large herbivores to meet economic, conservation and environmental objectives," *Journal of Applied Ecology*, vol. 41, no. 6, pp. 1021–1031, 2004.

[35] P. J. Weisberg, M. B. Coughenour, and H. Bugmann, "Modelling of large herbivore-vegetation interactions in a landscape context," in *Large Herbivore Ecology, Ecosystem Dynamics and Conservation*, K. Danell, P. Duncan, R. Bergström, and J. Pastor, Eds., pp. 348–382, Cambridge University Press, Cambridge, UK, 2006.

[36] P. J. Weisberg and M. B. Coughenour, "Model-based assessment of aspen responses to Elk herbivory in rocky mountain national park, USA," *Environmental Management*, vol. 32, no. 1, pp. 152–169, 2003.

[37] K. Kramer, J. M. Baveco, R. J. Bijlsma et al., "Landscape forming processes and diversity of forested landscapes-description and application of the model FORSPACE," Alterra Report 216, Wageningen, The Netherlands, 2001.

[38] K. Kramer, G. W. T. A. Groot Bruinderink, and H. H. T. Prins, "Spatial interactions between ungulate herbivory and forest management," *Forest Ecology and Management*, vol. 226, no. 1–3, pp. 238–247, 2006.

[39] D. F. Greene and C. Calogeropoulos, "Measuring and modelling seed dispersal of terrestrial plants," in *Dispersal Ecology*, J. M. Bullock, R. E. Kenward, and R. S. Hails, Eds., pp. 3–23, Blackwell, Oxford, UK, 2002.

[40] P. Vittoz and R. Engler, "Seed dispersal distances: a typology based on dispersal modes and plant traits," *Botanica Helvetica*, vol. 117, no. 2, pp. 109–124, 2007.

[41] P. J. Weisberg and H. Bugmann, "Forest dynamics and ungulate herbivory: from leaf to landscape," *Forest Ecology and Management*, vol. 181, no. 1-2, pp. 1–12, 2003.

[42] R. A. Stillman, "MORPH-An individual-based model to predict the effect of environmental change on foraging animal populations," *Ecological Modelling*, vol. 216, no. 3-4, pp. 265–276, 2008.

[43] R. N. Wright and D. V. Westerhoff, *New Forest SAC Management Plan, English Nature*, Lyndhurst, NJ, USA, 2001.

[44] A. C. Newton, E. Cantarello, G. Myers, S. Douglas, and N. Tejedor, "The condition and dynamics of New Forest woodlands," in *Biodiversity in the New Forest*, A. C. Newton, Ed., pp. 132–147, Pisces, Newbury Park, Calif, USA, 2010.

[45] G. F. Peterken and C. R. Tubbs, "Woodland regeneration in the New Forest, Hampshire, Since 1650," *Journal of Applied Ecology*, vol. 2, no. 1, pp. 159–170, 1965.

[46] R. J. Putman, R. M. Pratt, J. R. Ekins, and P. J. Edwards, "Food and feeding behaviour of cattle and ponies in the New Forest, Hampshire," *Journal of Applied Ecology*, vol. 24, no. 2, pp. 369–380, 1987.

[47] E. P. Mountford and G. F. Peterken, "Long-term change and implications for the management of wood-pastures: experience over 40 years from Denny Wood, New Forest," *Forestry*, vol. 76, no. 1, pp. 19–40, 2003.

[48] E. P. Mountford, G. F. Peterken, P. J. Edwards, and J. G. Manners, "Long-term change in growth, mortality and regeneration of trees in Denny Wood, an old-growth wood-pasture in the New Forest (UK)," *Perspectives in Plant Ecology, Evolution and Systematics*, vol. 2, no. 2, pp. 223–272, 1999.

[49] R. M. A. Gill, "The influence of large herbivores on tree recruitment and forest dynamics," in *Large Herbivore Ecology, Ecosystem Dynamics and Conservation*, K. Danell, P. Duncan, R. Bergström, and J. Pastor, Eds., pp. 170–202, Cambridge University Press, Cambridge, UK, 2006.

[50] C. R. Tubbs, *The New Forest: History, Ecology and Conservation*, New Forest Ninth Century Trust, Lyndhurst, NJ, USA, 2001.

[51] H. Ellenberg, *Vegetation Ecology of Central Europe*, Cambridge University Press, Cambridge, UK, 4th edition, 1988.

[52] M. J. Grant and M. E. Edwards, "Conserving idealized landscapes: past history, public perception and future management in the New Forest (UK)," *Vegetation History and Archaeobotany*, vol. 17, no. 5, pp. 551–562, 2008.

[53] CAB International, *Forestry Compendium*, CAB International, Wallingford, UK, 2003.

[54] M. O. Hill, C. D. Preston, and D. B. Roy, *PLANTATT, Attributes of British and Irish Plants: Status, Size, Life History, Geography and Habitats For Use in Connection With the New Atlas of the British and Irish Flora*, Centre for Ecology and Hydrology, Peterborough, UK, 2004.

[55] O. Johnson and D. More, *Tree Guide, Harper*, Collins, London, UK, 2004.

[56] A. Mitchell, *Trees of Britain and Northern Europe*, Harper Collins, London, UK, 1978.

[57] U. Niinemets and F. Valladares, "Tolerance to shade, drought, and waterlogging of temperate northern hemisphere trees and shrubs," *Ecological Monographs*, vol. 76, no. 4, pp. 521–547, 2006.

[58] P. Thomas, *Trees: Their Natural History*, Cambridge University Press, Cambridge, UK, 2000.

[59] R. M. A. Gill, "A review of damage by mammals in north temperate forests: 1. Deer," *Forestry*, vol. 65, no. 2, pp. 145–169, 1992.

[60] R. M. A. Gill, *The Impact of Deer on Woodland Biodiversity*, vol. 36 of *Information Note*, Forestry Commission, Edinburgh, Scotland, 2000.

[61] R. M. A. Gill and V. Beardall, "The impact of deer on woodlands: the effects of browsing and seed dispersal on vegetation structure and composition," *Forestry*, vol. 74, no. 3, pp. 209–218, 2001.

[62] R. Harmer, A. Kiewitt, G. Morgan, and R. Gill, "Does the development of bramble (*Rubus fruticosus* L. agg.) facilitate the growth and establishment of tree seedlings in woodlands by reducing deer browsing damage?" *Forestry*, vol. 83, no. 1, pp. 93–102, 2010.

[63] R. J. Putman, *Competition and Resource Partitioning in Temperate Ungulate Assemblies*, Chapman and Hall, London, UK, 1996.

[64] R. J. Putman, P. J. Edwards, J. C. E. Mann, R. C. How, and S. D. Hill, "Vegetational and faunal changes in an area of heavily grazed woodland following relief of grazing," *Biological Conservation*, vol. 47, no. 1, pp. 13–32, 1989.

[65] A. F. M. Van Hees, A. T. Kuiters, and P. A. Slim, "Growth and development of silver birch, pedunculate oak and beech as affected by deer browsing," *Forest Ecology and Management*, vol. 88, no. 1-2, pp. 55–63, 1996.

8

A Communal Sign Post of Snow Leopards (*Panthera uncia*) and Other Species on the Tibetan Plateau, China

Juan Li,[1] **George B. Schaller,**[2] **Thomas M. McCarthy,**[3] **Dajun Wang,**[1]
Zhala Jiagong,[4] **Ping Cai,**[5] **Lamao Basang,**[6] **and Zhi Lu**[1,4]

[1] *Center for Nature and Society, College of Life Sciences, Peking University, Beijing 100871, China*
[2] *Panthera and Wildlife Conservation Society, 8 West 40th Street, 18th Floor, New York, NY 10018, USA*
[3] *Snow Leopard Program, Panthera, 8 West 40th Street, 18th Floor, New York, NY 10018, USA*
[4] *Shan Shui Conservation Center, Beijing 100871, China*
[5] *Wildlife Conservation and Management Bureau, Qinghai Forestry Department, Xining 810008, Qinghai, China*
[6] *Sanjiangyuan National Nature Reserve, Qinghai Forestry Department, Xining 810008, Qinghai, China*

Correspondence should be addressed to Juan Li; lijuan924@gmail.com

Academic Editor: Rafael Riosmena-Rodríguez

The snow leopard is a keystone species in mountain ecosystems of Central Asia and the Tibetan Plateau. However, little is known about the interactions between snow leopards and sympatric carnivores. Using infrared cameras, we found a rocky junction of two valleys in Sanjiangyuan area on the Tibetan Plateau where many mammals in this area passed and frequently marked and sniffed the site at the junction. We suggest that this site serves as a sign post to many species in this area, especially snow leopards and other carnivores. The marked signs may also alert the animals passing by to temporally segregate their activities to avoid potential conflicts. We used the Schoener index to measure the degree of temporal segregation among the species captured by infrared camera traps at this site. Our research reveals the probable ways of both intra- and interspecies communication and demonstrates that the degree of temporal segregation may correlate with the degree of potential interspecies competition. This is an important message to help understand the structure of animal communities. Discovery of the sign post clarifies the importance of identifying key habitats and sites of both snow leopards and other species for more effective conservation.

1. Introduction

The snow leopard (*Panthera uncia*), an endangered species listed in IUCN Red List [1], occurs in remote and rugged mountains across Central Asia and the Tibetan Plateau [2]. As a keystone species, this cat has an important role in these mountain ecosystems [3]. But with only an estimated population of 3500–7000 remaining, the snow leopard generally occurs at low densities. Poaching, retaliatory killing due to livestock-snow leopard conflict, decimation of prey, and habitat degradation have all contributed to its rarity [2]. Most studies of snow leopard have focused on its habitat [4, 5] and food habits [6–8], whereas little is known about the interaction of snow leopards and their sympatric species.

Olfactory and visual signals are known as crucial mediators for intraspecies communication. Mammals, especially solitary species, use these signals to advertise reproductive status, recognize sex, identify individuals, establish territory and so on [9, 10]. Communal sign posts have, for example, been found in snow leopard [11], giant panda (*Ailuropoda melanoleuca*) [12], and American black bear (*Ursus americanus*) [13] habitats. However, previous studies did not describe interspecies interactions at such locations.

In this study, we discovered a site sniffed and marked by snow leopards and several other carnivores, a sign post. White et al. (2003) used the term "bulletin board" to describe such a site used by pandas [14]. There, one animal leaves a message (such as feces or urine), followed by another animal which sees or smells this message. Such communication could help different species avoid direct confrontation and may lead to temporal segregation [15].

To identify the interaction between snow leopards and sympatric species, we recorded the date, time, and corresponding behavior of the animals passing by this site with infrared camera traps. The analysis of these behaviors showed that snow leopards and other carnivores frequently marked and sniffed at this site, indicating the site functions as a sign post. Our data also showed that the sympatric carnivores were temporally segregated in both annual and daily activity patterns. The degree of temporal segregation may be related to the degree of competition among the species.

2. Materials and Methods

2.1. Study Area. The field study was conducted in Suojia Township in Yushu Prefecture (Figure 1), the core area of the Sanjiangyuan National Nature Reserve, Qinghai Province, China. This region has a high density of snow leopards [6, 16]. The climate is typically windy and dry. The average annual temperature is $-4°C$ (range from $-12°C$ to $-20°C$ in January and $8°C$ to $-10°C$ in July), and the average annual precipitation is 150–420 mm [17]. Alpine meadow is the major vegetation type, consisting of mainly a layer up to 30 cm thick of turf densely covered with sedge (*Kobresia* spp.). In addition to the snow leopard, the main mammal species in this area include the gray wolf (*Canis lupus*), the Tibetan brown bear (*Ursus arctos*), the Eurasian lynx (*Lynx lynx*), the red fox (*Vulpes vulpes*), the Tibetan fox (*Vulpes ferrilata*), Pallas's cat (*Otocolobus manul*), the stone marten (*Martes foina*), the blue sheep (*Pseudois nayaur*), the Himalayan marmot (*Marmota himalayana*), the plateau pika (*Ochotona curzoniae*), the large-eared pika (*Ochotona macrotis*), and various small rodents [17].

2.2. Camera Trapping. From January 1, 2011 to December 31, 2011, one Reconyx (HC500 HyperFireTM) passive infrared camera (Reconyx Inc., USA) was installed 35 cm above ground beside an overhanging rock with several snow leopard scrapes and carnivore feces (site SJ001 in Figure 1). The camera worked without delay between each consecutive exposure. Every time animals triggered the camera, ten continuous images were captured. The camera was checked about every ten days to ensure the angle remained consistent and batteries were charged. No bait or scent lure was used. In the same way, another 15 infrared cameras were placed at 15 sites from late April 2011 onward. Unfortunately, only cameras on five sites worked properly for around six months, and these data are included here (site SJ002, SJ004, SJ008, SJ010, and SJ012 in Figure 1). Other cameras were lost or damaged.

The behaviors of carnivores were divided into three categories: sniffing (receiving information), marking (sending information), and others. Sniffing behavior was defined as smelling the ground or rock. And the marking behavior included behaviors such as scraping, cheek rubbing, back rubbing, spraying, urinating, or defecating. Dogs were included only when they were out on their own, not when they followed people.

2.3. Data Analysis. Camera trap data are usually analyzed by grouping detections occurring within ten or thirty minutes into one capture [18, 19]. However, this method has two flaws. The first flaw is that the time frame used to group the detections is artificial. Neither ten minutes nor thirty minutes has a clear biological explanation. The second flaw is that grouping detections occurring within a period of time into one capture decrease data accuracy, a method less suitable for analysis of daily activity patterns. Therefore, we define a single capture as one individual (including every individual within a group) passing by the camera once. To build the daily activity curves, we averaged the number of captures within the adjacent two hours as the number of captures for each species. For example, the capture number at 3:00 is the 2-hour average of captures from 2:00 to 3:59. The annual and daily overlap indexes between two species were calculated according to the formula of Schoener [20].

To summarize, we used the following formulas in our study.

(1) Activity index of time point A = (captures at this time point)/(captures at the whole time frame).

(2) Capture = $\sum_{i=1}^{n}$ Individuals (n = number of times animal passes by the camera).

(3)

$$\alpha_{ij} = 1 - \frac{1}{2}\sum_{a=1}^{n}\left|p_{ia} - p_{ja}\right|, \qquad (1)$$

where α_{ij} is the overlap index of temporal niche between species i and species j, p_{ia} and p_{ja} are the proportions of ath month (for annual overlap index) or hour (for daily overlap index) used by species i and species j, respectively.

(4) Total overlap index = Annual overlap index × Daily overlap index.

We employed the software R to do hierarchical cluster analysis of the pairwise total overlap index with average linkage method and generated a cluster dendrogram [21].

We did not include marmot, pika, and stone marten in these analyses because they were too small for our camera to detect with its 35 cm height above ground, or too infrequently recorded for analysis.

3. Results

3.1. Camera Trap Captures. In 365 trap days from January 1, 2011 through December 31, 2011, we obtained 485 captures of carnivores and 1100 captures of herbivores at the site SJ001 (Figure 1), including snow leopards (145 captures), gray wolves (106), domestic dogs (73), brown bears (14), red foxes (81), Tibetan foxes (44), Pallas's cats (16), stone martens (6), blue sheep (891), and livestock (209) (Table 1, Figure 2). Snow leopards were the most frequently photographed carnivores,

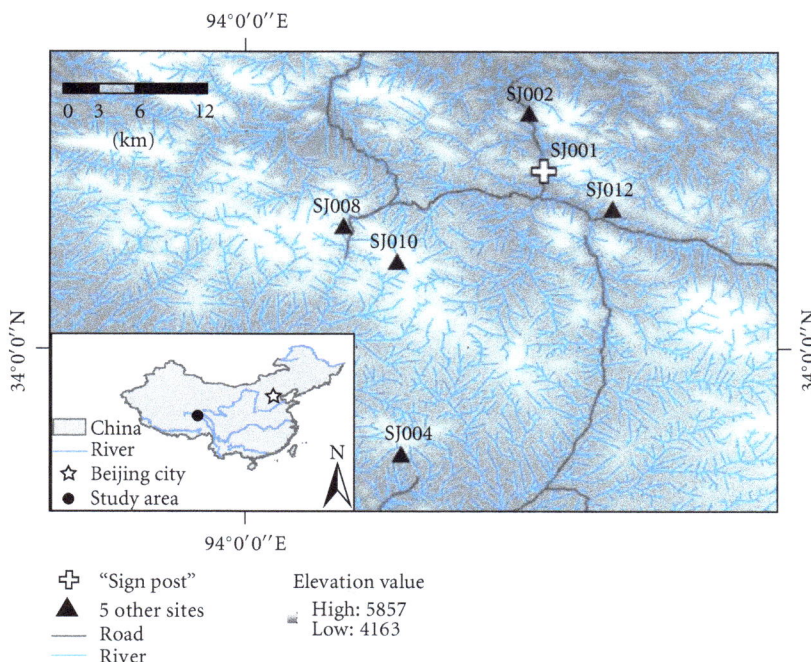

FIGURE 1: Locations of camera traps in Suojia Township in Yushu Prefecture, Qinghai Province, China.

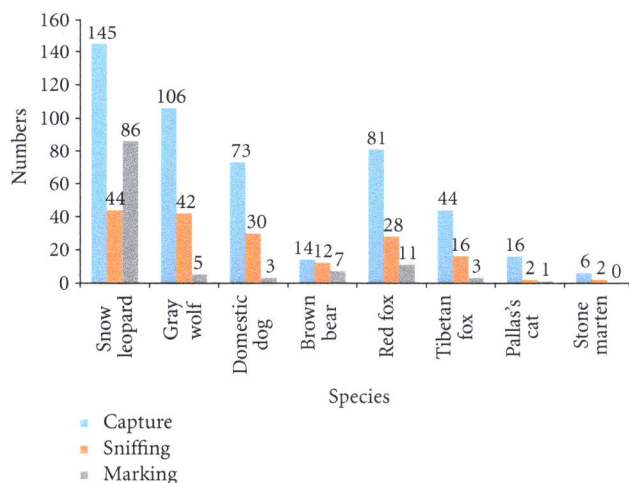

FIGURE 2: Camera trap captures of each carnivore species with sniffing and marking behaviors on the communal sign post.

and we identified six adults and one cub. As for sympatric carnivores, no fewer than five brown bears and six wolves were identified. We were unable to discriminate between the individuals of red foxes (including one young) or Tibetan foxes. The species photographed represented almost all the carnivores found in this area except the lynx and other small mustelids. The captures of camera traps from 5 other sites will be used for comparison in discussion section.

3.2. Behavior. Most carnivores did not just pass by the site SJ001 but spent time sniffing and marking it (Table 1, Figure

2). Snow leopards marked most actively, doing so in 86 out of 145 captures. Wolves marked approximately ten meters from this camera, but from October to December they began to mark the ground near by it. Dogs also sniffed here frequently. Brown bears sniffed this site almost every time they passed and often rubbed their back. Red and Tibetan foxes both seemed to prefer to sniff more than to mark.

3.3. Activity Patterns. Among the large carnivores, snow leopards visited this site (SJ001) throughout the year, their annual activity curve peaking in February (Figure 3(a)). In contrast, brown bears occurred mainly in May and October, and wolves appeared to be most active at this site in February and October. Domestic dogs showed a peak in winter (Figure 3(a)). As for daily activity patterns, snow leopards were active at this site throughout the night with a slightly crepuscular trend. Wolves passed by at noon and dusk, whereas brown bears only appeared at the rock before dawn (Figure 3(b)).

Among the small carnivores, the annual activity pattern of Tibetan foxes was very similar to that of brown bears, while red foxes passed by throughout the year (Figure 4(a)). Daily patterns showed that red foxes typically visited only at night, whereas Tibetan foxes and Pallas's cats came here throughout the day (Figure 4(b)).

The herbivores photographed at this site were livestock (only domestic yaks at this site) and blue sheep. Livestock passed by this site after 8:00 AM and returned before 2:00 PM, their activity pattern depending on the herders (Figure 5(a)). Blue sheep appeared to avoid livestock by appearing at this site before 10:00 AM and after 2:00 PM, but never at night (Figure 5(b)).

TABLE 1: Numbers of captures, sniffing, and marking behavior of each species captured by camera trapping at the communal sign post.

Group	English name	Latin name	IUCN[1]	CN[2]	Capture	Sniffing	Marking
Large carnivores	Snow leopard	*Panthera uncia*	EN	I	145	44	86
	Gray wolf	*Canis lupus*	LC	II	106	42	5
	Domestic dog	*Canis lupus*			73	30	3
	Brown bear	*Ursus arctos*	LC	II	14	12	7
Small carnivores	Red fox	*Vulpes vulpes*	LC	—	81	28	11
	Tibetan fox	*Vulpes ferrilata*	LC	—	44	16	3
	Pallas's cat	*Otocolobus manul*	NT	II	16	2	1
	Stone marten	*Martes foina*	LC	II	6	2	0
Herbivores	Blue sheep	*Pseudois nayaur*	LC	II	891	NA	NA
	Domestic yak	*Bos mutus*			209	NA	NA

[1] IUCN endangered species category (Red List): EN: endangered, NT: near threatened, LC: least concern.

[2] Category of protected wildlife species under China's Wildlife Protection Law, —: means not listed.

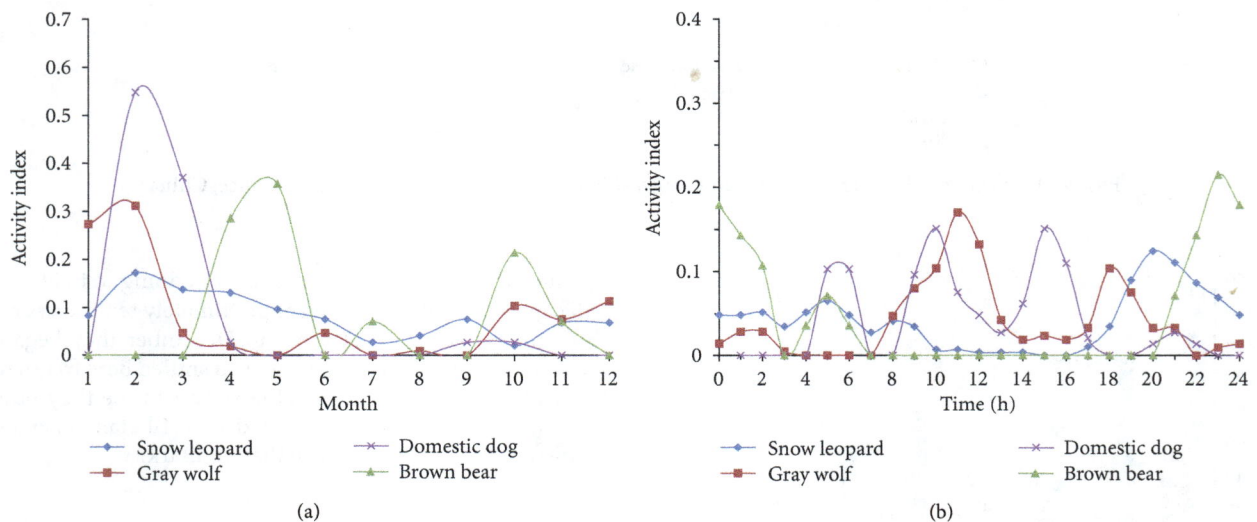

FIGURE 3: (a) Annual activity patterns and (b) daily activity patterns of large carnivores on the communal sign post. Snow leopard ($n = 145$), gray wolf ($n = 106$), domestic dog ($n = 73$), and brown bear ($n = 14$) are included.

4. Discussion

4.1. Sign Post. Compared with the captures of 5 other cameras placed nearby (Figure 1), the number of snow leopards captured at site SJ001 was much greater (see in Table 2). This difference indicates that this site had special relevance for intraspecies communication in this area. Additionally, the diversity of carnivore species photographed was the highest, implying that this site might also serve as a focal point for interspecies communication (see in Table 2). Previous studies have showed that mammals select conspicuous objects at major traffic hubs to communicate with each other. For example, snow leopards prefer to mark at mountain passes, cliff bases, confluence of rivers, and other conspicuous places [4, 22]. Brown bears tend to mark along travel routes [23], whereas wolves prefer crossroads [24, 25]. This site is located at a juncture of two major valleys, with an overhanging rock at the base of a cliff. Animals traveling along these valleys pass by here, and the rock is at an appropriate height for

depositing scent and rubbing. Whereas, other 5 sites are not located at crossroads. Therefore, it is not surprising that this site is favored by so many carnivore species and serves as a communal sign post.

4.2. Temporal Segregation. Many mammals shared this sign post, especially the carnivores which sniffed and marked it frequently. In addition, carnivores prey on herbivores, and they also prey on each other; and age, size, and patterns of grouping largely determine the outcome of such interactions [26, 27]. Temporal segregation is an important mechanism for these species to coexist and avoid direct confrontation [28]. To understand this temporal segregation, we first analyzed the probable factors affecting annual and daily activity patterns of the different species and then combined the pairwise annual and daily overlap indexes into a cluster dendrogram to show their interrelationship.

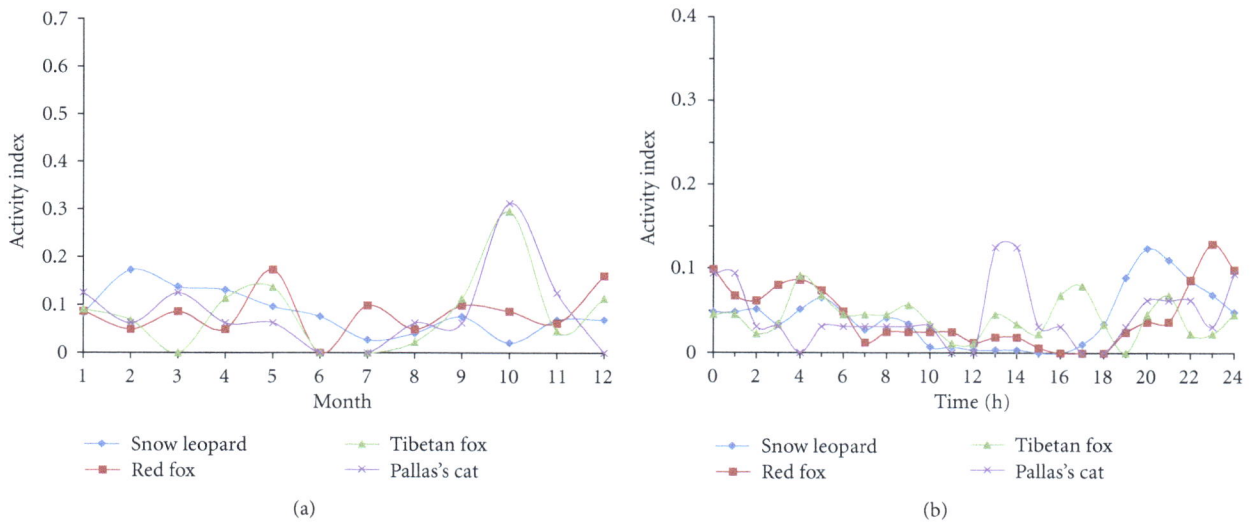

FIGURE 4: (a) Annual activity patterns and (b) daily activity patterns of snow leopard ($n = 145$) and small carnivores on the communal sign post. Red fox ($n = 81$), Tibetan fox ($n = 44$), and Pallas's cat ($n = 16$) are included.

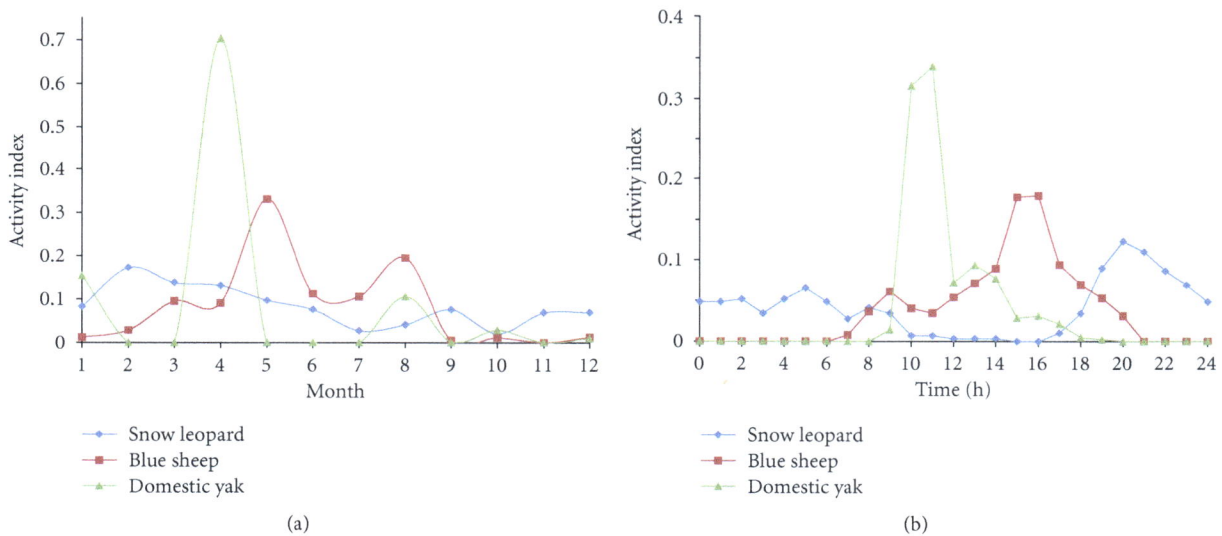

FIGURE 5: (a) Annual activity patterns and (b) daily activity patterns of the snow leopard ($n = 145$) and herbivores on the communal sign post. Blue sheep ($n = 891$) and livestock ($n = 209$) are included.

Annual activity patterns are, to a large extent, determined by the natural cycles and somewhat different habitat preferences of species. Snow leopards, for example, remain in steep rocky terrains the whole year, but they are slightly more active in the February mating season when they may roam widely to find a mate. Wolves prefer to live in less precipitous terrains [29]. The activity peak in February may be due to their mating season, but we have no explanation for the October one. Bears apparently mate mainly in May and June and hibernate from about late October to early March, according to our preliminary data. We found several caves in which bears had hibernated near this communal sign post. Brown bears are said to be much more active before hibernation to lay on fat [30]. Consequently, the annual activity curve of these bears shows peaks in May and in October.

Compared to annual activity patterns, the daily ones appeared to be more flexible. Affected by various factors, among them hunting and other human activities, animals may become more nocturnal or seek different habitats. The interactions among the various species may also affect it. To be specific, the snow leopard is active mainly at night, and its prey, the blue sheep which is active around this site only during the day, showed a conspicuous difference of their activity patterns. Wolves tend to be nocturnal, with activity

Cluster dendrogram

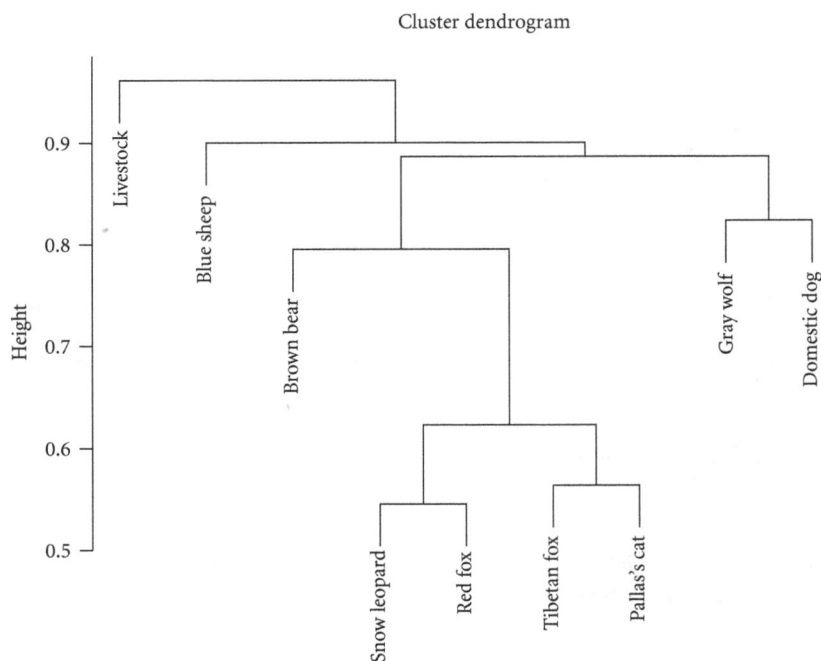

FIGURE 6: Cluster analysis dendrogram of pairwise total overlap indexes. It shows the interrelationship of the nine species' activity patterns at the communal sign post. Height represents the distance linkage between clusters.

TABLE 2: Number of captures of each species on six sites from May 1 to October 31, 2011.

Group	Species	SJ001[1]	SJ002	SJ004	SJ008[2]	SJ010	SJ012
Large carnivores	Snow leopard	49	14	8	6	7	36
	Gray wolf	17	2	0	0	0	1
	Domestic dog	4	0	0	1	0	0
	Brown bear	9	0	0	0	3	1
Small carnivores	Red fox	41	5	14	46	7	2
	Tibetan fox	25	8	0	0	0	0
	Pallas's cat	8	25	7	9	2	1
	Stone marten	2	0	1	2	0	0
Herbivores	Blue sheep	677	188	365	242	10	69
	Livestock	28	81	46	184	0	0
Total		860	323	441	490	29	110

[1] The communal sign post.

[2] The camera did not work from July 28 to August 9 because batteries were dead.

levels peaking at dawn and dusk in other areas [31, 32], whereas their activity patterns were diurnal at the communal sign post. This could be due to avoidance between wolves and snow leopards, causing one species shift its activity pattern into an opposite phase [28]. Brown bears only passed by this site at night with peaks at 23:00 and 5:00 as shown in our few records. We found that bears would follow snow leopards to scavenge on their kills, and this is perhaps one reason their daily activity patterns are similar.

To measure the degree of temporal segregation between any two species at this communal sign post, we introduced the total overlap index (see Section 2), which represents the overall effect of annual and daily temporal segregation. The dendrogram concisely shows the result of hierarchical clustering of the complex pairwise total overlap indexes (Figure 6). The species within the same group tend to have similar activity patterns. As shown in Figure 6, the nine species were clustered into 3 large branches, namely, livestock, blue sheep, and the carnivores. As expected, herbivores avoid predation by large carnivores by staggering their presence at this site. Then there are two apparent clusters under the carnivore branch, one composed of brown bears, Pallas's cats, snow leopards, and the two fox species; and the other of gray wolves and domestic dogs. Wolves occasionally kill dogs, but they were clustered together. The three main large carnivores, those with the greatest potential conflict—snow leopard, wolf, and brown bear—were grouped into different clusters, showing that they might have developed a mechanism to avoid conflicts, especially the wolf and the snow leopard. Thus, the degree of segregation may relate to the position of species in the food chain, from herbivores, small carnivores to large carnivores, as well as to body size.

5. Conclusions

In summary, using camera traps, we identified an important marking site or sign post used by snow leopards and several sympatric species for intra- and interspecies communication. We introduced the Schoener index and clustering dendrogram to measure and represent the degree of temporal segregation between species photographed and found that they were temporally segregated more or less at this location, presumably to avoid direct confrontations. Such a communal sign post provided preliminary knowledge of the finer-scaled habitat use of snow leopards and of niche segregation between species, an important consideration for protecting key habitats of snow leopards and other sympatric mammals. Future research could locate and investigate other such sites in order to learn how intraspecies and interspecies communications are conducted, as well as understanding the community structures and role of snow leopards in the ecosystem.

Conflict of Interest

We certify that there is no conflict of interest in this paper.

Acknowledgments

This study was a preliminary effort of a larger snow leopard research and conservation project jointly conducted by Qinghai Forestry Department, Shan Shui Conservation Center, the Center for Nature and Society of Peking University, and the Panthera and the Snow Leopard Trust. The authors thank Tudengzhaxi, Angye, Nangcaicicheng, Naobowending, Bairang, and others from Sub-Gongsa Monastery; and Lan Wu, Lingyun Xiao, and Meiqi Liu from Peking University for their assistance in the field work. Others contributed to the project, among them Hang Yin, Haiyuan Ma, Dawajiangcai, and Zhaxiduojie. They also extend their special appreciation to Sandan Li, Li Zhang, Yu Zhang of the Qinghai Forestry Department and Ruofan Li, Renzeng, and others from Qinghai Sanjiangyuan Nature Reserve.

References

[1] R. M. Jackson, D. Mallon, T. McCarthy, R. A. Chundaway, and B. Habib, "*Panthera uncia,*" in *IUCN, 2011: IUCN Red List of Threatened Species*, Version 2011. 2, 2008.

[2] T. M. McCarthy and G. Chapron, *Snow Leopard Survival Strategy*, ISLT and SLN, Seattle, Wash, USA, 2003.

[3] Y. V. Bhatnagar, V. B. Mathur, and T. McCarthy, "A regional perspective for snow leopard conservation in the Indian Trans-Himalaya," in *Contributed Papers to the Snow Leopard Survival Strategy Summit*, T. McCarthy and J. Weltzin, Eds., pp. 25–47, International Snow Leopard Trust, Seattle, Wash, USA, 2002.

[4] R. M. Jackson and G. Ahlborn, "A preliminary habitat suitability model for the snow leopard (*Panthera uncia*)," in *International Pedigree Book of Snow Leopards*, vol. 4, pp. 43–52, 1984.

[5] M. Wolf and S. Ale, "Signs at the top: habitat features influencing snow leopard *Uncia uncia* activity in sagarmatha national park, nepal," *Journal of Mammalogy*, vol. 90, no. 3, pp. 604–611, 2009.

[6] G. B. Schaller, R. Junrang, and Q. Mingjiang, "Status of the snow Leopard *Panthera uncia* in Qinghai and Gansu Provinces, China," *Biological Conservation*, vol. 45, no. 3, pp. 179–194, 1988.

[7] S. Bagchi and C. Mishra, "Living with large carnivores: predation on livestock by the snow leopard (*Uncia uncia*)," *Journal of Zoology*, vol. 268, no. 3, pp. 217–224, 2006.

[8] T. Sangay and K. Vernes, "Human-wildlife conflict in the Kingdom of Bhutan: patterns of livestock predation by large mammalian carnivores," *Biological Conservation*, vol. 141, no. 5, pp. 1272–1282, 2008.

[9] K. Rails, "Mammalian scent marking," *Science*, vol. 171, no. 3970, pp. 443–449, 1971.

[10] J. F. Eisenberg and D. G. Kleiman, "Olfactory communication in mammals," *Annual Review of Ecology and Systematics*, vol. 3, pp. 1–32, 1972.

[11] R. M. Jackson, *Home range, movements and habitat use of snow Leopard (Uncia uncia) in Nepal [Ph.D. thesis]*, University of London, London, UK, 1996.

[12] R. R. Swaisgood, D. G. Lindburg, A. M. White, Z. Hemin, and Z. Xiaoping, "Chemical communication in giant pandas: experimentation and application," in *Giant Pandas: Biology and Conservation*, D. G. Lindburg and K. Baronga, Eds., pp. 106–120, Oxford University Press, Oxford, UK, 2004.

[13] T. L. Burst, "Black bear mark trees in the Smoky Mountains," in *Proceedings of the International Conference on Bear Research and Management*, vol. 5, pp. 45–53, 1983.

[14] A. M. White, R. R. Swaisgood, and H. Zhang, "Chemical communication in the giant panda (*Ailuropoda melanoleuca*): the role of age in the signaller and assessor," *Journal of Zoology*, vol. 259, no. 2, pp. 171–178, 2003.

[15] T. W. Schoener, "Resource partitioning in ecological communities," *Science*, vol. 185, no. 4145, pp. 27–39, 1974.

[16] Y. Liao, "The geographical distribution of ounces in Qinghai Province," *Acta Theriologica Sinica*, vol. 5, pp. 183–188, 1985.

[17] D. P. Mallon, "Research, survey and biodiversity planning on the Tibet-Qinghai Plateau, China," Final Report For Darwin Initiative For the Survival of Species, 2004.

[18] S. Li, W. J. Mcshea, D. Wang, L. Shao, and X. Shi, "The use of infrared-triggered cameras for surveying phasianids in Sichuan Province, China," *Ibis*, vol. 152, no. 2, pp. 299–309, 2010.

[19] T. G. O'Brien, M. F. Kinnaird, and H. T. Wibisono, "Crouching tigers, hidden prey: Sumatran tiger and prey populations in a tropical forest landscape," *Animal Conservation*, vol. 6, no. 2, pp. 131–139, 2003.

[20] T. W. Schoener, "The Anolis lizards of Bimini: resource partitioning in a complex fauna," *Ecology*, vol. 49, pp. 704–726, 1968.

[21] R Development Core Team, *R: A language and Environment for Statistical Computing*, R Foundation for Statistical Computing, Vienna, Austria, 2011.

[22] G. B. Schaller, *Mountain Monarchs: Wild Sheep and Goats of the Himalaya*, University of Chicago Press, Chicago, Ill, USA, 1977.

[23] G. I. Green and D. J. Mattson, "Tree rubbing by yellowstone grizzly bears *Ursus arctos*," *Wildlife Biology*, vol. 9, no. 1, pp. 1–9, 2003.

[24] R. P. Peters, "Scent-marking in wolves: radio-tracking of wolf packs has provided definite evidence that olfactory sign is used for territory maintenance and may serve for other forms of communication within the pack as well," *American Scientist*, vol. 63, pp. 628–637, 1975.

[25] I. Barja, F. J. De Miguel, and F. Bárcena, "The importance of crossroads in faecal marking behaviour of the wolves (*Canis lupus*)," *Naturwissenschaften*, vol. 91, no. 10, pp. 489–492, 2004.

[26] G. A. Polis, C. A. Myers, and R. D. Holt, "The ecology and evolution of intraguild predation: potential competitors that eat each other," *Annual review of ecology and systematics*, vol. 20, pp. 297–330, 1989.

[27] F. Palomares and T. M. Caro, "Interspecific killing among mammalian carnivores," *American Naturalist*, vol. 153, no. 5, pp. 492–508, 1999.

[28] N. Kronfeld-Schor and T. Dayan, "Partitioning of time as an ecological resource," *Annual Review of Ecology, Evolution, and Systematics*, vol. 34, pp. 153–181, 2003.

[29] L. D. Mech and L. Boitani, *Wolves: Behavior, Ecology, and Conservation*, University of Chicago Press, Chicago, Ill, USA, 2003.

[30] J. Naves, A. Fernández-Gil, C. Rodrìguez, and M. Delibes, "Brown bear food habits at the border of its range: a long-term study," *Journal of Mammalogy*, vol. 87, no. 5, pp. 899–908, 2006.

[31] S. B. Merrill and L. David Mech, "The usefulness of GPS telemetry to study wolf circadian and social activity," *Wildlife Society Bulletin*, vol. 31, no. 4, pp. 947–960, 2003.

[32] M. Unit, "Daily patterns and duration of wolf activity in the Bialowieza forest, Poland," *Journal of Mammalogy*, vol. 84, pp. 243–253, 2003.

9

Status, Distribution, and Diversity of Birds in Mining Environment of Kachchh, Gujarat

Nikunj B. Gajera, Arun Kumar Roy Mahato, and V. Vijay Kumar

Gujarat Institute of Desert Ecology, Mundra Road, P.O. Box 83, Bhuj, Kachchh, Gujarat 370001, India

Correspondence should be addressed to Nikunj B. Gajera; gajeranikunj@gmail.com

Academic Editor: Rafael Riosmena-Rodríguez

Opencast mining is one of the major reasons for the destruction of natural habitats for many wildlife including birds. The Kachchh region belongs to the arid part of India and is one of the rich areas of mineral resources in the country. In the recent time and after the 2001 earthquake, mining and other developmental activities are increased, and as a result, the natural habitats of birds are disturbed and fragmented. So, this study was conducted to assess the impact of mining and associated activities on the diversity and distribution of birds. Birds were studied by surveying 180 transects along 9 zones around three selected major mines, and each zone is made in every 2 km radius from the mine. Based on the record, it was found that the density and diversity of birds are highest in zone 5 and lowest in zone 1 and zone 2, respectively. The result indicates that the diversity and abundance of birds were less in zones which are located close to the mines in comparison to the zones far from the mines. In conclusion, mining and its associated activities have some impacts on the diversity and distribution of birds in Kachchh region in India.

1. Introduction

Mining and its related activities are one of the major causes for the destruction of natural habitats for wildlife. As these activities increased in context of rapid use of natural resources to meet the demand of the market and the development of region and country, the pressure of threat to the wildlife is increasing day by day. In addition to mining, large parts of forests and wildlife habitats are being cleared for agriculture, industry, roads and railways network, and human habitation leading to degradation, fragmentation, and loss of habitat contributing to the overall loss of biological diversity. The loss of habitat and biological diversity by means of mining and its associated activities creates imbalance in ecological equilibrium.

Kachchh is one of the rich avifaunal diversity areas of India and also falls under one of the migratory route of the avifauna of this country. The list of birds of Kachchh district was made by some workers from time to time; notably among them were Lester [1], Ali [2], Ripley [3], Himmatsinhji [4], and Maharao [5]. The Kachchh district supports around 303

species of birds as recorded by Tiwari [6] and 370 bird species as reported by Sen [7] with some species including raptors, waterfowl, waders, and larks which are commonly found in this district. Similar to that, Kachchh is also a rich area of minerals including limestone, bentonite, and lignite. After the 2001 earthquake in Kachchh, many steps like industrialization and mining activities were increased for the development of this region. The consequence of the above activities results in the loss and fragmentation of habitat for wildlife. The major impacts on avifaunal diversity by means of the above development in postearthquake period are not being studied and assessed. In addition to that, so far no studies were made on the status, distribution, and diversity of birds in relation to the impact of mining and other developmental activities.

The bird species are widely distributed among various habitats in western Kachchh region. Various factors associated with different habitat types had a distinct impact on bird species. Likewise, various land use activities especially opencast mining that is being done at a mass scale in the region also affect the bird populations considerably, so we tested the following research hypotheses.

Ho: There is an impact of mining on the distribution and diversity of bird species.

2. Materials and Methods

2.1. Study Area. The study was conducted in three major mining areas, namely, Pandhro, Mata-nu-madh, and Jadva. The mining activities are carried out for extraction of minerals including limestone, bentonite, and lignite. All the three mines are opencast mines results into that major loss of habitats by extraction of minerals and dumping of waste materials in open ground. These mines are situated in the westernmost part of Kachchh district, Gujarat, and are very close to the Narayan Sarovar Wildlife Sanctuary, which is the only habitat in Kachchh for Indian gazelle (*Gazella benetti*) and some other wildlife. This area falls in the arid zone of the country and is characterized by presence of an admixture of distinguishable habitats such as *Acacia* forest (AF), *Euphorbia salvadora* forest (ES), mixed thorn forest (MTF), *Prosopis* forest (PF), dense grassland (DG), and sparse grassland (SG). Some seasonal and perennial wetlands and agricultural fields were also interspersed in and around the mining areas. The habitats found in the study area were broadly divided into seven habitat types which are described below.

Acacia Forest (AF). In this habitat, *Acacia senegal* and *Acacia nilotica* were dominant species with negligible distribution of *P. juliflora*.

Mixed Thorn Forest. This habitat is a complex distribution of species like *Acacia nilotica*, *A. senegal*, *Euphorbia caducifolia*, *Grewia* spp. *Commiphora wightii*, and *Salvadora* spp. with some individuals of *P. juliflora*.

Euphorbia salvadora Forest. This habitat, dominated by *Euphorbia caducifolia*, *Salvadora persica*, and *Salvadora oleoides*.

Prosopis Forest. This habitat dominated by the *P. Juliflora* (an Invasive alien species) with few other species like *Grewia tenax*, *G. villosa*, and *Capparis deciduas* which were also found.

Grassland Forest. Some areas were flat to gentle undulate terrain covered by annual or perennial grasses, and absence of shrub or tree species which were categorized under grassland. Grassland has a main component of the member of Gramineae family. *Cymbopogon spp.* and *Dichanthium spp.* are the major grass species.

Agriculture Field. This habitat included agriculture field (fallow or current agriculture) in which annually single cropping system was existed. The agricultural fields were also invaded by *P. juliflora*, *Capparis deciduas*, or *Salvadora spp.*

Wetland. Wetland habitat included man-made and natural water bodies with the presence of water in the whole year or some parts of the year. Most of the wetlands of the study area were man-made due to low precipitation rate.

2.2. Sampling Method. Prior to field surveys, the study area was stratified into various distinguishable habitats such as *Acacia* forest (AF), *Euphorbia salvadora* forest (ES), mixed thorn forest (MTF), *Prosopis* forest (PF), dense grassland (DG), and sparse grassland (SG). The above habitats were evenly distributed in the whole study area. After stratification, the entire area was divided into 5 × 5 km grids using survey of India's georeferenced coordinate system. Those grid cells were further subdivided into 1 × 1 km smaller grids, and a total of 180 transects were laid down randomly within the grids (Figure 1). The length of each transect was 1.1 km and was laid down randomly to cover each kind of habitats. To understand the impact of mining on birds, the diversity and distribution pattern of bird was recorded by dividing the area into nine circular zones (zone 1 to zone 9) with an interval of 2 km from the centre of mine (Figure 1). Care was taken to have adequate sampling in each of the topographical features across lateral and vertical gradients like altitudinal range and terrain, and spatially within each habitat type.

Bird survey was done using direct count methods which include (1) point count method and (2) area search method. In point count method [8], birds were recorded in four circular plots with 25 m radius in every 300 m distance along transects. In area search method [9], survey was made in 3 m wide belt along transect in between the circular plot. The time taken for survey varied depending on the terrain and topography. Surveys were conducted in the morning hours (6.30 a.m. to 9.30 a.m.) and evening hours (3.30 p.m. to 6.30 p.m.) by a single observer. Samplings were made in seasonal basis for the period of three years (2008 to 2010) using same transect and time. Observations were carried out with the aid of 8 × 40 binoculars, and field characteristics were noted down on special ornithological data sheet which included species, number of individuals, activity, microhabitat, threats with other minor details. The birds were identified with the help of Ali and Ripley [10], Ali [11], and Grimmett et al. [12]. Birds sighted during our survey were categorized based on their migratory nature including resident breeder (RB), resident (R), and migrant (M) according to Ali [11] and also categorized into their trophic guilds like insectivore, nectarivore, omnivore, scavenger, frugivore, carnivore, piscivore, and granivore according to Willis [13], Robinson et al. [14], and Anjos [15].

2.3. Data Analysis. The data collected during the whole study period were analyzed in PAST statistical software [16] to calculate species diversity, density, richness, and species composition by using the following formulae.

(A) Shannon Wiener diversity index (H') $H' = \sum P_i \times \ln(P_i)$, where

H = index of species diversity,

P_i = proportion of total sample belonging to the ith species,

$\ln$ = natural logl.

(B) Density = no. of individuals/total area surveyed (in hectare).

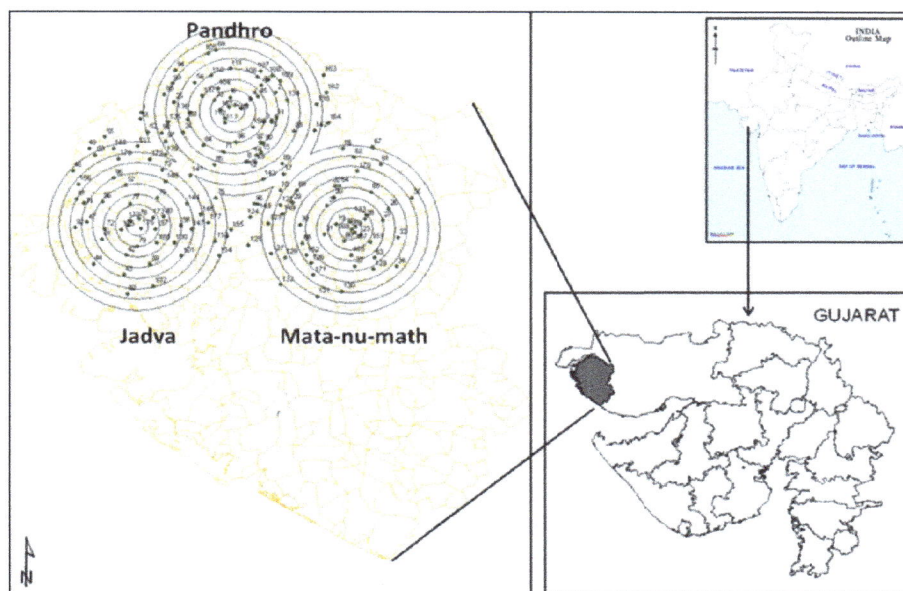

FIGURE 1: Location of the study area and transects (spots) and various zones (in circles) made around the three selected mines (Pandhro, Mata-na-madh, and Jadva) of western Kachchh for bird study.

(C) Menhinick richness index = the ratio of the number of taxa to the square root of sample size.

(D) Buzas and Gibson's evenness: e^H/S.

The rarefaction analysis was carried by using Biodiversity Pro software, 1997, to cope up the problem in comparing diversity among various land cover or habitat categories evaluated during the present study.

2.4. Hypothesis Testing. The null hypotheses tested for significance of differences was the following.

Ha: There is no impact of mining on the distribution and diversity of bird species.

Analysis of variance (ANOVA) was used to test the hypotheses, and Tukey's HSD post hoc analysis was carried to identify specific variables that differ significantly. The variables used for testing the hypotheses included the number of species, density, Shannon diversity index, and Menhinick species richness index. These variables representing bird species diversity were evaluated from each mine zone from Z1 to Z9.

Further, cluster analysis was made to quantify the resemblance among bird communities inhabiting various mine zones. Cluster analysis was performed on bird species. Abundance data were obtained from transect survey using minimum variances technique, known as Ward's method, which has been recognized as the best way to classify the ecological communities and to identify community structure. The cluster analysis generated a dendrogram providing hierarchically nested groups or clusters representing distinct bird communities, which were represented by subclusters.

3. Results

On surveying the mining and surrounding peripheral areas of the study area including all kinds of habitat, 252 species of birds (see Table 2) were recorded in western Kachchh. Various zones from zone 1 to zone 9 made for this study were located in and around the mines as shown in Figure 1. Zone 1 and zone 2 are overlapping with mines and are very close to the mines while zone 9 is the farthest from the mines. These zones were described as Z1, Z2, Z3, Z4, Z5, Z6, Z7, Z8, and Z9. The total number of bird species, mean number of species/transects, and their density recorded in various zones are shown in Table 1. Among these zones, Z1 recorded the least number of species and Z5 recorded the highest number of species of birds, while mean number of species/transects recorded its minimum value in Z2 and its maximum in Z7. Among the zones, Larks were the most abundant group of birds in Z1, Z3, and Z9, common coot was the most abundant in Z2, Z5, and Z8, while lesser short-toed lark, common crane, and white pelican were the most abundant in Z4, Z6, and Z7, respectively. The density or mean number of individual/ha recorded its highest value in Z5 (322) and its lowest in Z1 (204) in comparison to the other zones.

The overall diversity of the birds was rich within each zone of mining areas (Figure 2) with the highest value in Z8 (H' = 4.47) and the lowest in Z1 (H' = 4.10) and Z4 (H' = 4.10), while diversity of bird species/transects recorded the highest value in Z9 (H' = 3.40) and the lowest in Z2 (H' = 2.71). The species richness in various zones and per transect within each zone was calculated (Figure 3), and found that the Menhinick species richness index/zone was recorded the highest value in Z1 (2.92) and the lowest in Z4 (1.18), whereas Menhinick species richness index/transect recorded the highest value in Z9 (2.65) and the lowest in Z2 (1.96). Similarly, the species

TABLE 1: Bird species diversity and density in various zones of mining areas.

Zone	Total number of species	No. of species/transects	Mean density/Ha
Z1	114	31.66	204
Z2	153	31.06	250
Z3	126	34.33	271
Z4	139	32.73	222
Z5	195	37.89	322
Z6	149	39.54	254
Z7	175	40.42	303
Z8	154	35	270
Z9	157	42.04	248

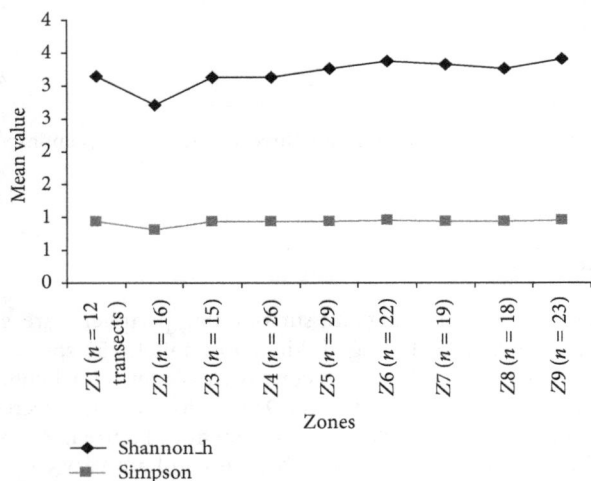

FIGURE 3: Species richness and equitability of birds in various zones made around the mines in the study area.

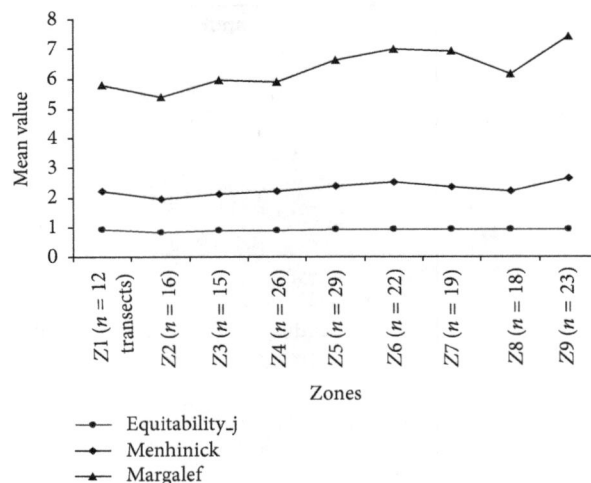

FIGURE 2: Species diversity of birds in various zones made around the mines in the study area.

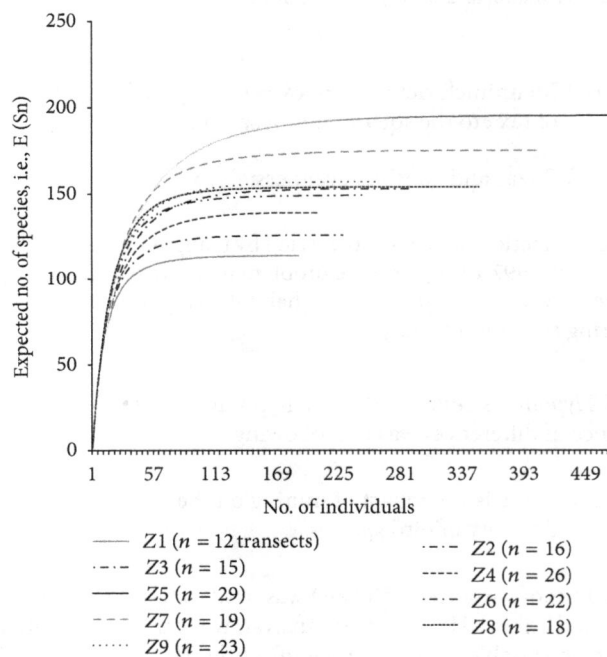

FIGURE 4: Rarefaction curve for expected number of bird species in different zones of mining environment of Kachchh, Gujarat.

evenness values of each zone and transect within were also calculated (Figure 3), and we found that the highest species evenness index was recorded in Z8 (0.57) and the lowest in Z5 (0.34), while species evenness index per transect was recorded more or less similar with the highest value in Z8 (0.78) and the lowest in Z2 (0.66).

On analysis of the data on various aspects of birds diversity in and around the mining areas based on various zones, it was found that diversity (Shannon diversity: $F = 2.984$, df = 8, and $P < 0.05$) and richness (Menhinik species richness: $F = 2.403$, df = 8, and $P < 0.05$) of birds in between zones differ significantly. Further, a Tukey HSD post hoc analysis results suggest that Z2 differs significantly from Z5 ($P = 0.018$), Z6 ($P = 0.003$), Z7 ($P = 0.015$), Z8 ($P = 0.046$), and Z9 ($P = 0.001$) in terms of mean diversity, and Z2 differs significantly from Z9 ($P = 0.012$) in terms of species richness. Rarefaction analysis was done for the expected number of species in each zone of the mining areas to standardize unequal sampling sizes (Figure 4). The expected number of species recorded its highest value in Z5 and its lowest in Z1. On cluster analysis of the data, abundance of birds revealed

great similarities among sampling points belonging to the same kind of habitat. Six distinct groups, namely, Z1, Z3; Z4, Z6; Z9; Z8; Z2; Z7, Z5 were identified (Figure 5), and we found that Z4 and Z6 have maximum similarity in abundance and next to that were Z1 and Z3, while they have very less similarity of clusters with Z5.

TABLE 2: Checklist of birds recorded during the study.

Sl. no.	Order/family	Species scientific name	Common name	MS	IUCN 2010	Feeding habit
	Anseriformes					
1	Anatidae	*Anas acuta*	Northern pintail	M	—	A
2		*Anas clypeata*	Northern shoveller	M	—	A
3		*Anas penelope*	Eurasian wigeon	M	—	A
4		*Anas platyrhynchos*	Mallard	RM	—	A
5		*Anas poecilorhyncha*	Spot-billed duck	RM	—	A
6		*Anas querquedula*	Garganey	M	—	A
7		*Anas strepera*	Gadwall	M	—	A
8		*Aythya ferina*	Common pochard	M	—	A
9		*Aythya fuligula*	Tufted pochard	M	—	A
10		*Sarkidiornis melanotos*	Comb duck	R	—	A
11	Dendrocygnidae	*Dendrocygna javanica*	Lesser whistling duck	R	—	A
	Apodiiformes					
12	Apodidae	*Apus affinis*	House(little) swift	R	—	I
	Ciconiiformes					
13	Accipitridae	*Accipiter badius*	Shikra	R	—	C
14		*Accipiter nisus*	Eurasian sparrow-hawk	R	—	C
15		*Aquila heliaca*	Imperial eagle	R	VU	C
16		*Aquila nipalensis*	Steppe eagle	M	—	C
17		*Aquila pomarina*	Lesser spotted eagle	M	—	C
18		*Aquila rapax*	Tawny eagle	R	—	C
19		*Butastur teesa*	White-eyed buzzard	R	—	C
20		*Buteo buteo*	Long-legged buzzard	R	—	C
21		*Circaetus gallicus*	Short-toad snake eagle	R	—	C
22		*Circus aeruginosus*	Eurasian marsh harrier	M	—	C
23		*Circus macrourus*	Pallid harrier	M	NT	C
24		*Circus melanoleucos*	Pied harrier	RM	—	C
25		*Circus pygargus*	Montagu's harrier	M	—	C
26		*Elanus caeruleus*	Black-shouldered kite	R	—	C
27		*Gyps bengalensis*	Indian white-backed vulture	R	CR	C
28		*Gyps indicus*	Long-billed vulture	R	CR	C
29		*Haliaeetus leucogaster*	White-bellied sea-eagle	R	—	C
30		*Haliaeetus leucoryphus*	Pallas's fish-eagle	RM	VU	C
31		*Haliastur Indus*	Brahminy kite	R	—	C
32		*Hieraaetus fasciatus*	Bonelli's eagle	M	—	C
33		*Hieraaetus pennatus*	Booted eagle	R	—	C
34		*Milvus migrans*	Black kite	R	—	C
35		*Pernis ptilorhynchus*	Oriental honey-buzzard	R	—	C
36		*Spilornis cheela*	Crested serpent eagle	R	—	C
37	Ardeidae	*Ardea alba*	Great egret	RM	—	A
38		*Ardea cinerea*	Grey heron	RM	—	A
39		*Ardea purpurea*	Purple heron	RM	—	A
40		*Ardeola grayii*	Indian pond heron	R	—	A
41		*Bubulcus ibis*	Cattle egret	R	—	I
42		*Butorides striatus*	Little green heron	RM	—	A
43		*Egretta garzetta*	Little egret	R	—	A
44		*Egretta gularis*	Western reef-egret	R	—	A
45		*Mesophoyx intermedia*	Intermediate egret	R	—	A
46	Burhinidae	*Burhinus oedicnemus*	Eurasian thick-knee	R	—	A

TABLE 2: Continued.

Sl. no.	Order/family	Species scientific name	Common name	MS	IUCN 2010	Feeding habit
47	Charadriidae	*Calidris alpine*	Dunlin	M	—	A
48		*Charadrius alexandrinus*	Kentish plover	RM	—	A
49		*Charadrius mongolus*	Lesser sand plover	RM	—	A
50		*Charadrius dubius*	Common ring plover	RM	—	A
51		*Esacus magnirostris*	Greater sand-plover	R	—	A
52		*Himantopus himantopus*	Black-winged stilt	R	—	A
53		*Pluvialis squatarola*	Grey plover	M	—	A
54		*Rostratula benghalensis*	Painted snipe	R	—	A
55		*Tringa erythropus*	Spotted redshank	M	—	A
56		*Vanellus indicus*	Red-wattled lapwing	R	—	I
57		*Vanellus malabaricus*	Yellow-wattled lapwing	R	—	I
58		*Calidris minuta*	Little stint	M	—	A
59	Ciconiidae	*Ephippiorhynchus asiaticus*	Black necked stork	R	NT	P
60		*Mycteria leucocephala*	Painted stork	RM	NT	A
61	Falconidae	*Falco chicquera*	Red-necked falcon	R	—	C
62		*Falco jugger*	Lagger falcon	R	NT	C
63		*Falco naumanni*	Lesser kestrel	R	VU	C
64		*Falco subbuteo centralasiae*	Eurasian hobby	M	—	C
65		*Falco tinnunculus*	Common kestrel	RM	—	C
66	Glareolidae	*Cursorius coromandelicus*	Indian courser	R	—	I
67	Laridae	*Calidris alba*	Sanderling	RM	—	P
68		*Chlidonias leucopterus*	White winged black tern	RM	—	P
69		*Larus brunnicephalus*	Brown-headed gull	RM	—	P
70		*Larus cachinnans*	Yellow-legged gull	M	—	P
71		*Larus fuscus*	Lesser black backed gull	M	—	P
72		*Sterna acuticauda*	Black-bellied tern	M	NT	P
73		*Sterna aurantia*	River tern	R	—	A
74		*Sterna caspia*	Caspian tern	RM	—	P
75	Pelecanidae	*Pelecanus onocrotalus*	Great white-pelican	RM	—	P
76	Phalacrocoracidae	*Phalacrocorax carbo*	Great cormorant	R	—	A
77		*Phalacrocorax fuscicollis*	Indian cormorant	R	—	A
78		*Phalacrocorax niger*	Little cormorant	R	—	A
79	Phoenicopteridae	*Phoenicopterus minor*	Lesser flamingo	RM	NT	A
80		*Phoenicopterus ruber*	Greater flamingo	RM	—	A
81	Podicipedidae	*Podiceps cristatus*	Great crested grebe	RM	—	A
82		*Podiceps nigricollis*	Black-necked grebe	M	—	A
83		*Tachybaptus ruficollis*	Little grebe	R	—	A
84	Pteroclididae	*Pterocles alchata*	White-bellied sandgrouse	M	—	G
85		*Pterocles exustus*	Chestnut-bellied sandgrouse	R	—	G
86		*Pterocles indicus*	Painted sandgrouse	R	—	G
87	Scolopacidae	*Actitis hypoleucos*	Common sandpiper	R	—	A
88		*Limosa limosa*	Black-tailed godwit	M	NT	A
89		*Numenius phaeopus*	Whimbrel	M	—	A
90		*Tringa glareola*	Wood sandpiper	M	—	A
91		*Tringa nebularia*	Common greenshank	M	—	A
92		*Tringa ochropus*	Green sandpiper	M	—	A
93		*Tringa stagnatilis*	Marsh sandpiper	M	—	A
94	Threskiornithidae	*Platalea leucorodia*	Eurasian spoonbill	RM	—	A
95		*Plegadis falcinellus*	Glossy ibis	RM	—	I
96		*Pseudibis papillosa*	Black ibis	R	—	I
97		*Threskiornis melanocephalus*	Black-headed ibis	RM	NT	A

Table 2: Continued.

Sl. no.	Order/family	Species scientific name	Common name	MS	IUCN 2010	Feeding habit
	Columbiformes					
98	Columbidae	*Columba livia*	Rock pigeon	R	—	G
99		*Streptopelia chinensis*	Spotted dove	R	—	G
100		*Streptopelia decaocto*	Eurasian collared dove	R	—	G
101		*Streptopelia orientalis*	Oriental turtle-dove	RM	—	G
102		*Streptopelia senegalensis*	Laughing dove	R	—	G
103		*Streptopelia tranquebarica*	Red-collared dove	R	—	G
	Coraciiformes					
104	Alcedinidae	*Alcedo Hercules*	Common kingfisher	R	—	P
105	Cerylidae	*Ceryle rudis*	Lesser pied kingfisher	R	—	P
106	Coraciidae	*Coracias benghalensis*	Indian roller	R	—	I
107		*Coracias garrulus*	European roller	RM	NT	I
108	Dacelonidae	*Halcyon smyrnensis*	White-throated	R	—	P
109	Meropidae	*Merops leschenaulti*	Chestnut-headed bee-eater	R	—	I
110		*Merops orientalis*	Green bee-eater	R	—	I
111		*Merops persicus*	Blue-cheeked bee-eater	RM	—	I
112		*Merops philippinus*	Blue-tailed bee-eater	RM	—	I
	Cuculiformes					
113	Centropodidae	*Centropus sinensis*	Greater coucal	R	—	O
114	Cuculidae	*Cuculus canorus*	Pied-crested cuckoo	R	—	I
115		*Eudynamys scolopacea*	Asian koel	R	—	F
116		*Phaenicophaeus leschenaultii*	Sirkeer cuckoo	R	—	O
	Galliformes					
117	Phasianidae	*Coturnix coromandelica*	Rain quail	R	—	G
118		*Coturnix coturnix*	Common quail	R	—	G
119		*Francolinus francolinus*	Black francolin	R	—	G
120		*Francolinus pictus*	Painted francolin	R	—	G
121		*Francolinus pondicerianus*	Grey francolin	R	—	G
122		*Pavo cristatus*	Indian peafowl	R	—	G
123		*Perdicula asiatica*	Jungle bush quail	R	—	G
	Gruiformes					
124	Gruidae	*Grus grus*	Common crane	M	—	O
125		*Grus virgo*	Demoiselle crane	M	—	O
126	Otididae	*Ardeotis nigriceps*	Great indian bustard	R	EN	G
127	Rallidae	*Fulica atra*	Common coot	R	—	A
128		*Gallinula chloropus*	Common moorhen	R	—	A
	Passeriformes					
129	Alaudidae	*Alauda arvensis*	Eurasian skylark	R	—	G
130		*Alauda gulgula*	Oriental skylark	R	—	G
131		*Ammomanes phoenicurus*	Rufous-tailed lark	R	—	G
132		*Calandrella brachydactyla*	Greater short-toed lark	M	—	G
133		*Calandrella rufescens*	Lesser short-toed lark	RM	—	G
134		*Eremopterix grisea*	Ashy-crowned sparrow-lark	R	—	G
135		*Galerida cristata*	Crested lark	R	—	G
136		*Galerida deva*	Sykes's crested lark	R	—	G
137		*Galerida malabarica*	Malabar lark	RM	—	G
138		*Mirafra affinis*	Jerdon's bushlark	RM	—	G
139		*Mirafra cantillans*	Singing bushlark	R	—	G
140		*Mirafra erythroptera*	Indian bushlark	R	—	G
141		*Mirafra erythroptera*	Red-winged bush-lark	R	—	G

TABLE 2: Continued.

Sl. no.	Order/family	Species scientific name	Common name	MS	IUCN 2010	Feeding habit
142	Certhiidae	Salpornis spilonotus	Spotted creeper	R	—	I
143	Cisticolidae	Orthotomus sutorius	Tailor bird	R	—	I
144		Prinia buchanani	Rufous-fronted prinia	R	—	I
145		Prinia hodgsonii	Grey-breasted prinia	R	—	I
146		Prinia inornata	Plain prinia	R	—	I
147		Prinia socialis	Ashy prinia	R	—	I
148		Prinia sylvatica	Jungle prinia	R	—	I
149	Corvidae	Aegithina nigrolutea	Marshall's iora	R	—	I
150		Coracina macei	Large cuckoo shrike	R	—	I
151		Corvus splendens	House crow	R	—	O
152		Dendrocitta vagabunda	Rufous treepie	R	—	I
153		Dicrurus caerulescens	White-bellied drongo	R	—	I
154		Dicrurus macrocercus	Black drongo	R	—	I
155		Pericrocotus cinnamomeus	Small minivet	R	—	I
156		Rhipidura aureola	White-browed fantail	R	—	I
157		Tephrodornis gularis	Large wood shrike	R	—	I
158		Tephrodornis pondicerianus	Common woodshrike	R	—	I
159	Fringillidae	Emberiza buchanani	Grey-necked bunting	M	—	G
160		Emberiza cia	Rock bunting	M	—	G
161		Emberiza melanocephala	Ortolan bunting	M	—	G
162		Emberiza striolata	House bunting	R	—	G
163		Melophus lathami	Crested bunting	R	—	G
164	Hirundinidae	Hirundo concolor	Dusky crag-martin	R	—	I
165		Hirundo daurica	Red-rump swallow	R	—	I
166		Hirundo fluvicola	Streak-throated swallow	R	—	I
167		Hirundo rupestris	Eurasian crag-martin	R	—	I
168		Hirundo rustica	Barn swallow	RM	—	I
169		Hirundo smithii	Wire-tailed swallow	R	—	I
170	Laniidae	Lanius collurio	Red-backed shrike	R	—	C
171		Lanius cristatus	Brown shrike	M	—	C
172		Lanius isabellinus	Rufous-tailed shrike	RM	—	C
173		Lanius meridionalis	Southern grey shrike	R	—	C
174		Lanius schach	Long-tailed shrike	R	—	C
175		Lanius schach canipes	Rufous-backed shrike	R	—	C
176		Lanius vittatus	Bay-backed shrike	R	—	C
177	Muscicapidae	Cercomela fusca	Brown rock chat	R	—	I
178		Cercotrichas galactotes	Rufous chat	M	—	I
179		Copsychus saularis	Oriental magpie robin	R	—	I
180		Culicicapa ceylonensis	Grey-headed flycatcher	R	—	I
181		Ficedula parva	Red-throated flycatcher	R	—	I
182		Ficedula superciliaris	Ultramarine flycatcher	R	—	I
183		Muscicapa striata	Spotted flycatcher	RM	—	I
184		Oenanthe deserti	Desert wheatear	RM	—	I
185		Oenanthe isabellina	Isabelline wheatear	RM	—	I
186		Oenanthe picata	Variable wheatear	M	—	I
187		Phoenicurus erythronotus	Rufous-backed redstart	RM	—	I
188		Phoenicurus ochruros	Black redstart	RM	—	I
189		Saxicola caprata	Pied bush chat	R	—	I
190		Saxicola jerdoni	Pied chat	R	—	I
191		Saxicola macrorhyncha	Stoliczka's bushchat	RM	VU	I

TABLE 2: Continued.

Sl. no.	Order/family	Species scientific name	Common name	MS	IUCN 2010	Feeding habit
192		*Saxicola torquata*	Common stonechat	RM	—	I
193		*Saxicoloides fulicata*	Indian robin	R	—	I
194		*Turdus naumanni*	Dusky thrush	RM	—	I
195		*Turdus obscurus*	Eyebrowed thrush	RM	—	I
196		*Zoothera citrina*	Orange-headed thrush	R	—	I
197	Nectariniidae	*Aethopyga siparaja*	Crimson sunbird	R	—	N
198		*Dicaeum agile*	Thick-billed flowerpecker	R	—	N
199		*Nectarinia asiatica*	Purple sunbird	R	—	N
200		*Nectarinia zeylonica*	Purple-rumped sunbird	R	—	N
201	Paridae	*Parus major*	Great tit	R	—	I
202		*Parus nuchalis*	Pied tit	R	VU	I
203	Passeridae	*Anthus campestris*	Tawny pipit	RM	—	G
204		*Anthus godlewski*	Blyth's pipit	RM	—	G
205		*Anthus rufulus*	Paddyfield pipit	R	—	G
206		*Anthus similis jerdoni*	Brown rock pipit	M	—	G
207		*Anthus trivialis*	Tree pipit	RM	—	G
208		*Dendronanthus indicus*	Forest wagtail	RM	—	I
209		*Lonchura malabarica*	Indian silverbill	R	—	G
210		*Lonchura striata*	White-rumped munia	R	—	G
211		*Motacilla alba*	White wagtail	RM	—	I
212		*Motacilla cinerea*	Grey wagtail	RM		I
213		*Motacilla citreola*	Citrine wagtail	RM	—	I
214		*Motacilla flava*	Yellow wagtail	RM	—	I
215		*Passer domesticus*	House sparrow	R	—	G
216		*Petronia xanthocollis*	Chestnut-shouldered petronia	R	—	G
217		*Ploceus philippinus*	Baya weaver	R	—	G
218	Pycnonotidae	*Pycnonotus cafer*	Red-vented bulbul	R	—	I
219		*Pycnonotus leucotis*	White-eared bulbul	R	—	I
220	Sturnidae	*Acridotheres ginginianus*	Bank myna	R	—	O
221		*Acridotheres tristis*	Common myna	R	—	I
222		*Sturnus pagodarum*	Brahminy starling	R	—	O
223		*Sturnus roseus*	Rosy starling	M	—	G
224	Sylviidae	*Acrocephalus aedon*	Thick-billed warbler	M	—	I
225		*Acrocephalus dumetorum*	Blyth's reed-warbler	RM	—	I
226		*Acrocephalus stentoreus*	Indian great reed-warbler	RM	—	I
227		*Chaetornis striatus*	Bristled grass-warbler	M	VU	I
228		*Chrysomma sinense*	Yellow-eyed babbler	R	—	I
229		*Cisticola juncidis*	Streak fantail warbler	R	—	I
230		*Hippolais caligata*	Booted warbler	RM	—	I
231		*Locustella naevia*	Grasshopper warbler	R	—	I
232		*Phylloscopus inornatus*	Desert warbler	R	—	F
233		*Phylloscopus magnirostris*	Large-billed leaf-warbler	M	—	I
234		*Phylloscopus neglectus*	Plain-leaf warbler	R	—	F
235		*Phylloscopus trochiloides*	Greenish leaf-warbler	M	—	I
236		*Sylvia communis*	Greater whitethroat	R	—	I
237		*Sylvia curruca*	Lesser whitethroat	M	—	I
238		*Sylvia hortensis*	Orphean warbler	M	—	I
239		*Turdoides caudatus*	Common babbler	R	—	G
240		*Turdoides malcolmi*	Large grey babbler	R	—	G
241		*Turdoides striatus*	Jungle babbler	R	—	G
242	Zosteropidae	*Zosterops palpebrosus*	Oriental white-eye	RM	—	F

TABLE 2: Continued.

Sl. no.	Order/family	Species scientific name	Common name	MS	IUCN 2010	Feeding habit
	Piciformes					
243	Picidae	*Dendrocopos mahrattensis*	Yellow-fronted pied woodpecker	R	—	I
244		*Jynx torquilla*	Eurasian wryneck	R	—	I
	Psittaciformes					
245	Psittacidae	*Psittacula cyanocephala*	Plum-headed parakeet	R	—	F
246		*Psittacula krameri*	Rose-ringed parakeet	R	—	F
	Strigiformes					
247	Strigidae	*Bubo bubo*	Eurasian eagle owl	R	—	C
248		*Otus bakkamoena*	Collared scops-owl	R	—	I
249	Caprimulgidae	*Caprimulgus asiaticus*	Indian nightjar	R	—	I
250		*Caprimulgus europaeus*	Eurasian nightjar	M	—	I
251	Strigidae	*Athene brama*	Spotted owlet	R	—	C
	Upupiformes					
252	Upupidae	*Upupa epops*	Common hoopoe	RM	—	I

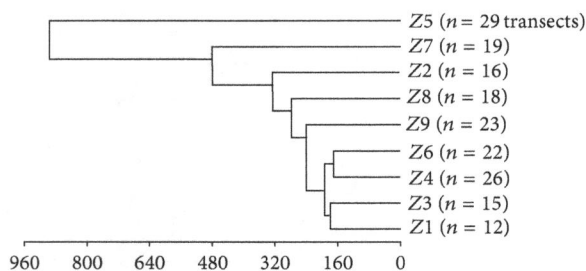

FIGURE 5: Clusters of zones based on abundance of bird species in mining areas of Kachchh, Gujarat.

4. Discussion

Ecosystem disturbance is one of the major phenomena in recent times which alters the relationship of organisms and their habitat in time and space. The extraction of mineral resources through mining activities in wildlife areas is one of the major factors for ecosystem disturbances and habitat fragmentation results impact on the survival of precious wildlife. Large-scale denudation of forest cover, scarcity of water, pollution of air, water, and soil, and degradation of agricultural lands are some of the conspicuous environmental implications of mining in western Kachchh. The result of the study reveals that the whole mining areas and its surrounding areas are rich in terms of avifaunal diversity and abundance. It's also found that the species diversity, richness, and abundance were less in the zones which are close to the mines within 4 km radius to mines. Interestingly, the species diversity was found lowest in zone 2 in comparison to zone 1 which is probably due to the frequency of mining activities which were more in zone 2 in comparison to zone 1. The mining activity is completed in zone 1 and as a result some aquatic water bodies were developed in this zone which attracts aquatic birds to it. Salovarov and Kuznetsova [17] also found similar type of results in coal mining areas of Angara region, Russia. At the present time, mining activities are more frequent in zone 2 as a result of bird species diversity, and abundance recorded less in comparison to the other zones. The species diversity, abundance, and richness were found highest in zones beyond zone 5 to zone 9 which were located in between 8 km to 18 km from the mines. The expected number of species in zone 5 was found to be the highest and zone 2 was the lowest. Some of the species of birds may adapt to the human habitation and mining environment as it provides easy food and habitat. Smith et al. [18] found that mining has no impact on birds, but it helps in supporting some breeding birds. The species of birds which naturally survive in certain kind of habitat in this area will get more threat from the mining activities as the mining in this areas is opencast type.

5. Conclusion

These results indicate that mining and its associated activities have some impacts on the diversity and distribution of birds in Kachchh region of Gujarat. The enhancement of mining and other developmental activities after the earthquake and in recent times has worsened the situation which will further increase the rate of habitat destruction and their avifaunal diversity. Apart from the above, western Kachchh is the gateway of large number of migratory birds to India which will also be affected if the rate of mining activities and industrial development will go on like today.

Acknowledgments

The authors are grateful to the, Director of the Gujarat Institute of Desert Ecology, Bhuj, for his support and for providing facility. The authors are also grateful to scientists and researchers of the Terrestrial Ecology Division, GUIDE, for their help and encouragement.

References

[1] C. D. Lester, *The Birds of Kutch*, Kutch Darba, Bhuj, Kutch, 1904.

[2] S. Ali, *The Birds of Kutch*, Oxford University Press, Bombay, India, 1st edition, 1945.

[3] S. D. Ripley, "Review: the birds of Kutch," *The Auk*, vol. 65, no. 1, p. 148, 1948.

[4] M. K. Himmatsinhji, "More bird notes from Kutch," *Journal of the Bombay Natural History Society*, vol. 55, no. 3, pp. 575–576, 1959.

[5] K. Maharao, "Some bird records from Kutch," *Journal of the Bombay Natural History Society*, vol. 65, no. 1, p. 225, 1968.

[6] J. K. Tiwari, "Checklist of birds of Kachchh," 2011, http://www.kolkatabirds.com/gujarat/gujaratclist.htm.

[7] S. K. Sen, "Birds of Kachchh, Gujarat," 2012, http://www.kolkatabirds.com/gujarat/gujaratclist.htm.

[8] C. J. Bibby, N. D. Burgess, and D. A. Hill, *Bird Census Techniques*, Academic Press, London, UK, 1992.

[9] J. S. Dieni and S. L. Jones, "A field test of the area search method for measuring breeding bird populations," *Journal of Field Ornithology*, vol. 73, no. 3, pp. 253–257, 2002.

[10] S. Ali and S. D. Ripley, *A Pictorial Guide to the Birds of the Indian Subcontinent*, Bombay Natural History Society. Oxford University Press, Bombay, India, 1983.

[11] S. Ali, *Book of Indian Birds*, Bombay Natural History Society, 2002.

[12] R. Grimmett, C. Inskipp, and T. Inskipp, *Pocket Guide To the Birds of the Indian Sub-Continent*, Oxford University Press, New Delhi, India, 2006.

[13] E. O. Willis, "The composition of avian communities in reminiscent woodlots in southern Brazil," *Papéis Avulsos De Zoologia*, vol. 33, no. 1, pp. 1–25, 1979.

[14] S. K. Robinson, J. G. Blake, and R. O. Bierregaard Jr., "Birds of Four Neotropical Forests," in *Four Neotropical Rainforest. New Haven*, A. H. Fentry, Ed., pp. 237–269, Yale University Press, London, UK, 1990.

[15] L. Anjos, "Bird communities in five Atlantic forest fragments in southern Brazil," *Ornithologia Neotropical*, vol. 12, pp. 11–27, 2001.

[16] Ø. Hammer, D. A. T. Harper, and P. D. Ryan, "Past: paleontological statistics software package for education and data analysis," *Palaeontologia Electronica*, vol. 4, no. 1, 2001.

[17] V. O. Salovarov and D. V. Kuznetsova, "Impact of coal mining on bird distribution in Upper Angara Region," *Izvestiia Akademii nauk. Seriia biologicheskaia*, no. 2, pp. 248–251, 2006.

[18] A. C. Smith, J. A. Virgl, D. Panayi, and A. R. Armstrong, "Effects of a diamond mine on tundra-breeding birds," *Arctic*, vol. 58, no. 3, pp. 295–304, 2005.

10

An Infectious Disease and Mortality Survey in a Population of Free-Ranging African Wild Dogs and Sympatric Domestic Dogs

G. Flacke,[1,2] P. Becker,[3,4] D. Cooper,[5] M. Szykman Gunther,[3,6] I. Robertson,[7] C. Holyoake,[1] R. Donaldson,[1] and K. Warren[1]

[1] Conservation Medicine Program, College of Veterinary Medicine, School of Veterinary and Life Sciences, Murdoch University, Perth, WA 6150, Australia
[2] School of Animal Biology, University of Western Australia, 35 Stirling Highway, Crawley, WA 6009, Australia
[3] Center for Species Survival, Smithsonian Conservation Biology Institute, National Zoological Park, Front Royal, VA 22630, USA
[4] Centre for Wildlife Management, University of Pretoria, Pretoria 0002, South Africa
[5] Ezemvelo KZN Wildlife, Queen Elizabeth Park, Pietermaritzburg 3202, South Africa
[6] Department of Wildlife, Humboldt State University, Arcata, CA 95521, USA
[7] Veterinary Epidemiology Programme, College of Veterinary Medicine, School of Veterinary and Life Sciences, Murdoch University, Perth, WA 6150, Australia

Correspondence should be addressed to G. Flacke; gflacke@hotmail.com

Academic Editor: Antonio Terlizzi

Disease can cause declines in wildlife populations and significantly threaten their survival. Recent expansion of human and domestic animal populations has made wildlife more susceptible to transmission of pathogens from domestic animal hosts. We conducted a pathogen surveillance and mortality survey for the population of African wild dogs (Lycaon pictus) in KwaZulu-Natal (KZN), South Africa, from January 2006–February 2007. Samples were obtained from 24 wild dogs for canine distemper virus (CDV) and canine parvovirus (CPV) serological testing. Data were collected on the presence of CDV, CPV, and rabies virus in the KZN domestic dog (Canis familiaris) population from 2004–06. The presence of these pathogens was confirmed in domestic dogs throughout KZN. Wild dogs exhibited 0% and 4.2% prevalence for CDV and CPV antibodies, respectively. In 2006 the largest wild dog pack in KZN was reduced from 26 individuals to a single animal; disease due to rabies virus was considered the most probable cause. This study provides evidence that CDV, CPV and rabies are potential threats to African wild dog conservation in KZN. The most economical and practical way to protect wild dogs from canine pathogens may be via vaccination of sympatric domestic dogs; however, such programmes are currently limited.

1. Introduction

The impacts of infectious disease on wildlife populations and the importance of disease surveillance for endangered species conservation and management programmes have been repeatedly demonstrated [1–5]. The KwaZulu-Natal (KZN) African Wild Dog Reintroduction and Conservation Programme, in the South African province of KwaZulu-Natal, has incorporated infectious pathogen and disease surveillance as one of many components within their long-term population monitoring strategy. The African wild dog is the most endangered carnivore in southern Africa, with

just 3000–5700 individuals remaining [6, 7]. Across the continent, the total number of free-ranging African wild dogs is estimated at less than 8000 individuals, surviving in only 14 of their original 39 range countries [6–9]. In South Africa, the number of individuals is estimated to be only 300–400 [6, 7, 10], and Kruger National Park currently contains the largest population with nearly half of the country's African wild dogs living within its boundaries.

The long-term goal of the KZN African Wild Dog Reintroduction and Conservation Programme is to establish a self-sustaining population of wild dogs in South Africa, specifically within KZN protected areas. As of 2006-2007 when this study was conducted, the KZN population of 80 to 90 wild dogs accounted for approximately a quarter of South Africa's total population of this species [10]. Although the KZN wild dogs occur primarily within Hluhluwe-iMfolozi Park (HiP), the establishment of a viable population with adequate gene flow for outbreeding in this species requires the dispersal of individuals for maintenance of population-wide genetic diversity [11–13]. Due to land-use practices in KZN province, such dispersal requires movement between and among many small, separate protected areas and subpopulations. Disease risks are heightened during such dispersal events as contact with domestic dogs (Canis familiaris) may enable transmission of key threatening diseases to dispersing wild dogs [5, 14] and subsequently between their subpopulations. Unfortunately, long-term consequences of infectious pathogens for persistence of free-ranging African wild dog populations remain poorly understood and difficult to evaluate, due in part to a variety of interacting factors affecting disease dynamics [15].

Pathogens previously identified as a threat to African wild dogs, and thus of particular concern to the KZN African wild dog population, are canine distemper virus (CDV), canine parvovirus (CPV), and rabies virus [15]. The first of these, CDV, of the genus Morbillivirus, is the etiological agent of a contagious disease to which all carnivore species are susceptible, causing respiratory and central nervous system derangements [16]. CDV infection is usually more severe in immunologically naïve individuals and can reach epidemic proportions in immunologically naïve populations [17]. A number of documented outbreaks of CDV in African wild dogs across the continent indicate that the disease is of primary concern for the species, reducing long-term viability and increasing extinction risk for already small populations [18–21]. A single outbreak in a pack of wild dogs in South Africa's Tswalu Kalahari Reserve in 2005 resulted in the loss of the entire pack [22].

The second pathogen of concern, CPV, causes one of the most common and highly contagious infectious diseases of the domestic dog with clinical signs including haemorrhagic diarrhoea, emesis, dehydration, hypoproteinemia, and septicaemia [23]. Clinical disease resulting from CPV infection is often fatal, particularly in young pups, and all canids are susceptible [23]. Exposure followed by mild to no clinical signs of illness is more common in individuals exposed to the virus as adults; these animals still shed live viral particles and thus serve as a source of infection and environmental contamination [23]. Although CPV has not been considered as important a disease threat as CDV to endangered wild carnivores [24], it poses a major risk to small populations due to the potential for significant early pup mortality [25] and can negatively affect all age classes in immunologically naïve populations [26]. Seroprevalence to CPV among African wild dogs has been demonstrated across southern Africa [5], indicating that pathogen exposure does occur, albeit with unknown population-level impacts.

Rabies is an acute, fatal infection of the central nervous system by Lyssavirus spp. All mammals are considered susceptible, and while clinical signs of rabies can be extremely variable, they classically include acute behavioural alterations and pansystemic neurological dysfunction ranging from hyperesthesia, ataxia, paresis, and paralysis to tremors, seizures, and convulsions [27]. Rabies is considered a significant disease threat to the African wild dog across the continent and has been linked to the loss of entire packs, including the 1991 disappearances of all wild dog study packs from the Serengeti ecosystem in Tanzania [27–29]. Major confirmed rabies outbreaks in free-ranging African wild dogs also include those reported in the Masai Mara Reserve in Kenya [30, 31], Madikwe Game Reserve in South Africa [32, 33], and Etosha National Park in Namibia [34]. For the 1991 Serengeti outbreak, virus isolation and molecular diagnostics pointed to domestic dogs as the most probable source of rabies virus infection for the wild dogs [32]; diagnostic evidence collected during the 1989 Masai Mara outbreak suggested that the virus originated from domestic dogs living on the periphery of the reserve [30, 31].

With human populations increasing in rural areas of KZN and throughout South Africa, more people are moving closer to the boundaries of game reserves, bringing with them various domestic animals. Across the continent, disease risks to wild dogs have been shown to increase dramatically in areas close to human habitation, as domestic dogs have the potential to serve as either a reservoir host or as a transmission source for many canine pathogens [35]. For the purposes of this study a reservoir is defined as a primary host that harbors the pathogen, demonstrates little to no ill effects, serves as a means of sustaining a pathogen in the environment, and serves as a source of infection for others. In Zimbabwe, Butler et al. [36] demonstrated that domestic dogs pose a direct risk for pathogen transmission to various wild carnivore species, including African wild dogs. In Kenya, Woodroffe et al. [14] followed 19 wild dog packs and found that exposure to rabies and CPV (but not CDV) was associated with contact with domestic dogs. In KZN, domestic dogs in rural community areas are unlikely to be vaccinated [37]. Since African wild dogs often cross out of protected areas during natural dispersal events and domestic dogs have been witnessed to range inside protected areas along park boundaries, there exists a substantial risk of contact between the two species. Wildlife poachers will also frequently bring domestic dogs into protected areas for hunting purposes.

The potential for contact between the KZN wild dogs and domestic dogs is of heightened concern because small populations of carnivore species living in fragmented habitats are particularly susceptible to population declines due to transmission of pathogens from larger populations of reservoir hosts [38, 39]. In such situations infectious pathogens can result in disease epidemics from which small, fragmented populations are unable to recover. For African wild dogs these population-level effects are exacerbated by pack structures and the highly social nature of the species. Further, even diseases with relatively low mortality rates or decreases in fecundity may provide sufficient additional loss to promote extinctions in small, declining, or fragmented populations [12]. Despite the apparent risks posed by infectious diseases, prior to our study period in 2006, there was no previous information available concerning the disease exposure status of African wild dogs in KZN province. The goals of this study were (1) to investigate the CDV and CPV status of the KZN wild dogs by serologically screening a representative proportion of the population (approximately 30%), (2) to investigate the occurrence of CDV, CPV, and rabies in the KZN domestic dog population, and (3) to use this information to help develop management recommendations concerning the risks of these pathogens to the KZN African wild dog population.

2. Materials and Methods

2.1. Study Area, African Wild Dog Population, and Sample Collection. Serum samples and mortality information were collected from seven African wild dog packs in the KZN population ($n = 24$; from a total population of approximately 70 individuals in 2006) between January 2006 and February 2007. Sampled individuals came from five packs in Hluhluwe-iMfolozi Park (28°4′59″S, 32°4′59″E), one pack in Mkuze Game Reserve (27°37′59″S, 32°15′E), and one pack in Thanda Private Game Reserve (27°24′S, 32°9′E). Demographic data were available for all animals sampled in this study through the broader KZN African Wild Dog Reintroduction and Conservation Programme. This programme undertakes capture and chemical immobilization of wild dogs for telemetry collar attachment/removal and translocation between and within reserves; blood samples for this study were collected opportunistically during these immobilization procedures. Samples were collected in serum separator tubes and centrifuged within 24 hours of collection. The serum was frozen at −20°C until analysis was conducted.

2.2. African Wild Dog Disease Analysis. All serological testing was conducted at the Department of Veterinary Tropical Diseases Laboratory, University of Pretoria, Onderstepoort campus. Serum antibodies reactive to canine distemper virus (CDV) and canine parvovirus (CPV-2) were detected by means of an indirect fluorescent antibody technique as previously described [40]. Serum specimens were either screened at a serum dilution of 1 : 20 to minimize nonspecific fluorescence or titrated by testing serial twofold dilutions of sera. The CDV strain used in the preparation of the capture antigen slides was the Onderstepoort strain, while the CPV-2 strain was a field strain isolated from a clinically ill domestic dog in South Africa. Rabies serological assays were not conducted as part of this study, because all founding members of the KZN wild dog population and most wild dogs immobilized for management reasons had been vaccinated against rabies and because a positive rabies virus titre in an unvaccinated dog in the absence of clinical signs of rabies is extremely uncommon [41, 42].

2.3. Domestic Dog Data Collection. While the collection of blood samples from domestic dogs in KZN rural communities was outside the limits of this study, we used a variety of indirect methods to investigate the incidence or presence of CDV, CPV, and rabies in domestic dogs in KZN. For CDV, we conducted 75 community disease surveys in rural Zulu communities near KZN protected areas and in dispersal corridors used by wild dogs to investigate for the presence

of clinical signs of CDV. These surveys involved interviews conducted in Zulu, with unlabelled photographs of domestic dogs with clinical signs of the respiratory form of CDV for visual reference. Survey participants were asked if any of their household dogs had been observed in the past year with clinical signs similar to those in the photographs. Data from these surveys were used as a broad indication of disease prevalence of CDV in Zulu community dogs. The surveys also questioned interviewees about the number of dogs they owned, whether they used veterinary services, and the causes of death (if known) of any of their dogs which had died within the past year. CPV and rabies were not included in the surveys, as these two diseases do not present with "classical" clinical signs readily identifiable via photographs.

To further investigate for the presence of CDV and CPV in domestic dogs in areas near KZN protected areas, data on proportional disease rates of CDV and CPV in pet domestic dogs presenting to the practice during 2006 was obtained from five veterinarians with practices in towns surrounding the KZN protected areas. The proportional disease rate data is based on the number of cases seen at the practices, as established via presenting clinical signs and standard veterinary diagnostic methods. Although disease rates in these pet dogs were not expected to be exactly representative of disease status in rural Zulu community dogs, which are generally not vaccinated nor taken to private veterinarians, positive responses by local practitioners would confirm the presence of these diseases among the domestic dog population in areas covered by their practices.

To investigate rabies in domestic dogs, we obtained retrospective disease data from the KZN State Veterinary Authority (SVA), an organization which maintains data regarding all cases of reportable animal diseases. The data was reviewed to determine the number of reported canine rabies cases and estimated prevalence of canine rabies for 2004–2006 in the regions surrounding protected areas and in areas where wild dogs were known to disperse. The SVA also provided rabies vaccination information and domestic dog census data from 2001 onwards. These data on rabies cases and vaccine coverage in KZN would both represent minimums because, although the Veterinary Authority holds annual rabies vaccine campaigns in rural communities in KZN, many areas are considered too difficult or dangerous to access. In addition, most community dogs are free-roaming semiferal animals which receive no veterinary attention outside of the vaccine campaigns and in which diseases usually go undetected and unreported.

2.4. African Wild Dog Mortality. We investigated age- and cause-specific mortality for the KZN population of African wild dogs from 01 January 2006 to 31 December 2006. An overall population mortality was calculated, along with age- and cause-specific mortalities, using data from all mortalities occurring during 2006. In addition to confirmed mortalities, any individual which disappeared from its pack and was not sighted within six months was considered deceased. All wild dogs are sighted frequently as part of the broader KZN Reintroduction and Conservation Programme, and to date there have been no individuals resighted after an absence of greater than six months. Given pack social structures and the likelihood that any wild dog dispersing into the rural, densely populated areas around parks would be reported, it was considered very unlikely that any disappearances were due to dispersal and ongoing survival outside KZN protected areas.

3. Results

3.1. African Wild Dog Serology. All sampled wild dogs ($n = 24$) exhibited negative titres for CDV, indicating 0% seroprevalence (95% CI 0–11.7). One (1) of the 24 wild dogs exhibited a positive titre for CPV, representing 4.17% seroprevalence (95% CI 0.1–21.1).

3.2. Community Surveys for CDV and Mortalities in Domestic Dogs. Of the 65 Zulu households around HiP and in northern KZN which responded to interviews, 31% ($n = 20$) households reported having 1-2 dogs, 35% ($n = 23$) reported 3-4 dogs, and 34% ($n = 22$) reported having more than five dogs. All interviewed households reported owning at least one dog. Very few households (9%; $n = 6$) reported having taken their dogs to a private veterinarian.

Of interviewed households, 58% ($n = 38$) reported the death of at least one of their dogs within the last year, with 26% of deaths attributed to illness, 21% to vehicle strike, 12% to attacks by other dogs, and 41% to unknown causes. When shown photographs of domestic dogs exhibiting clinical signs of the respiratory form of CDV, 8% ($n = 5$) of households reported that at least one of their dogs had exhibited such clinical signs within the last year.

3.3. CDV and CPV in Domestic Dogs Presenting to Veterinary Clinics. In domestic dogs seen at the five private veterinary practices surrounding KZN protected areas during 2006, 1.7% (95% CI 1.5–2.0%) presented with CDV, and 1.7% (95% CI 1.5–2.0%) presented with CPV (D. Baxter—Vryheid; H. Kohrs—Pongola; N. Meunier—Hluhluwe; C. Pryke—Eshowe; T. Viljoen—Mtubatuba).

3.4. Rabies Cases and Vaccination Coverage from State Veterinary Authority Data. Annual observed incidence of rabies in domestic dogs in northern KZN, as contained in State Veterinary Authority records, was reported to range 17.4–18 (per 100000 animals) for 2004–2006 (Table 1) (KZN SVA, unpublished data). The SVA confirmed rabies-positive cases via indirect fluorescent antibody testing. Rabies vaccination coverage ranged 27.4%–30.8% over the same three years (Table 1), which represents between 100,000 and 120,000 dogs vaccinated among an estimated population of 365,000 to 390,000 dogs for the area in question (KZN SVA, unpublished data).

3.5. African Wild Dog Mortality. During 2006 the KZN wild dog population lost 53 of 70 individuals (including confirmed mortalities and permanent disappearances). The overall mortality was 75.7% during 2006 (Table 2). The highest proportion of mortalities was attributed to "suspected disease" (37.7% of cases; Table 2), followed by snaring

TABLE 1: Rabies prevalence and rabies vaccination coverage data for domestic dogs in northern KZN for 2004, 2005, and 2006, from State Veterinary Authority records. Annual observed incidence is presented as a combined value per 100,000 animals for all districts for each year.

Year	Annual observed incidence (per 100,000 animals)	Average vaccination coverage
2004	17.4/year	30.8%
2005	18.0/year	28.9%
2006	18.0/year	27.4%

(13.2% of cases). In 32.1% of cases the cause of death was unknown, although this included at least one individual strongly suspected to have been killed by lions or hyenas, having disappeared overnight from its pack, with fresh tracks of multiple lions and hyenas observed the following day at the pack's overnight resting site. The "Other" category included two individuals apparently killed by crocodiles (they both disappeared immediately after chasing a prey animal into the Hluhluwe River, which is densely populated by crocodiles) and one individual which died secondary to an anaesthetic reaction during a routine immobilization procedure.

3.6. Loss of the iMfolozi Pack: Observations Suggest Disease as the Cause. Twenty of 26 wild dogs from KZN's largest pack, the iMfolozi pack, were lost in a period of less than 2 months in August/September of 2006 (11 adults/yearlings and all nine of the pack's pups). By mid-2007, five of the remaining six animals had disappeared, and only the alpha female remained. No carcasses could be recovered for diagnostic testing because the pack's only radiocollared animal was the second animal to disappear, and his carcass was severely decomposed when retrieved. However, several pieces of anecdotal evidence support disease as a primary factor associated with the pack's disappearance: (1) the short time in which such a large number of animals were lost is highly consistent with disease and highly inconsistent with other common causes of death in wild dogs (i.e., snaring or predation); (2) the pack denned on the south-western boundary of the reserve, a location where wild dogs were often observed to cross in and out of the park; (3) domestic dogs were observed inside the park in this area; (4) a domestic dog was observed feeding with the iMfolozi pack on a carcass in July 2006 (this dog was destroyed, but no disease testing was conducted); (5) within one month of the observation of the domestic dog feeding with the iMfolozi pack, a yearling member of the pack was observed exhibiting clinical signs consistent with either rabies or the neurological form of CDV (emaciated, weakness, and ataxia), and within two months a large majority of the pack had disappeared; (6) over the same two-month period there was a focal epidemic of rabies in domestic dogs near the south-western boundary of the park where the iMfolozi pack denned including a confirmed case less than 20 km from the den site where the pack was known to cross in and out of reserve boundaries [43]; (7) the surviving alpha female was one of only two pack members to have ever received a rabies vaccination (six years previously);

TABLE 2: Cause-specific mortality (%) for 12 pups and 41 adults/yearlings in the KZN population of African wild dogs for 2006. The overall population mortality for 2006 is also presented. The "unknown" category includes both animals confirmed to have died and animals which disappeared and were never resighted (minimum 6 months).

2006	Pups ($n = 12$)	Adults/yearlings ($n = 41$)	Total ($n = 53$)
Overall mortality (53/70)			75.71%
Snaring	0.0%	17.07% ($n = 7$)	13.20%
Lion predation/attack	16.67% ($n = 2$)	4.88% ($n = 2$)	7.55%
Old age	0.0%	4.88% ($n = 2$)	3.77%
Suspected disease	75.0% ($n = 9$)	26.83% ($n = 11$)	37.74%
Other	0.0%	7.32% ($n = 3$)	5.66%
Unknown	8.33% ($n = 1$)	39.02% ($n = 16$)	32.08%

the other previously vaccinated animal was the radiocollared alpha male.

4. Discussion

The loss of 53 African wild dogs in a single year from a population of only around 70 individuals is clearly unsustainable and is markedly higher than mortality rates reported for other African wild dog populations [24, 44–46]. We note that the loss of the iMfolozi pack contributed substantially to the mortality total and that this loss was suspected to be linked to a disease epidemic. Given the lack of other confirmed or suspected disease epidemics during 10 years of demographic research at KZN, the 2006 mortality rate does not reflect the average annual mortality rate for the population. However, even when excluding the iMfolozi pack losses from the mortality tally ($n = 20$ by the end of 2006), a loss of 33 of 70 animals in a single year is still highly unsustainable for a small population.

The unsustainable mortality rate lends extra significance to a large number of deaths suspected to be caused by disease in KZN wild dogs. In addition to the 37.7% of deaths linked directly to the suspected rabies outbreak (the loss of the iMfolozi pack), disease cannot be ruled out as an underlying factor in other wild dog deaths in 2006. Several wild dogs were confirmed predator kills, and multiple "unknown" and "other" mortalities were strongly suspected to be due to hyena, lion, and crocodile attacks. Although not listed as a known cause of mortality for 2006, trauma secondary to motor vehicle collision is also a relatively frequent cause of mortality for the KZN wild dog population due to the presence of the Corridor Road, a two-lane paved commercial road bisecting HiP. Wild dogs will be more likely to be subject to predation or vehicle collisions when they are weakened by an underlying subclinical or ongoing chronic disease.

Consistent with Prager et al. [5], which included samples shared from our study, we found a low to zero prevalence of antibodies to CDV and CPV in the animals sampled. We caution that these findings are concerning given the presence of these pathogens in sympatric domestic dogs and the severe consequences of these infectious diseases for small, immunologically naïve populations. The wild dog which tested positive for CPV antibodies in this study is likely to have been exposed to CPV before arriving in KZN, as the animal was translocated to HiP in February 2006 from

Limpopo province and was sampled upon arrival in HiP. Excluding this new arrival, the prevalence of CDV and CPV antibodies among sampled individuals originating from KZN was 0%. In domestic dogs, serum antibody titres to CPV are known to remain high for years after exposure, even in the absence of reexposure [23], and while it is not known whether nondomestic canids have a similar immunological response, a detectable CPV titre would be expected in adult pack members who had been exposed to CPV as pups or juveniles. The negative results may therefore indicate either a lack of CPV exposure to date or 100% mortality for all exposed/infected animals for this population.

There are several reasons to remain vigilant in the face of these CDV and CPV results: (1) although no CDV exposure was detected, the 95% confidence intervals for CDV seroprevalence indicate that the actual prevalence could still be as high as 11.73% in the total population; (2) the finding of 0% CDV seroprevalence could indicate that exposed wild dogs do not survive infection, which may be the case if the viral strain is highly pathogenic, particularly since immunologically naïve animals are more susceptible to severe disease; (3) private veterinarians confirmed the presence of both CDV and CPV among pet dogs presenting to clinics in areas around KZN protected areas; (4) communal area surveys indicated that 26% of dog deaths in the past year were attributed by their owners to disease and that 8% of owners reported one or more dogs with clinical signs consistent with CDV infection; (5) African wild dogs and domestic dogs are both confirmed to cross boundaries between protected and community areas in KZN, and direct contact between individuals of both species was observed in KZN in 2006 by one of this paper's authors (P. Becker).

Reports from other populations of African wild dogs demonstrate the capacity of CDV to cause rapid extinction of packs or local populations. In Tswalu Kalahari Reserve (Northern Cape Province, South Africa) a confirmed CDV outbreak resulted in 100% mortality in a pack of nine wild dogs over a one-month period in 2005 [22]. In Chobe National Park, Botswana, CDV similarly resulted in the local extinction of a small pack of resident African wild dogs in 1994. As with all infectious pathogens, CDV risks are particularly high for small, fragmented populations [38, 47]. A population of African wild dogs from the Okavango Delta, Botswana, showed high seroprevalence of CDV without associated reduction in pup survivorship or increased adult mortality [15], possibly indicating exposure to a CDV strain of lower virulence and resulting in long-lasting immunity for this population. Interestingly, in 1996 this same Okavango wild dog population also experienced a substantial loss of individuals (five packs over a 4-week period), similar to that which occurred for the iMfolozi pack in our study; CDV or rabies was hypothesized to be the ultimate cause, but no carcasses could be recovered for confirmation [15].

CPV infections among wild dogs in Selous Game Reserve (Tanzania) were hypothesized to be responsible for significant reductions in litter size before den emergence [24]. The introduction of CPV into immunologically naïve populations, such as the KZN population, could potentially increase pup mortality during subsequent denning seasons, precipitating population decline. Indeed, Barker and Parrish [26] warn that, for small, isolated populations, the establishment of CPV as an endemic disease with ongoing reductions in pup recruitment would have substantial impacts on population survival. Another concern for CPV is the highly persistent nature of this pathogen in the environment, with viral particles remaining infectious for months to years depending on environmental conditions [23]. Prager et al. [5] demonstrated high prevalence of CPV antibodies in wild dogs with limited to not known domestic dog contact, an indication that due to prolonged pathogen persistence in the environment CPV may become self-perpetuating in small wildlife populations even in the absence of a domestic dog reservoir.

Rabies has repeatedly shown to pose a significant threat to wild dog conservation [28, 30, 31, 47]. Although not directly investigated via serology for African wild dogs in this study, the presence of rabies in domestic dogs in northern KZN during 2004–2006 was confirmed by State Veterinary Authority records, including in communities bordering protected areas inhabited by wild dogs. There were likely many undocumented rabies cases among domestic dogs in KZN during this time period, given the fact that less than one-third of communal areas were visited during annual rabies vaccination campaigns and only 9% of surveyed households reported taking their dogs to veterinarians. Prevalence of rabies is described to be grossly underestimated in KZN due to a high level of underreporting of cases, both in domestic animals and humans [37].

The domestic dog rabies epidemic in the Ntambanana district at the south-western border of HiP during 2006 occurred over the same 2-month period when 20 out of 26 wild dogs were lost from the iMfolozi pack just inside the park, including confirmation of death for the only radiocollared pack member. Clinical signs consistent with rabies were observed in another animal before it disappeared, within one month of the pack being observed feeding on a kill with a domestic dog inside park borders at the start of the rabies outbreak. These observations are consistent with rabies being the probable cause of mortalities for the iMfolozi pack, particularly given the highly contagious nature of the virus, the ease with which rabies could be transmitted between pack members due to close social interactions [47], and the rapid mortality rates observed in other African wild dog packs infected with rabies [30, 32, 48]. The case for disease as the

cause of mortality is strengthened by the lack of a convincing alternative explanation, given that the other known common causes of mortality for KZN wild dogs (snaring, predation, and vehicle strike) would be highly unlikely to result in the loss of so many pack members in such a short time frame. Additionally, it is of interest that the sole surviving pack member identified in mid-2007 was one of only two pack members that had been previously vaccinated against rabies. However, the vaccine was administered six years earlier, and the individual never received a booster vaccine; the efficacy of a single rabies vaccination in preventing mortality after exposure to the virus has been variable in previous reports [32, 33, 48, 49].

Vaccination of African wild dogs against infectious disease is usually not considered a viable management option given that most vaccines need to be administered parenterally on an annual or triannual basis. Hofmeyr et al. [33] reported that although rabies vaccination can have a beneficial impact in the face of a disease epidemic, multiple doses may be needed to ensure protection. Vial et al. [50] advocate that rabies vaccination of a "core" population of 30–40% of wild dogs in small populations every 1-2 years would theoretically be effective at controlling rabies and thus ensuring wild dog persistence. However, the efficacy and duration of immunity of domestic dog vaccine products in African wild dogs is largely unknown [50]. Inactivated CDV vaccines generally do not stimulate adequate seroconversion or protective immunity in nondomestic carnivore species [51]. At least one study has demonstrated that modified-live CDV vaccines designed for use in domestic dogs can induce seroconversion in captive African wild dogs [52], but seroconversion does not necessarily guarantee protective immunity against disease. There are several instances in which modified-live CDV vaccines have been used in this species without complications [52–54]; however, modified-live CDV vaccines also have the potential to cause mortality in this species, most probably through reversion to virulence and as a result of host immune system differences [55–57].

Given these various limitations, complexities, and concerns about vaccination in wildlife populations as a disease control strategy, the most economical and practical way to protect wild dogs from exposure to key canine pathogens may be via vaccination and control of domestic dog populations in areas adjacent to protected areas where wild dogs reside. It is generally reported that a vaccine coverage threshold of 70% is required in order to eliminate rabies as a disease threat from any given population [58–61]. In northern KZN the overall rabies vaccination coverage for domestic dogs is far below this threshold at approximately 30%. In contrast to many other parts of southern Africa, KZN does not have a significant wildlife reservoir of rabies; rather the primary rabies host species is the domestic dog [43, 62]. Therefore control or eradication of rabies in domestic dogs, including semiferal community dogs, has the potential to control or eliminate rabies as a threat to African wild dogs and other wildlife in the region [14]. A recent study reported that vaccination of domestic dogs in Tanzania is a feasible option for canine rabies control and thus also rabies control in sympatric wildlife [61].

Unfortunately, there are many reasons why domestic dog vaccination campaigns in northern KZN do not achieve adequate vaccine coverage in rural community areas, including difficulty accessing many communities due to poor road infrastructure and safety concerns, lack of funding for staff and supplies for larger-scale vaccine campaigns, opinions of local people regarding administration of pharmaceuticals to their animals, and political pressures placed on state veterinary authorities to concentrate efforts on other high-profile domestic livestock diseases. These challenges are not unique to KZN; Alexander et al. [15] as well as Prager et al. [39] outline the many complications inherent in attempting to manage disease risks in African wild dog populations by focussing on controlling disease in the reservoir host(s). In situations where it is not feasible to manage disease in the domestic dog pathogen host, the vaccination of African wild dogs may be the only viable option to limit the risk of widespread disease. Pathogen control via vaccination of key pack members and more vulnerable packs (e.g., those located in edge habitats) could function as an alternative control strategy [39, 50] and could be considered for the KZN African wild dog population.

In conclusion, the data and information presented indicate that infectious diseases, especially those with broad host ranges, such as CDV, CPV, and rabies, are a conservation concern for the African wild dog both in South Africa and across the continent. Mortality data and field observations in this study strongly suggest that infectious disease was the major cause of mortality in the KZN African wild dog population for 2006. Additionally, CDV, and CPV were found to be present in the geographically sympatric domestic dog population and thus could pose a threat in the future, especially as previous exposure (as measured via serology) and hence immune system competence against these diseases were low to nonexistent in the KZN African wild dog population. To reduce infectious disease risk for this population, wildlife managers need to further explore bolstering sympatric domestic dog vaccination campaigns and consider vaccinating key packs or a strategic number of key pack members. Pathogen management strategies must be implemented in combination with management of other threats to wild dog survival, including continued habitat fragmentation and landscape-associated factors limiting natural dispersal events.

Abbreviations

CDV: Canine distemper virus
CI: Confidence interval
CPV: Canine parvovirus
KZN: KwaZulu-Natal
n: Sample size.

Acknowledgments

The authors wish to thank the management of Hluhluwe-iMfolozi Park, the Isimangaliso (Greater St. Lucia) Wetland Park Authority, Mkuze Game Reserve, and Thanda Private Game Reserve for permission to conduct this research and for logistical support. They also thank the Ezemvelo KZN Wildlife Game Capture Unit for use of laboratory and darting equipment. They are grateful to the private and state veterinarians who kindly provided data regarding disease and vaccinations in domestic dogs: Drs. G. Archibald, R. Bagnall, D. Baxter, H. Kohrs, N. Meunier, C. Pryke, and T. Viljoen. They are also grateful to Gus van Dyk, wildlife manager of the Tswalu Kalahari Reserve, for information regarding CDV in African wild dogs. They thank Dr. Moritz van Vuuren and the laboratory at Department of Veterinary Tropical Diseases, Onderstepoort, for conducting the serology for this project. Finally, they wish to thank Sboniso "Zama" Zwane for assistance with field monitoring and for conducting disease surveys in Zulu communities and Thadaigh Baggallay for assistance with field monitoring. This project was supported by the National Geographic Society Conservation Trust (Grant no. C91-06) and the Conservation Medicine Small Research Grants Program at Murdoch University.

References

[1] P. Daszak, A. A. Cunningham, and A. D. Hyatt, "Anthropogenic environmental change and the emergence of infectious diseases in wildlife," *Acta Tropica*, vol. 78, no. 2, pp. 103–116, 2001.

[2] D. Spielman, "The roles of contagious diseases in natural populations, endangered populations, captive populations, and in wildlife breeding, translocation and rehabilitation programmes," in *Veterinary Conservation Biology Wildlife Health and Management in Australasia, Proceedings of the International Joint Conference of World Association of Wildlife Veterinarians*, A. Martin and L. Vogelnest, Eds., pp. 205–210, Wildlife Disease Association: Australasian Section, Australian Association of Veterinary Conservation Biologists and Wildlife Society of the New Zealand Veterinary Association, Australasian Veterinary Association, Sydney, Australia, 2001.

[3] L. Munson and W. Karesh, "Disease monitoring for the conservation of terrestrial animals," in *Conservation Medicine: Ecological Health In Practice*, A. A. Aguirre, G. M. Tabor, M. C. Pearl, R. S. Ostfeld, and C. House, Eds., pp. 95–102, Oxford University Press, Oxford, UK, 2002.

[4] D. A. Randall, J. Marino, D. T. Haydon et al., "An integrated disease management strategy for the control of rabies in Ethiopian wolves," *Biological Conservation*, vol. 131, no. 2, pp. 151–162, 2006.

[5] K. C. Prager, J. A. K. Mazet, L. Munson et al., "The effect of protected areas on pathogen exposure in endangered African wild dog (*Lycaon pictus*) populations," *Biological Conservation*, vol. 150, pp. 15–22, 2012.

[6] R. Woodroffe, J. W. McNutt, and M. G. L. Mills, "African wild dog," in *Foxes, Wolves, Jackals and Dogs: Status Survey and Conservation Action Plan*, C. Sillero-Zubiri and D. W. MacDonald, Eds., pp. 174–183, IUCN, Gland, Switzerland, 2nd edition, 2004.

[7] IUCN/SSC, *Regional Conservation Strategy for the Cheetah and African Wild Dog in Southern Africa*, I.S.S. Commission, Gland, Switzerland, 2007.

[8] Kenya Wildlife Service, *Proposal for Inclusion of Species on the Appendices of the Convention of the Conservation of Migratory Species of Wild Animals*, Kenya Wildlife Service, Nairobi, Kenya, 2010, http://www.cms.int/bodies/COP/cop9/proposals/Eng/II_5_Lycaon_pictus_KEN_E.pdf.

[9] IUCN/SSC, *Regional Conservation Strategy for the Cheetah and Wild Dog in Eastern Africa*, I.S.S. Commission, Gland, Switzerland, 2008.

[10] H. T. Davies-Mostert, M. G. L. Mills, and D. W. MacDonald, "A critical assessment of South Africa's managed metapopulation recovery strategy for African wild dogs," in *ReIntroduction of Top-Order Predators*, M. W. Hayward and M. J. Somers, Eds., pp. 10–42, Wiley-Blackwell, London, UK, 2009.

[11] T. K. Fuller, M. G. L. Mills, M. Borner, M. K. Laurenson, and P. W. Kat, "Long distance dispersal by African wild dogs in East and South Africa," *Journal of African Zoology*, vol. 106, pp. 535–537, 1992.

[12] S. Creel and N. M. Creel, "Six ecological factors that may limit African wild dogs, *Lycaon pictus*," *Animal Conservation*, vol. 1, no. 1, pp. 1–9, 1998.

[13] K. A. Leigh, K. R. Zenger, I. Tammen, and H. W. Raadsma, "Loss of genetic diversity in an outbreeding species: small population effects in the African wild dog (*Lycaon pictus*)," *Conservation Genetics*, vol. 13, pp. 767–777, 2012.

[14] R. Woodroffe, K. C. Prager, L. Munson, J. Conrad, E. J. Dubovi, and J. A. K. Mazet, "Contact with domestic dogs increases pathogen exposure in endangered African wild dogs (*Lycaon pictus*)," *PLoS One*, vol. 7, pp. 1–9, 2012.

[15] K. A. Alexander, J. W. McNutt, M. B. Briggs et al., "Multi-host pathogens and carnivore management in southern Africa," *Comparative Immunology, Microbiology and Infectious Diseases*, vol. 33, no. 3, pp. 249–265, 2010.

[16] S. L. Deem, L. H. Spelman, R. A. Yates, and R. J. Montali, "Canine distemper in terrestrial carnivores: a review," *Journal of Zoo and Wildlife Medicine*, vol. 31, no. 4, pp. 441–451, 2000.

[17] C. E. Greene and M. J. G. Appel, "Canine distemper virus," in *Infectious Diseases of the Dog and Cat*, C. E. Green, Ed., pp. 25–41, Elsevier, St. Louis, Miss, USA, 3rd edition, 2006.

[18] K. A. Alexander and M. J. Appel, "African wild dogs (*Lycaon pictus*) endangered by a canine distemper epizootic among domestic dogs near the Masai Mara National Reserve, Kenya," *Journal of wildlife diseases*, vol. 30, no. 4, pp. 481–485, 1994.

[19] K. A. Alexander, P. W. Kat, L. A. Munson, A. Kalake, and M. J. G. Appel, "Canine distemper-related mortality among wild dogs (*Lycaon pictus*) in Chobe National Park, Botswana," *Journal of Zoo and Wildlife Medicine*, vol. 27, no. 3, pp. 426–427, 1996.

[20] M. W. G. Van De Bildt, T. Kuiken, A. M. Visee, S. Lema, T. R. Fitzjohn, and A. D. M. E. Osterhaus, "Distemper outbreak and its effect on African wild dog conservation," *Emerging Infectious Diseases*, vol. 8, no. 2, pp. 211–213, 2002.

[21] K. V. Goller, R. D. Fyumagwa, V. Nikolin et al., "Fatal canine distemper infection in a pack of African wild dogs in the Serengeti ecosystem, Tanzania," *Veterinary Microbiology*, vol. 146, no. 3-4, pp. 245–252, 2010.

[22] G. van Dyk, *Wildlife Manager of the Tswalu Kalahari Reserve*, personal communication, 2007.

[23] D. McCaw and J. Hoskins, "Canine viral enteritis," in *Infectious Diseases of the Dog and Cat*, C. E. Greene, Ed., pp. 63–73, Elsevier, St. Louis, Miss, USA, 3rd edition, 2006.

[24] S. Creel, N. M. Creel, L. Munson, D. Sanderlin, and M. J. G. Appel, "Serosurvey for selected viral diseases and demography of African wild dogs in Tanzania," *Journal of Wildlife Diseases*, vol. 33, no. 4, pp. 823–832, 1997.

[25] L. D. Mech and S. M. Goyal, "Effects of canine parvovirus on gray wolves in Minnesota," *Journal of Wildlife Management*, vol. 59, no. 3, pp. 565–570, 1995.

[26] I. K. Barker and C. R. Parrish, "Parvovirus infections," in *Infectious Diseases of Wild Mammals*, E. S. Williams and I. K. Barker, Eds., pp. 131–146, Iowa State Press, Ames, Iowa, USA, 3rd edition, 2001.

[27] C. E. Rupprecht, K. Stoehr, and C. Meredity, "Rabies," in *Infectious Diseases of Wild Mammals*, S. Williams and I. K. Barker, Eds., pp. 3–36, Iowa State Press, Ames, Iowa, USA, 3rd edition, 2001.

[28] S. C. Gascoyne, M. K. Laurenson, S. Lelo, and M. Borner, "Rabies in African wild dogs (*Lycaon pictus*) in the Serengeti region, Tanzania," *Journal of Wildlife Diseases*, vol. 29, no. 3, pp. 396–402, 1993.

[29] D. W. Macdonald, M. Artois, M. Aubert et al., "Cause of wild dog deaths," *Nature*, vol. 360, pp. 633–634, 1992.

[30] K. A. Alexander, J. D. Richardson, L. Munson, and P. W. Kat, "An outbreak of rabies among African wild dogs (*Lycaon pictus*) in the Masai Mara, Kenya," in *Proceedings of the American Association of Zoo Veterinarians*, p. 340, St. Louis, Miss, USA, 1993.

[31] P. W. Kat, K. A. Alexander, J. S. Smith, and L. Munson, "Rabies and African wild dogs in Kenya," in *Proceedings of the Royal Society of London B Biological Sciences*, vol. 262, pp. 229–233, 1995.

[32] M. Hofmeyr, J. Bingham, E. P. Lane, A. Ide, and L. Nel, "Rabies in African wild dogs (*Lycaon pictus*) in the Madikwe Game Reserve, South Africa," *Veterinary Record*, vol. 146, no. 2, pp. 50–52, 2000.

[33] M. Hofmeyr, D. Hofmeyr, L. Nel, and J. Bingham, "A second outbreak of rabies in African wild dogs (*Lycaon pictus*) in Madikwe Game Reserve, South Africa, demonstrating the efficacy of vaccination against natural rabies challenge," *Animal Conservation*, vol. 7, no. 2, pp. 193–198, 2004.

[34] J. L. Scheepers and K. A. E. Venzke, "Attempts to reintroduce African wild dogs *Lycaon pictus* into Etosha National Park, Namibia," *South African Journal of Wildlife Research*, vol. 25, pp. 138–140, 1995.

[35] R. Woodroffe and J. R. Ginsberg, "Past and future causes of wild dogs' population decline," in *The African Wild Dog: Status, Survey, and Conservation Action Plan*, R. Woodroffe, J. R. Ginsberg, and D. W. MacDonald, Eds., pp. 58–74, IUCN, Gland, Switzerland, 1997.

[36] J. R. A. Butler, J. T. Du Toit, and J. Bingham, "Free-ranging domestic dogs (*Canis familiaris*) as predators and prey in rural Zimbabwe: threats of competition and disease to large wild carnivores," *Biological Conservation*, vol. 115, no. 3, pp. 369–378, 2004.

[37] N. Meunier, *State Veterinarian, Hlabisa District*, Personal Communication, 2006.

[38] S. M. Funk, C. V. Fiorello, S. Cleaveland, and M. E. Gompper, "The role of disease in carnivore ecology and conservation," in *Carnivore Conservation*, J. L. Gittleman, S. M. Funk, D. W. MacDonald, and R. K. Wayne, Eds., pp. 443–466, Cambridge University Press, Cambridge, UK, 2001.

[39] K. C. Prager, R. Woodroffe, A. Cameron, and D. T. Haydon, "Vaccination strategies to conserve the endangered African wild dog (*Lycaon pictus*)," *Biological Conservation*, vol. 144, no. 7, pp. 1940–1948, 2011.

[40] M. Van Vuuren, "Serological studies of bovine respiratory syncytial virus in feedlot cattle in South Africa," *Journal of the South African Veterinary Association*, vol. 61, no. 4, pp. 168–169, 1990.

[41] J. F. Bell, M. A. Gonzales, A. M. Diaz, and G. J. Moore, "Nonfatal rabies in dogs—experimental studies and results of a survey," *American Journal of Veterinary Research*, vol. 32, pp. 2049–2058, 1971.

[42] M. Fekudo and J. H. Shaddock, "Peripheral distribution of virus in dogs inoculated with two strains of rabies virus," *American Journal of Veterinary Research*, vol. 45, pp. 724–729, 1984.

[43] R. Bagnall, *KZN State Veterinary Authority*, Personal Communication, 2006.

[44] J. Van Heerden, M. G. Mills, M. J. Van Vuuren, P. J. Kelly, and M. J. Dreyer, "An investigation into the health status and diseases of wild dogs (*Lycaon pictus*) in the Kruger National Park," *Journal of the South African Veterinary Association*, vol. 66, no. 1, pp. 18–27, 1995.

[45] J. R. Ginsberg, K. A. Alexander, S. Creel, P. W. Kat, J. W. McNutt, and M. G. Mills, "Handling and survivorship of African wild dog (*Lycaon pictus*) in five ecosystems," *Conservation Biology*, vol. 9, no. 3, pp. 665–674, 1995.

[46] R. Woodroffe, H. Davies-Mostert, J. Ginsberg et al., "Rates and causes of mortality in Endangered African wild dogs *Lycaon pictus*: lessons for management and monitoring," *Oryx*, vol. 41, no. 2, pp. 215–223, 2007.

[47] M. G. Mills, "Social systems and behaviour of the African wild dog *Lycaon pictus* and the spotted hyaena *Crocuta crocuta* with special reference to rabies," *Onderstepoort Journal of Veterinary Research*, vol. 60, no. 4, pp. 405–409, 1993.

[48] R. Woodroffe, "Assessing the risks of intervention: immobilization, radio-collaring and vaccination of African wild dogs," *Oryx*, vol. 35, no. 3, pp. 234–244, 2001.

[49] D. L. Knobel, A. Liebenberg, and J. T. Du Toit, "Seroconversion in captive African wild dogs (*Lycaon pictus*) following administration of a chicken head bait/SAG-2 oral rabies vaccine combination," *Onderstepoort Journal of Veterinary Research*, vol. 70, no. 1, pp. 73–77, 2003.

[50] F. Vial, S. Cleaveland, G. Rasmussen, and D. T. Haydon, "Development of vaccination strategies for the management of rabies in African wild dogs," *Biological Conservation*, vol. 131, no. 2, pp. 180–192, 2006.

[51] R. J. Montali, C. R. Bartz, J. A. Teare, J. T. Allen, M. J. Appel, and M. Bush, "Clinical trials with canine distemper vaccines in exotic carnivores," *Journal of the American Veterinary Medical Association*, vol. 183, no. 11, pp. 1163–1167, 1983.

[52] J. van Heerden, J. Bingham, M. van Vuuren, R. E. J. Burroughs, and E. Stylianides, "Clinical and serological response of wild dogs (*Lycaon pictus*) to vaccination against canine distemper, canine parvovirus infection and rabies," *Journal of the South African Veterinary Association*, vol. 73, no. 1, pp. 8–12, 2002.

[53] J. Spencer and R. Burroughs, "Antibody response to canine distemper vaccine in African wild dogs," *Journal of Wildlife Diseases*, vol. 28, no. 3, pp. 443–444, 1992.

[54] J. Van Heerden, W. H. Swart, and D. G. A. Meltzer, "Serum antibody levels before and after administration of live canine distemper vaccine to the wild dog *Lycaon pictus*," *Journal of the South African Veterinary Association*, vol. 51, no. 4, pp. 283–284, 1980.

[55] A. E. McCormick, "Canine distemper in African hunting dogs (*Lycaon pictus*)—possibly vaccine induced," *The Journal of Zoo Animal Medicine*, vol. 14, pp. 66–71, 1983.

[56] J. Van Heerden, N. Bainbridge, R. E. Burroughs, and N. P. Kriek, "Distemper-like disease and encephalitozoonosis in wild dogs (*Lycaon pictus*)," *Journal of Wildlife Diseases*, vol. 48, pp. 19–21, 1989.

[57] B. Durchfeld, W. Baumgärtner, W. Herbst, and R. Brahm, "Vaccine-associated canine distemper infection in a litter of African hunting dogs (*Lycaon pictus*)," *Zentralblatt für Veterinärmedizin Reihe B*, vol. 37, no. 3, pp. 203–212, 1990.

[58] P. G. Coleman and C. Dye, "Immunization coverage required to prevent outbreaks of dog rabies," *Vaccine*, vol. 14, pp. 185–186, 1996.

[59] WHO, "WHO Expert Consultation on Rabies," Technical Report Series 931, World Health Organization, 2005.

[60] J. F. Reece and S. K. Chawla, "Control of rabies in Jaipur, India, by the sterilisation and vaccination of neighbourhood dogs," *Veterinary Record*, vol. 159, no. 12, pp. 379–383, 2006.

[61] M. C. Fitzpatrick, K. Hampson, S. Cleaveland, L. Ancel Meyers, J. P. Townsend, and A. P. Galvani, "Potential for rabies control through dog vaccination in wildlife-abundant communities of Tanzania," *PLoS Neglected Tropical Diseases*, vol. 6, no. 8, Article ID e1796, 2012.

[62] C. T. Sabeta, J. Bingham, and L. H. Nel, "Molecular epidemiology of canid rabies in Zimbabwe and South Africa," *Virus Research*, vol. 91, no. 2, pp. 203–211, 2003.

11

Vegetation Structure and Composition across Different Land Uses in a Semiarid Savanna of Southern Zimbabwe

Patience Zisadza-Gandiwa,[1] **Cheryl T. Mabika,**[2] **Olga L. Kupika,**[2]
Edson Gandiwa,[1] **and Chrispen Murungweni**[3]

[1] *Scientific Services, Gonarezhou National Park, Parks and Wildlife Management Authority, Private Bag 7003, Chiredzi, Zimbabwe*
[2] *Department of Wildlife and Safari Management, Chinhoyi University of Technology, Private Bag 7724, Chinhoyi, Zimbabwe*
[3] *Department of Animal Production and Technology, Chinhoyi University of Technology, Private Bag 7724, Chinhoyi, Zimbabwe*

Correspondence should be addressed to Patience Zisadza-Gandiwa; patience.gandiwa@gmail.com

Academic Editor: Antonio Terlizzi

We compared the structure and composition of vegetation communities across different land uses in the northern Gonarezhou National Park and adjacent areas, southeast Zimbabwe. Vegetation data were collected from 60 sample plots using a stratified random sampling technique from April to May 2012. Stratification was by land use, and sample plots in all three strata occurred on predominantly siallitic soils. Our results show that the communal area had higher woody plant species diversity ($H' = 2.66$) than the protected area ($H' = 1.78$). However, the protected area had higher grass species richness per plot than the communal area and resettlement area. Overall, the protected area had more structural and compositional diversity than the other land use areas. These findings suggest that the areas adjacent to protected areas contribute to plant diversity in the greater ecosystem; hence conservation efforts should extend beyond the boundaries of protected areas. We recommend that protected area management should engage community-based institutions in neighbouring areas for effective monitoring of woody vegetation structure and composition.

1. Introduction

Wildlife conservation in today's world is increasingly confronted by the challenges of understanding the dynamics shaping vegetation cover and species diversity as wildlife habitat straddles across the land use divide [1, 2]. One of the assumptions which have not been adequately tested is the protection of wildlife habitat in areas of different land uses surrounding protected areas. The International Union for Conservation of Nature defines a protected area as "a clearly defined geographical space, recognized, dedicated and managed through legal or other effective means to achieve the long-term conservation of nature with associated ecosystem services and cultural values" [3]. Moreover, the world commission on protected areas recently estimated that there are over 100,000 protected areas ranging from areas that strictly limit human activity to those that allow for sustainable human use [4]. Despite their prevalence in both developed and developing countries, there have been surprisingly few assessments on the ecological effectiveness of protected areas [5] and evaluation of vegetation structure and composition inside the protected areas and adjacent areas.

It is assumed that biodiversity is best managed in protected areas and other areas where land has not been fragmented due to human population pressure [6, 7]. Biodiversity conservation outside protected areas is increasingly taking centre stage in global conservation discourse [8–10]. Although it is seldom the focus of scientific investigations, wildlife habitat loss has alarmed conservationists because of its potential implications for native biodiversity [11]. However, little is currently known about the ecological consequences of the increasing demographic pressure of human and livestock populations to terrestrial wildlife habitat in areas of different land uses, yet some conservationists suggest that it may result in biodiversity loss [12, 13].

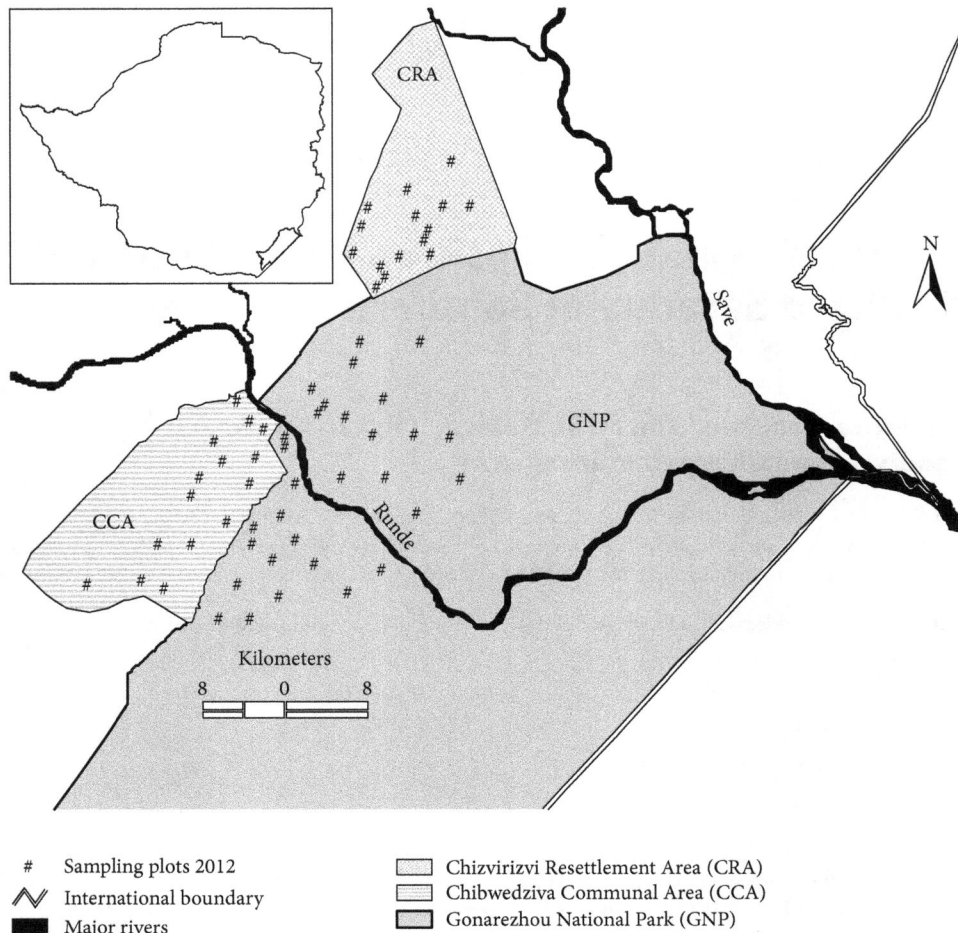

FIGURE 1: Location of study sites and sample plots in northern Gonarezhou National Park and adjacent areas in southern Zimbabwe.

In Zimbabwe, the assumption that the Communal Areas Management Programme for Indigenous Resources (CAMP-FIRE) surrounding mostly protected areas expands the habitat of the core wildlife area, forming a buffer around the protected area [14], needs to be continuously investigated. Such assumptions assume a land use gradient exists for biodiversity protection [15]. Therefore, the aim of this study was to investigate the current vegetation status in and around a large state protected area in a semiarid savanna landscape of southeast Zimbabwe. Our main objective was to determine the structure and composition of plant species across different land uses adjacent to a state protected area.

2. Materials and Methods

2.1. Study Area.
Our study focussed on the northern Gonarezhou National Park (GNP), Chibwedziva Communal Area (CCA) which is a CAMPFIRE area, and Chizvirizvi Resettlement Area (CRA)—a resettlement area (Figure 1). The entire GNP is about 5000 km² in extent whereas CCA and CRA are 315 km² and 240 km² in extent, respectively. All the selected sites are within the Great Limpopo Transfrontier Conservation Area in southeastern Zimbabwe, and wildlife

conservation is a recognised form of land use. The plant communities in the study area are typical of the savanna vegetation, comprised of a mosaic of trees and grass dominated by *Colophospermum mopane* and *Combretum apiculatum*.

Three climatic seasons can be recognized in the study area: hot and wet (from November to March), cool and dry (from April to July), and hot and dry (from August to October). The average annual precipitation ranges from 200 to 600 mm. Average monthly maximum temperatures are 25°C in July and 38°C in January. Average monthly minimum temperatures range between 11°C in June and 25°C in January [16]. The area is generally low-lying with a mean altitude of mostly 400 m above sea level [17].

2.2. Data Collection.
A stratified random sampling procedure was used in this study. Three strata were defined according to land use, namely, (i) strictly wildlife conservation, (ii) communal area, and (iii) resettlement area. Data collection was conducted from April to May 2012. The estimated variables of the woody vegetation (trees and shrubs) were plant species richness, plant height, and dead trees. Trees and shrubs were classified based on height; that is, rooted, woody, and self-supporting plants ≥ 3 m in height were

TABLE 1: Vegetation attributes for sample plots across different land use areas (mean ± standard error) and significant levels from one-way ANOVA with unequal sample size tests.

| Variable | Land use category | | | $F_{2,57}$ | P value |
	GNP	CRA	CCA		
Tree density ha^{-1}	842.78 ± 110.11[a]	416.67 ± 93.59[b]	407.78 ± 185.25[b]	5.41	**0.007**
Shrub density ha^{-1}	240.00 ± 56.73[a]	80.00 ± 17.68[a]	111.11 ± 48.87[a]	2.71	0.075
Sapling density ha^{-1}	694.44 ± 152.25[a]	204.44 ± 54.44[b]	147.78 ± 61.18[b]	5.36	**0.007**
Dead tree density ha^{-1}	27.77 ± 7.84[a]	28.88 ± 11.21[a]	42.22 ± 17.66[a]	0.44	0.647
Tree height (m)	4.56 ± 0.16[a]	5.63 ± 0.38[b]	7.42 ± 0.87[a]	11.13	**0.000**
Shrub height (m)	1.44 ± 0.15[a]	1.51 ± 0.22[a]	1.08 ± 0.25[a]	1.11	0.335
Woody species diversity (H')	1.78 ± 0.16[a]	1.54 ± 0.26[b]	2.66 ± 0.24[a]	6.37	**0.003**
Grass species richness per plot	11.27 ± 0.88[a]	6.80 ± 0.85[b]	7.47 ± 0.8[b]	7.68	**0.001**

GNP represents Gonarezhou National Park, CRA represents Chizvirizvi Resettlement Area, and CCA represents Chibwedziva Communal Area. Significant values are indicted in bold; values with different superscript letters within rows differ significantly (Tukey's HSD; $P < 0.05$).

classified as trees whereas rooted, woody, self-supporting, and multistemmed or single-stemmed plants greater than 1 m but < 3 m in height were classified as shrubs [16]. Herbaceous vegetation (forbs and herbs) species richness per plot was also recorded. A total of 60 plots (30 × 20 m^2) were sampled in the study sites, that is, 30 plots in northern GNP and 15 plots each in CCA and CRA (Figure 1). A 6 m graduated pole was used for measuring woody plant height, and a handheld Global Positioning System (GPS) was used to mark the location of each sampling plot.

2.3. Data Analyses. Collected data were summarised and tested for normality using the Kolmogorov-Smirnov test, and data for tree density, shrub density, sapling density, and tree height were found to be not normally distributed; hence, data were normalised using $\log_{10}(x + 1)$ transformation [18]. Species diversity in different land use areas was determined by calculating the Shannon-Weiner (H') diversity index [19]. Differences in vegetation structure and composition were tested using One-way Analysis of Variance (ANOVA), at 5% level of significance using the Statistical Package for Social Sciences (SPSS) version 19 for Windows (SPSS Inc., Chicago, IL, USA). *Post hoc* analysis for variables with significant differences was carried out using Tukey's Honestly Significant Difference (HSD). Furthermore, we performed a Principal Component Analysis (PCA) to determine the underlying patterns of the vegetation data using the 60 sample plots in CANOCO version 4.5 for Windows [20].

3. Results

3.1. Woody Vegetation Structure and Composition across Land Use. A total of 3670 woody plants (61% trees and 39% shrubs) were assessed, and 136 vegetation species were identified across all land uses. About 51% of the vegetation species were woody plant species whereas 49% were grass species. Vegetation structure and composition significantly differed across land use, particularly in the following variables: tree density, sapling density, tree height, woody species diversity, and grass species richness (Table 1). In contrast, there were no significant differences in densities of shrubs, dead trees, and shrub

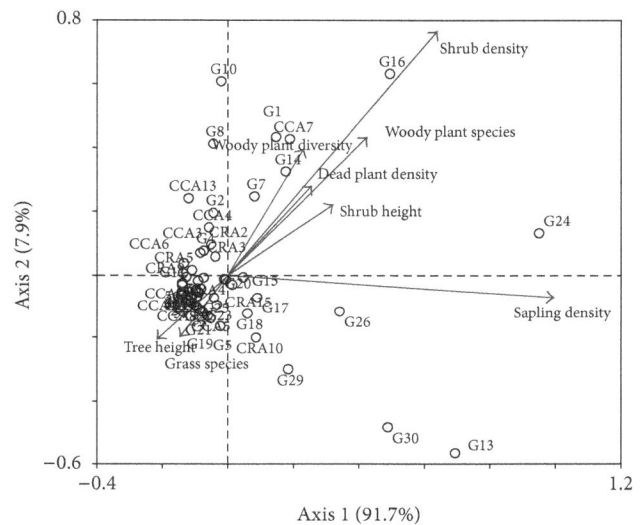

FIGURE 2: Principal Component Analysis biplot of measured vegetation variables from the 60 sample plots in northern Gonarezhou National Park and adjacent areas, southern Zimbabwe. G represents Gonarezhou National Park, CCA represents Chibwedziva Communal Area, and CRA represents Chizvirizvi Resettlement Area.

height. Woody vegetation community in the study area was dominated by *Colophospermum mopane*, *Acacia nigrescens*, *Combretum apiculatum*, *Dichrostachys cinerea*, *Kirkia acuminata*, *Spirostachys Africana*, and *Terminalia sericea*.

3.2. Patterns in Woody Vegetation Structure and Composition. Figure 2 shows a PCA biplot of sample plots and measured variables in the study area. Axis 1 explained 91.7% of the variance in vegetation data and defined a gradient from areas with taller trees and higher grass species richness to areas with a higher density of saplings, shrubs and higher woody vegetation diversity. Accordingly, the CCA, and CRA correlated negatively with Axis 1 whereas GNP correlated positively with Axis 1. Moreover, Axis 2 explained 7.9% of the vegetation data and defined a gradient from areas characterised with taller trees and higher grass species richness to areas with higher densities of shrubs and diversity of woody plants. GNP, CCA and CRA had a negative correlation with Axis 2 whereas mostly GNP and, to a lesser extend, CCA were positively correlated with Axis 2.

4. Discussion

The three land use areas examined in this study showed significant differences in structural and compositional attributes of vegetation. We find it interesting that tree species diversity was higher in CCA than in the protected area, that is, GNP. This finding is contrary to the widely accepted perception that diversity is poorly managed in areas settled by people. However, the perception is supported by our results on grass species richness where diversity was higher in the GNP than in CCA. This finding suggests that disturbance factors may have a significant effect on certain plant communities, their composition and functioning are important factors to consider when studying biodiversity [21, 22], and anthropogenic disturbances may be more pronounced outside protected areas [23, 24]. Similarities across the land use strata were in shrub height, shrub density, and dead tree density.

The vegetation structure and composition across different land uses suggest that the role of anthropogenic disturbance can have long-term effect in influencing habitat loss [11]. Despite topography, edaphic and moisture variation which is known to affect structure and composition in savannas [25, 26], the loss of woody vegetation due to herbivory, fires, droughts, frost, diseases, and human disturbances remains important in semiarid savanna ecosystem [27–29]. Our study confirms this finding and further suggests that human disturbance is likely to be a key factor in shaping woody vegetation communities in the southeastern Zimbabwe. This has implications on CAMPFIRE areas surrounding protected areas in Zimbabwe, as habitat availability affects distribution of wildlife [30]. Moreover, vegetation provides local communities with basic subsistence and economic resources [31]. Recent evidence of cattle grazing in the different land uses, including GNP, presents some important insights of habitat overlaps between wild and domesticated herbivores [32], which also leads to herbaceous layer changes due to human and livestock encroachments into protected areas.

Most communal areas in southeastern Zimbabwe are associated with human population increase, encroachments into wildlife areas, and increased dependency on natural resources for livelihood, which often results in habitat loss and degradation, thus influencing wildlife abundances and their distribution [32–35]. In the unprotected areas, vegetation losses can be a result of selective extraction of forest/woodland resources for purposes such as fuel wood, construction materials, and other nontimber forest products [36]. The varying levels of disturbance in the different land use categories have an effect on plant biodiversity. It has been reported that the structural complexity of an ecological community is positively correlated with the diversity of plant life [37]. Fully protected areas are often assumed to be the best way to conserve plant diversity and maintain intact woodland/forest composition and structure [38], that ultimately determines biodiversity at various scales, providing habitat for unique wildlife species that require unique and variable forage and cover opportunities or "niches" for survival and reproduction.

5. Conclusions

Our study provides some evidence that the protected areas are a more effective way to conserve diversity in grasses compared to nonprotected areas. However, how to improve diversity of trees inside protected areas or understanding what is causing less diversity of trees in these areas remains a puzzle. This study provides a reference baseline for monitoring changes in vegetation species diversity, which has, undoubtedly, important conservation implications requiring appropriate and timely management interventions if the direction of change is not desirable according to conservation objectives being pursued in the area. We, therefore, recommend regular monitoring of vegetation structure and composition in all areas surrounding protected areas and not restricting ecological monitoring effort within boundaries of protected areas. There is also a need for tapping into local ecological knowledge to understand the sociocultural issues surrounding the survival of some woody plant species in unprotected areas dominated by human activities.

Acknowledgments

This research was supported by the Gonarezhou Conservation Project, a conservation partnership between the Zimbabwe Parks and Wildlife Management Authority and Frankfurt Zoological Society. The authors are grateful to Hillary Madzikanda, Evious Mpofu, Daphine Madhlamoto, Tendai Chinho, Patience Nyabawa, and Gonarezhou National Park staff for rendering invaluable assistance during this study. Comments and suggestions from two anonymous reviewers helped improve this paper.

References

[1] N. Clerici, A. Bodini, H. Eva, J. M. Grégoire, D. Dulieu, and C. Paolini, "Increased isolation of two Biosphere Reserves and surrounding protected areas (WAP ecological complex, West Africa)," *Journal for Nature Conservation*, vol. 15, no. 1, pp. 26–40, 2007.

[2] R. DeFries, A. Hansen, A. C. Newton, and M. C. Hansen, "Increasing isolation of protected areas in tropical forests over the past twenty years," *Ecological Applications*, vol. 15, no. 1, pp. 19–26, 2005.

[3] International Union for Conservation of Nature, "1993 United Nations list of national parks and protected areas," WCMC and CNPPA. IUCN, Gland, Switzerland, 1994.

[4] L. Naughton-Treves, M. B. Holland, and K. Brandon, "The role of protected areas in conserving biodiversity and sustaining local livelihoods," *Annual Review of Environment and Resources*, vol. 30, pp. 219–252, 2005.

[5] K. J. Gaston, K. Charman, S. F. Jackson et al., "The ecological effectiveness of protected areas: the United Kingdom," *Biological Conservation*, vol. 132, no. 1, pp. 76–87, 2006.

[6] K. A. Brown, J. C. Ingram, D. F. B. Flynn, R. Razafindrazaka, and V. Jeannoda, "Protected area safeguard tree and shrub communities from degradation and invasion: a case study in eastern madagascar," *Environmental Management*, vol. 44, no. 1, pp. 136–148, 2009.

[7] R. Watson, K. H. Fitzgerald, and N. Gitahi, "Expanding options for habitat conservation outside protected areas in Kenya: the use of environmental easements," Technical Paper no. 2, African Wildlife Foundation, Kenya, 2010.

[8] M. Anyonge-Bashir and P. Udoto, "Beyond philanthropy: community nature-based enterprises as a basis for wildlife conservation," *The George Wright Forum*, vol. 29, no. 1, pp. 67–73, 2012.

[9] P. F. Langhammer, M. I. Bakarr, L. A. Bennun et al., *Identification and Gap Analysis of Key Biodiversity Areas: Targets for Comprehensive Protected Area Systems*, The World Conservation Union-IUCN, Gland, Switzerland, 2007.

[10] M. Niamir-Fuller, C. Kerven, R. Reid, and E. Milner-Gulland, "Co-existence of wildlife and pastoralism on extensive rangelands: competition or compatibility?" *Pastoralism: Research, Policy and Practice*, vol. 2, no. 8, 2012.

[11] A. J. Hansen, R. Rasker, B. Maxwell et al., "Ecological causes and consequences of demographic change in the new west," *BioScience*, vol. 52, no. 2, pp. 151–162, 2002.

[12] H. H. T. Prins, "The patoral road to extinction: competition between wildlife and traditional pastoralism in East Africa," *Environmental Conservation*, vol. 19, no. 2, pp. 117–123, 1992.

[13] W. D. Newmark, "Isolation of African protected areas," *Frontiers in Ecology and the Environment*, vol. 6, no. 6, pp. 321–328, 2008.

[14] P. G. H. Frost and I. Bond, "The CAMPFIRE programme in Zimbabwe: payments for wildlife services," *Ecological Economics*, vol. 65, no. 4, pp. 776–787, 2008.

[15] J. D. Maestas, R. L. Knight, and W. C. Gilgert, "Biodiversity across a Rural Land-Use gradient," *Conservation Biology*, vol. 17, no. 5, pp. 1425–1434, 2003.

[16] E. Gandiwa and S. Kativu, "Influence of fire frequency on *Colophospermum mopane* and *Combretum apiculatum* woodland structure and composition in northern Gonarezhou National Park, Zimbabwe," *Koedoe*, vol. 51, no. 1, 2009.

[17] Zimbabwe Parks and Wildlife Management Authority, "Gonarezhou National Park Management Plan: 2011–2021," Zimbabwe Parks and Wildlife Management Authority, Harare, Zimbabwe, 2011.

[18] J. H. McDonald, *Handbook of Biological Statistics*, Sparky House, Baltimore, Md, USA, 2nd edition, 2009.

[19] J. A. Ludwig and J. F. Reynolds, *Statistical Ecology: A Primer on Methods and Computing*, John Wiley & Sons, New York, NY, USA, 1988.

[20] C. J. F. Ter Braak and P. Šmilauer, "CANOCO reference manual and CanoDraw for Windows user's guide: software for Canonical Community Ordination," version 4.5, Microcomputer Power, Ithaca, NY, USA, 2002.

[21] J. P. Grime, *Plant Strategies and Vegetation Processes*, Wiley, Chichester, UK, 1979.

[22] P. Zisadza-Gandiwa, L. Mango, E. Gandiwa et al., "Variation in woody vegetation structure and composition in a semiarid savanna of Southern Zimbabwe," *International Journal of Biodiversity and Conservation*, vol. 5, no. 2, pp. 71–77, 2013.

[23] E. F. Lambin, B. L. Turner, H. J. Geist et al., "The causes of land-use and land-cover change: moving beyond the myths," *Global Environmental Change*, vol. 11, no. 4, pp. 261–269, 2001.

[24] W. Kperkouma, Y. W. Agbélessessi, B. Wiyao et al., "Assessment of vegetation structure and human impacts in the protected area of Alédjo, Togo," *African Journal of Ecology*, vol. 50, pp. 355–366, 2012.

[25] E. T. F. Witkowski and T. G. O'Connor, "Topo-edaphic, floristic and physiognomic gradients of woody plants in a semi-arid African savanna woodland," *Vegetatio*, vol. 124, no. 1, pp. 9–23, 1996.

[26] R. J. Williams, G. A. Duff, D. M. J. S. Bowman, and G. D. Cook, "Variation in the composition and structure of tropical savannas as a function of rainfall and soil texture along a large-scale climatic gradient in the Northern Territory, Australia," *Journal of Biogeography*, vol. 23, no. 6, pp. 747–756, 1996.

[27] C. D. Allen, A. K. Macalady, H. Chenchouni et al., "A global overview of drought and heat-induced tree mortality reveals emerging climate change risks for forests," *Forest Ecology and Management*, vol. 259, no. 4, pp. 660–684, 2010.

[28] E. Gandiwa, T. Magwati, P. Zisadza, T. Chinuwo, and C. Tafangenyasha, "The impact of African elephants on *Acacia tortilis* woodland in northern Gonarezhou National Park, Zimbabwe," *Journal of Arid Environments*, vol. 75, no. 9, pp. 809–814, 2011.

[29] C. Tafangenyasha, "Decline of the mountain acacia, *Brachystegia glaucescens* in Gonarezhou National Park, southeast Zimbabwe," *Journal of Environmental Management*, vol. 63, no. 1, pp. 37–50, 2001.

[30] M. Waltert, B. Meyer, and C. Kiffner, "Habitat availability, hunting or poaching: what affects distribution and density of large mammals in western Tanzanian woodlands?" *African Journal of Ecology*, vol. 47, no. 4, pp. 737–746, 2009.

[31] E. Gandiwa, "Importance of dry savanna woodlands in rural livelihoods and wildlife conservation in southeastern Zimbabwe," *Nature & Faune*, vol. 26, no. 1, pp. 60–66, 2011.

[32] E. Gandiwa, I. M. A. Heitkönig, P. Gandiwa, W. Matsvayi, H. Van Der Westhuizen, and M. M. Ngwenya, "Large herbivore dynamics in northern Gonarezhou National Park, Zimbabwe," *Tropical Ecology*, vol. 54, no. 3, pp. 343–352, 2013.

[33] P. Gandiwa, M. Matsvayi, M. M. Ngwenya, and E. Gandiwa, "Assessment of livestock and human settlement encroachment into northern Gonarezhou National Park, Zimbabwe," *Journal of Sustainable Development in Africa*, vol. 13, no. 5, pp. 19–33, 2011.

[34] W. Wolmer, J. Chaumba, and I. Scoones, "Wildlife management and land reform in southeastern Zimbabwe: a compatible pairing or a contradiction in terms?" *Geoforum*, vol. 35, no. 1, pp. 87–98, 2004.

[35] W. Wolmer, "Wilderness gained, wilderness lost: wildlife management and land occupations in Zimbabwe's southeast lowveld," *Journal of Historical Geography*, vol. 31, no. 2, pp. 260–280, 2005.

[36] K. A. Brown, S. Spector, and W. Wu, "Multi-scale analysis of species introductions: combining landscape and demographic models to improve management decisions about non-native species," *Journal of Applied Ecology*, vol. 45, no. 6, pp. 1639–1648, 2008.

[37] M. A. Huston, *Biological Diversity: The Coexistence of Species on Changing Landscapes*, Cambridge University Press, New York, NY, USA, 1996.

[38] T. Banda, M. W. Schwartz, and T. Caro, "Woody vegetation structure and composition along a protection gradient in a miombo ecosystem of western Tanzania," *Forest Ecology and Management*, vol. 230, no. 1-3, pp. 179–185, 2006.

12

Diversity of Millipedes in Alagar Hills Reserve Forest in Tamil Nadu, India

Periasamy Alagesan and Baluchamy Ramanathan

Post Graduate and Research Department of Zoology, Yadava College, Madurai, Tamil Nadu 625 014, India

Correspondence should be addressed to Periasamy Alagesan; maniragavi@rediffmail.com

Academic Editor: Curtis C. Daehler

Millipede diversity and abundance were analysed at sites lying between 250 and 650 meters above mean sea level in Alagar Hills of Eastern Ghats, Tamil Nadu, India. Millipede abundance and diversity peaked at midelevations influenced by favourable niche and food resources. Diversity of millipedes indicates the influence of local habitat and food resource availability. In the present study, millipede species, *Harpaphe haydeniana*, *Xenobolus carnifex*, *Arthrosphaera magna*, *Aulacobolus newtoni*, and *Spinotarsus colosseus*, are present at midelevation (450 MSL). Abundance of millipedes at 450 m elevation is due to moderate canopy and litter, which support understorey vegetation like herbs and shrubs.

1. Introduction

Biodiversity offers several direct and indirect economic benefits to humankind [1]. The interest in diversity especially over the past few years has focused on how diversity influences ecosystems and ecological processes [2]. Ninety-five percent of experimental studies support a positive relationship between diversity and ecosystem functioning [3–5]. The biotic diversity tends to play a significant role by enriching the soil, maintaining water and climatic cycles, and converting waste materials into nutrients. Soil macrofauna makes an important contribution to soil fertility by promoting the stability and productivity of forest ecosystems, mainly due to their influence on soil process such as litter decomposition and nutrient dynamics [6]. Furthermore, elevation is merely a surrogate for a suite of biotic and abiotic factors that influence species richness [7]. Therefore, identifying ecologically meaningful causal factors is essential in order to explain variation in species richness along elevation gradients [8]. Millipedes belong to class Diplopoda, a highly diverse group of terrestrial organisms with over 12,000 described species and an estimated 80, 000 species yet to be described [9, 10]. Among soil Arthropods, millipedes act as primary destructors of plant debris and play a crucial role in soil formation processes. Many millipedes can also serve as indicators of

environmental conditions and improve the structure content of organic matter and nutrient elements of soil [11, 12].

Gadagkar et al. [13] studied species richness and diversity of ant populations from different localities in Western Ghats, India. Diversity of forest litter-inhabiting ants along an elevation gradient in the Wayanad region of the Western Ghats is studied by Sabu et al. [14]. Bharti and Sharma [15] observed the diversity and abundance of ants along an elevational gradient in Jammu Kashmir, the Himalayas. Bubesh Guptha et al. [16] examined the preliminary observation on butterflies of Seshachalam biosphere reserve, Eastern Ghats, Andhra Pradesh, India. Diversity indices of tropical cockroach were reported by Bonsals [17] and Padmanaban [18]. Distribution, diversity, and population dynamics of chosen insects in Courtallam tropical evergreen forest were extensively studied by Edwin [19]. Anu et al. [20] studied litter arthropod diversity in an evergreen forest in the Wayanad region of Western Ghats, India. Biodiversity of spiders in Western Ghats of Tamil Nadu was also reported [21, 22]. Isabel [23] observed biodiversity of litter arthropod communities in Alagar Hills Reserve forest in Eastern Ghats, India. In India there is no proper information available on the identification, the diversity, and the role of millipedes in forest ecosystems. Hence, the broad objective of the present study is to identify and calculate the diversity indices of millipedes species along an elevation gradient within the Alagar Hills Reserve forest of Eastern Ghats, India.

2. Materials and Methods

2.1. Study Area. Alagar Hills, a biosphere evergreen reserve in Tamil Nadu, is located in Eastern Ghats ($10°0'$–$10°30'$ N and $75°55'$–$78°20'$ E), 20 km northeast of Madurai city, India. The vegetation is two layered, in which the ground vegetation is very poor. The canopy is open dry deciduous or evergreen vegetation. Five locations were selected at different altitudes at Alagar hills (250, 350, 450, 550, and 650 m), which were visited every month from July 2011 to May 2012. The lowest site was at 250 m elevation because that is where millipedes seemed to first occur, while 650 m is the highest elevation of Alagar hills reserve forest. Observation of millipedes was made through 6 quadrats (1 m × 1 m) in each study site and each sampling date, and the mean number of milli-pedes/quadrat was calculated. Millipedes were collected from the study area by hand picking, and species were identified by using various field guides and available literatures.

2.2. Data Analysis. Standard methods were used to calculate the richness and evenness of millipede species at different altitudes. The diversity indices were calculated using the Ludwig and Reyonlds software package [24]. Two indices are needed to compute Hill's diversity numbers: (a) Simpson's index and (b) Shannon's index. These were calculated by using the following formulae:

Simpson's index (λ):

$$\lambda = \sum_{i=1}^{s} \frac{n_i (n_i - 1)}{n (n - 1)}, \quad (1)$$

Shannon's index (H'):

$$H' = \sum_{i=1}^{s} \left[\left(\frac{n_i}{n} \right) \ln \left(\frac{n_i}{n} \right) \right]. \quad (2)$$

3. Results

The following five species of millipedes were identified in the study area.

(1) *Harphaphe haydeniana* (Wood, 1984) belongs to order Polydesmida and family Xystodesmidae. *H. haydeniana* is a yellow spotted millipede and reaches a length of 4-5 cm, width of 0.1 to 0.3 cm, and weight of 0.8 to 1.5 g. The body is black and is distinctively marked along the sides with patches of a yellowish colour. It consists of approximately 15–20 body seg-ments, bearing a total of 30 (male) or 31 (female) pairs of legs. It lives for 2-3 years. (Figure 1).

(2) *Arthrosphaera magna* (Attems, 1936) belongs to order Sphaerotheriida and family Sphaerotheriidae. Adults have exactly 13 segments (including collum and anal

FIGURE 1: Harpaphe haydeniana.

segments), and the juveniles are of dark olive green colour. The head of the adult is yellow brown or olive brown or olive green; the second segment is dark brown with a black band bordered with yellow colour, forming a narrow stripe. The average weight, length, and width of an adult millipede range from 4.5 to 12.5 g, from 3.5 to 6.5 cm, and from 1.5 to 2.5 cm, respectively, (Figure 2).

(3) *Xenobolus carnifex* (Fabricius, 1775) belongs to order Spirobolida and family Pachybolidae. The body seg-ments are black with a broad median band of dark pink colour. Adults have exactly 50 segments and attain an average length of 6 cm and width of 0.5 cm. The seventh segment in the male is without legs. The sexes are easily distinguishable from the fifth stadium based on their length and width. Adult males have an average body width of 0.8 cm and weight of 7.8 g, whereas the adult female has an average midsegment width of 0.9 cm. and weight of 8.5 g (Figure 3).

(4) *Aulacobolus newtoni* (Silvestri, 1916) belongs to order Spirobolida and family Pachybolidae. Both sexes exhibit dark greenish yellow colouration, and the maximum number of body segments is 50. Adult males have a width of 0.8 cm and an average weight of 7.8 g, whereas the adult female has an average weight of 8.5 g and midsegment width of 0.9 cm. The body length of both male and female ranges from 8.0 to 9.5 cm (Figure 4).

(5) *Spinotarsus colosseus* (Attems, 1928) belongs to order Spirostreptida and family Odontopygidae. Both males and females of *S. colosseus* are of darkblack colour. The length, width, and weight of the millipede range from 6.5–15.4 cm, 0.3–1.0 cm, and 5.5–15.7 g, respectively. The number of body segments in adult is 55–58. (Figure 5).

The maximum number of millipedes ($721/m^2$) was observed at 450 m and the minimum ($251/m^2$) was observed at 650 m elevation (Figure 6). The density of millipedes decreased in the order of 350 m ($393/m^2$), 550 m ($387 m^2$), and 250 m ($251/m^2$) elevations. Richness indices (R_1: 0.607; R_2: 0.186) and Simpson's index (0.207) were found to be lower in the midelevation (450 m) than in the other elevations (Table 1).

TABLE 1: Diversity of five different species of millipedes distributed at five elevations of Alagar hills reserve forest.

Indices		Elevations				
		250 m	350 m	450 m	550 m	650 m
Richness	R_1	0.720	0.669	0.607	0.671	0.723
	R_2	0.311	0.252	0.186	0.254	0.315
Diversity	Simpson's index	0.222	0.215	0.207	0.213	0.223
	Shannon's index	1.539	1.562	1.589	1.561	1.489
Evenness	E	0.955	0.970	0.981	0.969	0.924

FIGURE 2: Arthrosphaera magna.

FIGURE 4: Aulacobolus newtoni.

FIGURE 3: Xenobolus carnifex.

FIGURE 5: Spinotarsus colosseus.

Remarkably, a very high evenness (0.981) was noticed in 450 m elevation. With reference to Shannon's index, it was found to decrease from 1.589, 1.562, 1.561, 1.539, and 1.489 at 450, 350, 550, 250 and 650 m elevations, respectively.

4. Discussion

The present study, an inventory of millipedes, is the first of its kind in Alagar hills of Eastern Ghats. Also, this is the first report of the elevational distribution of millipedes in the Eastern Ghats of Alagar hills. Two diversity indices were calculated. Shannon's index (H') is sensitive to changes in the abundance of rare species in a community, whereas Simpson's index (λ) is sensitive to changes of the most abundant species in a community. These two diversity indices showed much difference in millipede distribution and diversity. Shannon index assumes that individuals are randomly sampled from an indefinitely large population [25]. This index also assumes

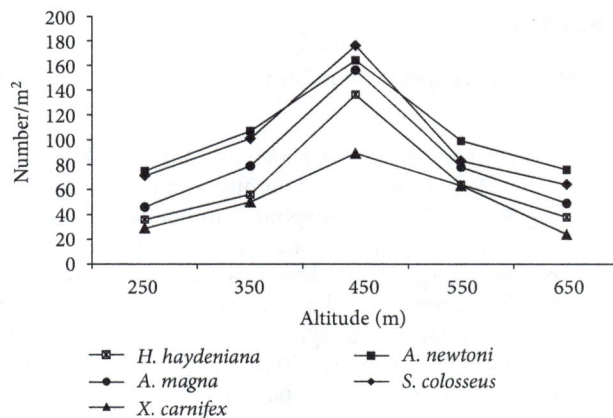

FIGURE 6: Density of different millipede species in the study sites of Alagar Hills.

that all millipedes are represented in the sample. Simpson's index, which gives the probability of two individuals drawn at random from a population belonging to the same species, increases with the decrease in diversity. Hence, it is understood that midelevation has the highest diversity, and low and highest elevations have the lowest diversity pattern.

Brown [26] reported that changes in life on earth are due to abundance and diversity of organisms along the earth major environmental gradients, including those of elevation.

The significant change of abiotic and biotic environmental variables along elevational gradients strongly affects patterns of abundance, distribution, and diversity of most organisms. As emphasized by Brown and Lomolino [27], lower elevational zones usually differ from higher altitude by the following: (1) a greater total amount of resources and population number; (2) more refugia and space for species with large home ranges; (3) greater habitat diversity; and (4) a greater potential for serving as a target for potential immigrants. The abundance of feeding guilds can strongly depend on habitat structure [28] which changes with increasing altitude, thereby differentially altering the available amount of food resources for feeding guilds and the vegetation structures required for foraging by different functional groups along the elevation gradient [29].

The present study reveals the fact that maximum millipede species diversity in midelevation, which confirms to the midelevational richness in species abundance is recorded from Philippines by Samson et al. [30] and from Madagascar by Fisher [31]. The higher Shannon's diversity index shown by middle elevation indicates that it provides more opportunities for survival in the form of ecological niches than those lower and higher elevations. The lower level of elevation is highly disturbed due to human activities and cattle grazing. The canopy cover is open, which results in nearly dry and barren land rendering fewer potential niches. The boulders, which are very common, have been used only as a shelter from extremes of temperature and humidity during summer. In high elevation, the closed canopy reduces the sunlight and the under storey vegetation which again results in a reduced number of potential niches. Besides, the high rate of transpiration by plants and nearby water streams makes the atmosphere very humid and cool. This could be the reason for moderate evenness and diversity indices shown by these elevations. But, the middle elevation with moderate canopy and litters supports understorey vegetation like herbs and shrubs and provides more potential niches and protection during periodic flooding. This elevation is occupied by all species of millipedes in fairly equal proportion, and this explains the high evenness and diversity indices observed in midelevation.

Cardelús et al. [32] emphasized that many species from lower and higher elevations overlap at midelevation, which generally provides the most suitable environment for arthropods. Sabu et al. [14] commented that peak litter ant abundance recorded at midelevations in the Wayanad region suggests that these are centres of the richest diversity and abundance. Rahbek [7] demonstrated that hump-shaped relationship with maximum species numbers at midelevations. Different elevational richness pattern may be due to (a)

an overall decline of species richness with increasing altitude [33], (b) a plateau of species being the richest at lower altitude then declining towards the highest elevations [34], and (c) a midelevation peak of species richness [35, 36]. Padmanaban [18] reported that cockroach diversity and abundance were the maximum in the middle altitudes in comparison with those in the higher and lower altitudes of Alagar hills reserve forest., Isabel [23] also emphasized that arthropod species diversity was higher in midelevations compared to that in lower and higher elevations of Alagar hills.

Krebs [37] consolidated six reasons out of the hypotheses proposed for variation in species diversity, namely, time available for speciation and dispersal, spatial heterogeneity, vegetation structure, competition, environmental stability, and productivity. These reasons can be attributed to the variation in diversity indices between elevations observed in the present study. From the present study, it is concluded that distinct variations in the diversity of millipedes at five different altitudes and peak diversity of millipedes recorded at midlevel elevations in the Alagar hills suggest that these are centres of the richest diversity and abundance that should be prioritised as areas for further intense conservation.

Acknowledgments

The authors are thankful to the University Grants Commission, New Delhi, India, for providing grants in the form of Major Research Project for this research. Editorial improvements by Cutis C. Daehler are gratefully acknowledged.

References

[1] P. R. Ehrlich and A. H. Ehrlich, "The value of biodiversity," *Ambio*, vol. 21, no. 3, pp. 219–226, 1992.

[2] D. Tilman, "The ecological consequences of changes in biodiversity: a search for general principles," *Ecology*, vol. 80, no. 5, pp. 1455–1474, 1999.

[3] A. Purvis and A. Hector, "Getting the measure of biodiversity," *Nature*, vol. 405, no. 6783, pp. 212–219, 2000.

[4] K. S.. McCann, "The diversity-stability debate," *Nature*, vol. 405, pp. 228–233, 2000.

[5] P. Jaen-Francois, "Interaction between soilfauna and their environment," *Soil Fauna Environment*, pp. 45–76, 1999.

[6] U. Irmler, "Changes in the fauna and its contribution to mass loss and N release during leaf litter decomposition in two deciduous forests," *Pedobiologia*, vol. 44, no. 2, pp. 105–118, 2000.

[7] C. Rahbek, "The elevational gradient of species richness: a uniform pattern?" *Ecography*, vol. 18, no. 2, pp. 200–205, 1995.

[8] R. Naniwadekar and K. Vasudevan, "Patterns in diversity of anurans along an elevational gradient in the Western Ghats, South India," *Journal of Biogeography*, vol. 34, no. 5, pp. 842–853, 2007.

[9] R. M. Shelley, "Taxonomy of extant Diplopoda (Millipeds) in the modern era: perspectives for future advancements and observations on the global diplopod community (Arthropoda: Diplopoda)," *Zootaxa*, no. 1668, pp. 343–362, 2007.

[10] P. Sierwald and J. E. Bond, "Current status of the myriapod class diplopoda (millipedes): taxonomic diversity and phylogeny," *Annual Review of Entomology*, vol. 52, pp. 401–420, 2007.

[11] G. Loranger-Merciris, D. Imbert, F. Bernhard-Reversat, J. F. Ponge, and P. Lavelle, "Soil fauna abundance and diversity in a secondary semi-evergreen forest in Guadeloupe (Lesser Antilles): influence of soil type and dominant tree species," *Biology and Fertility of Soils*, vol. 44, no. 2, pp. 269–276, 2007.

[12] J. Seeber, G. U. H. Seeber, R. Langel, S. Scheu, and E. Meyer, "The effect of macro-invertebrates and plant litter of different quality on the release of N from litter to plant on alpine pastureland," *Biology and Fertility of Soils*, vol. 44, no. 5, pp. 783–790, 2008.

[13] R. Gadagkar, K. Chandrasekaran, and D. M. Bhat, "Ant species richness and diversity in some selected localities in Western Ghats, India," *Hexpoda*, vol. 5, no. 2, pp. 79–94, 1993.

[14] T. K. Sabu, P. J. Vineesh, and K. V. Vinod, "Diversity of forest litter-inhabiting ants along elevations in the Wayanad region of the Western Ghats," *Journal of Insect Science*, vol. 8, article 69, 2008.

[15] H. Bharti and Y. P. Sharma, "Diversity and abundance of ants along an elevational gradient in Jammu—Kashmir Himalaya," *Halteres*, vol. 1, pp. 10–24, 2009.

[16] M. Bubesh Guptha, D. Chalapathi Rao, and D. Srinivas Reddy, "A preliminary observation on butterflies of Seshachalam Biosphere Reserve, Eastern Ghats, Andrapradesh, India," *World Journal of Zoology*, vol. 7, pp. 83–89, 2012.

[17] M. B. . Bonsals, "Domiciciary cockroach diversity in Ecuador," *Endomologist*, pp. 14431–14439, 1995.

[18] A. . Padmanaban, *Distribution, population dynamics and biodiversity of litter dwelling feral cockroaches in Alagar hill reserve forest of Eastern Ghats [Ph.D thesis]*, Madurai Kamaraj University, Madurai, India, 2002.

[19] J. Edwin, *Distribution, diversity and population dynamics of chosen insects in the Courtallam tropical evergreen forest [Ph.D. dissertation]*, Madurai Kamaraj University, Madurai, India, 1997.

[20] A. Anu, T. K. Sabu, and P. J. Vineesh, "Seasonality of litter insects and relationship with rainfall in a wet evergreen forest in south western Ghats," *Journal of Insect Science*, vol. 9, article 46, 2009.

[21] M. . Sugumaran, M. Ganesh Kumar, and K. Ramasamy, "Biodiversity of spiders in Western Ghats of Tamil Nadu," *Entomon*, vol. 30, pp. 157–163, 2005.

[22] H. Upamanyu and V. P. Uniyal, "Diversity and composition of spider assemblages in five vegetation types of the Terai Conservation Area, India," *Journal of Arachnology*, vol. 36, no. 2, pp. 251–258, 2008.

[23] W. Isabel, *Litter dynamics and biodiversity of litter Arthropod Communities [Ph.D thesis]*, Madurai Kamaraj University, Madurai, India, 2002.

[24] J. A. Ludwig and J. F. Reyonlds, *Statistical ecology. A primer on methods and computing richness evenness and species diversity*, John Wiley and Sons, New York, NY, USA, 1988.

[25] C. PeilouE, *Ecological Diversity*, Wiley Eastern, New York, NY, USA, 1975.

[26] J. H. Brown, "Mammals on mountainsides: elevational patterns of diversity," *Global Ecology and Biogeography*, vol. 10, no. 1, pp. 101–109, 2001.

[27] J. Brown and M. V. Lomolino, *Biogeography*, Sinauer, Sunderland, Mass, USA, 2nd edition, 1998.

[28] H. Brunner, "Vogelgemeinschaften an der oberen Waldgrenze unter dem Einfluss traditioneller und moderner Landnutzung im Nockgebiet (Karnten, steiermark), Carinthia, Jahrgang," pp. 533–544, 2001.

[29] J. G. Blake and B. A. Loiselle, "Diversity of birds along an elevational gradient in the Cordillera Central, Costa Rica," *The Auk*, vol. 117, no. 3, pp. 663–686, 2000.

[30] D. A. Samson, E. A. Rickart, and P. C. Gonzales, "Ant diversity and abundance along an elevational gradient in the Philippines," *Biotropica*, vol. 29, no. 3, pp. 349–363, 1997.

[31] B. L. Fisher, "Ant diversity pattern along an elevational gradient in the Reserve Speciale de Manongarivo, Madagascar," *Boissiera*, vol. 59, pp. 311–328, 2002.

[32] C. L. Cardelús, R. K. Colwell, and J. E. Watkins, "Vascular epiphyte distribution patterns: explaining the mid-elevation richness peak," *Journal of Ecology*, vol. 94, no. 1, pp. 144–156, 2006.

[33] F. Sergio and P. Pedrini, "Biodiversity gradients in the Alps: the overriding importance of elevation," *Biodiversity and Conservation*, vol. 16, no. 12, pp. 3243–3254, 2007.

[34] S. K. Herzog, M. Kessler, and K. Bach, "The elevational gradient in Andean bird species richness at the local scale: a foothill peak and a high-elevation plateau," *Ecography*, vol. 28, no. 2, pp. 209–222, 2005.

[35] M. Kessler, S. K. Herzog, J. Fjeldså, and K. Bach, "Species richness and endemism of plant and bird communities along two gradients of elevation, humidity and land use in the Bolivian Andes," *Diversity and Distributions*, vol. 7, no. 1-2, pp. 61–77, 2001.

[36] P. F. Lee, M. Kessler, and R. R. Dunn, "What drives elevational patterns of diversity? A test of geometric constraints, climate and species pool effects for pteridophytes on an elevational gradient in Costa Rica," *Global Ecology and Biogeography*, vol. 15, no. 4, pp. 358–371, 2006.

[37] C. J. Krebs, *The Experimental Analysis of Distribution and Abundance*, Harper and Row, New York, NY, USA, 1972.

13

Quantifying Effects of Spatial Heterogeneity of Farmlands on Bird Species Richness by Means of Similarity Index Pairwise

Federico Morelli

DiSTeVA, University of Urbino, Scientific Campus, 61029 Urbino, Italy

Correspondence should be addressed to Federico Morelli; morellius@libero.it

Academic Editor: Antonio Terlizzi

Many studies have shown how intensification of farming is the main cause of loss biodiversity in these environments. During the last decades, agroecosystems in Europe have changed drastically, mainly due to mechanization of agriculture. In this work, species richness in bird communities was examined on a gradient of spatial heterogeneity of farmlands, in order to quantify its effects. Four categories of farmland spatial heterogeneity were defined, based on landscape and landuse parameters. The impact of features increasing the spatial heterogeneity was quantified comparing the similarity indexes between bird communities in several farmlands of Central Italy. The effects of environmental variables on bird richness were analyzed using GLM. The results highlighted that landscape features surrogates of high nature values (HNVs) of farmlands can increase more than 50% the bird species richness. The features more related to bird richness were hedgerows, scattered shrubs, uncultivated patches, and powerlines. The results confirm that the approach based on HNV for evaluating the farmlands is also suitable in order to study birds' diversity. However, some species are more sensitive to heterogeneity, while other species occupy mainly homogeneous farmlands. As a consequence, different conservation methods must be considered for each farmland bird species.

1. Introduction

Agricultural intensification is one of the main drivers of biodiversity decline. During the last twenty-five years a rapid and large scale change of the agricultural landscape occurred in Europe, caused by the intensification and mechanization of agricultural activities [1, 2], and for this reason understanding the relations between biodiversity and land-use intensity is quite important to developed effective plans for habitat conservation [3]. Many studies about biodiversity in agroecosystems usually faced the problem that both management type and landscapes features can affect it [4–6], but a quantification of the real impact of each of these factors on the composition of animal communities is complicated. Biodiversity in farmlands is affected by land-use management at a small spatial scale (grazing intensity or crop rotation) and also at a large spatial scale [3]. All of this is related to the presence and/or distribution of landscape features that may reflect a low fragmentation of the habitat, which is often called "functional heterogeneity," and is very important in supporting biodiversity [7–11].

Many studies have shown that organic systems may enhance bird species diversity over nonorganic counterparts as a result of increased complexity in landscape features [12, 13], because in ecology the habitat diversity is associated with an increase of niche availability for the species [14]. Mainly for this reason farmers are increasingly being encouraged to conserve biodiversity through the maintenance of extensive farming systems, the preservation of seminatural landscape features, or the extension of intensive farming systems [15, 16]. However, at the moment, most of our knowledge of agricultural practices and farmland bird ecology comes from a few countries only, with 76% of papers in the Web of Science outsourcing from the UK, France, the Netherlands, Spain, Germany, and Sweden, and less than 10% from Central and Eastern European countries [17]. The results of many of these works show that the correlation between biodiversity and agricultural heterogeneity varies from positive to negative in different systems in Europe [9, 18, 19].

The spatial heterogeneity in an agroecosystems is measurable through landscape features that can be used also to assess bird species richness, as a surrogate of biodiversity. [20].

The first aim of the present study is examining the relationships between bird species richness and spatial heterogeneity of farmlands in Central Italy and quantifying the impact of spatial heterogeneity on biodiversity, through the use of an index easy to apply and useful for conservation strategies.

The second aim is identifying the main environmental characteristics associated with the heterogeneity of farmlands and investigating whether the species living in heterogeneous agricultural areas are also present also in more homogeneous farmlands, to help in this way to identify the bird species most related to landscape heterogeneity as possible indicators of HNV farmlands.

2. Material and Methods

2.1. Study Area. The study area was an agricultural area of the North Eastern Marches region (Central Italy, 43°45′47.30″N; 12°45′5.20″E) that was 2,500 ha wide, ranged from 0 to 800 m/a.s.l. This area was selected because it includes farmlands representative of the different farming practices in the region. Within the area 80 sites were selected randomly and surveyed. There sites were located uniformly and each far at least 500 m from one other.

2.2. Species and Environmental Data. To survey the bird community composition during the breeding period of the year 2010, the sites were visited in the morning within 06:00 AM and 10:00 AM, with sunny weather conditions, between mid-April and the end of July. Each visit lasted 10 min, and all birds detected visually and acoustically within a radius of 100 m from the observer were recorded [21] using single point counts. Species not breeding and diurnal or nocturnal raptors were excluded from subsequent analyses, because sometimes they require different survey methods. Also occasional or not breeding species were excluded from the statistical analysis, because their presence was very rare and therefore did not affect the total richness. The only exception was Circus pygargus that was considered strongly related to the studied habitats [22, 23] and then counted into bird richness but excluded from the further environmental analysis because, if compared with passeriform birds, the species could be considered a wide-ranged species.

Description *in situ* of the 3 ha (100 m radius area) around the point-count was made in order to quantify the land-use composition and the structural characteristics in the sampling sites and to obtain an updated information concerning crop types. The percentages of land-use composition estimated *in situ* were adjusted by checking aerial photographs and using a 100 m radius buffer overlapped with ArcGIS 10, in order to avoid the difficulties determined by the exclusive use of digital maps to study limited environments [24]. Also altitude of the point count was recorded, using a GPS.

2.3. Spatial Heterogeneity of Farmlands. In order to classify the farmland heterogeneity, an approach in line with that recently used for evaluating high nature value (HNV) farmland in Europe [25–28] was used. The presence and size

TABLE 1: Environmental parameters used to describe the farmlands in Central Italy. The spatial scale of measurement was 100 m radius around a single point count.

Parameters	Abbreviation	Level	Details
Altitude	alt	Landscape	Altitude of the point count (m/a.s.l.)
Terrain slope	slo	Landscape	No slope (less than 3 degrees): 0, slight slope (between 3 and 8 degrees): 1, mean slope (greater than 8 degrees): 2
Roads	roa	Landscape	Presence and type of roads (paved and unpaved)
Power lines	pow	Landscape	Number of electricity wires
Urban	urb	Land-use	%
Forest	for	Land-use	%
Shrubs	shr	Land-use	%
Uncultivated	unc	Land-use	%
Badland	bad	Land-use	%
Grassland	gra	Land-use	%
Hedgerows	hed	Land-use	%
Isolated trees	tre	Land-use	%
Vineyards	vin	Land-use	%
Olive	oli	Land-use	%
Cultivated total	cul	Land-use	Sum of all crop types, %
Forage	ofo	Crop type	%
Wheat	whe	Crop type	%
Alfalfa	alf	Crop type	%
Hay	hay	Crop type	%
Sunflower	sun	Crop type	%
Coriander	cri	Crop type	%
Sugarbeet	sug	Crop type	%
Corn	cor	Crop type	%
Garlic	gar	Crop type	%
Oats	oat	Crop type	%
Brassica sp	bra	Crop type	%
Lettuce	let	Crop type	%

of natural and seminatural farmland features (i.e., shrubs, uncultivated patches) or linear features (i.e., field margins, hedgerows) were analyzed, since they could increase the number of ecological niches in which wildlife can coexist with farming activities and create a heterogeneity gradient. The spatial features were analyzed at three different scale level: landscape diversity, land-use diversity, and crop type (Table 1).

2.3.1. Landscape Diversity. To define the landscape diversity the following parameters were used:

(1) road presence and type: paved, unpaved, or mixed;

(2) power lines: presence and number;

TABLE 2: Ranking of farmland spatial heterogeneity according to landscape and land-uses characteristics, where FSH is farmland spatial heterogeneity.

FSH description	FSH rank	Terrain slope	Marginal vegetation	Land-use diversity
Very heterogeneous	4	None, slight, mean	Linear + point	>1.0
Heterogeneous	3	None, slight, mean	Linear	0.8 < 1.0
Homogeneous	2	None, slight, mean	Point	0.6 < 0.8
Very homogeneous	1	None, slight	None	<0.6

(3) terrain slope, classified as no slope (less than 3 degrees): 0, slight slope (between 3 and 8 degrees): 1, and mean slope (greater than 8 degrees): 2;

(4) marginal vegetation level on cultivated, taking into account the presence (isolated or simultaneous) of natural and semi-natural features: max level (linear features (hedgerows) + point features (scattered shrubs)), mid-level (linear features (hedgerows)), low level (point features (scattered shrubs)), and min level (without hedgerows and scattered shrubs) (see Figure 1). The hedgerows were considered as a factor of greater level than scattered shrubs because in the study area they always resulted as larger features in relative terms, thus offering a greater edge effects on habitat fragmentation. The minimum radius size of the scattered shrubs considered in this study was 0.5 meter. The marginal vegetation is also used to define the HNV of farmlands [27, 28].

2.3.2. Land-Use Diversity.
The land-use diversity in each site was calculated using Shannon diversity index on land-uses, $H' = -\sum p_i \times \log p_i$, where p_i is the relative proportion of each land-use type i [30]. This index can express the fragmentation of land-uses as more fragmented land-use types have larger number of patches, each of a smaller average size within the sampled area.

2.3.3. Crop Type.
Crop diversity was estimated for each site, applying the Shannon diversity index on crop types into the cultivated category (Table 1). However, this feature was also analyzed as another level of farmland heterogeneity, because it appears in contrast compared to the land-use (the crop diversity was greater when the land-use diversity was lower, because of being dominated by cultivated typology).

Four categories of spatial heterogeneity of farmland were defined considering both landscape and land-use data collected *in situ*. The sampled sites were ranked in relation to a decreasing spatial heterogeneity value as follows: very heterogeneous mosaic: 4, heterogeneous mosaic: 3, homogeneous mosaic: 2, and very homogeneous mosaic: 1 (see details in Table 2). When a site could not be clearly categorized (considering all parameters in Table 2), it was classified in the closest category.

2.4. Data Analysis.
Biodiversity was estimated as bird species richness at each sampled site and was estimated for each farmland heterogeneity level too. Total biodiversity was the sum of all bird species recorded in the entire study area. To

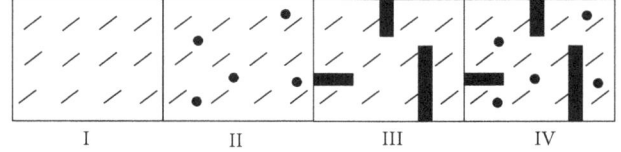

FIGURE 1: Scheme of the criteria used to classify the level of marginal vegetation of farmlands in Central Italy. The point features represent scattered shrubs while linear features represent hedgerows.

test the effects of environmental parameters (associated with farmland spatial heterogeneity) on biodiversity, bird richness in different spatial heterogeneity categories was compared using ANOVA (after doing the normality and homoscedasticity tests). Similarity between bird communities in different heterogeneity categories was explored using the Sorensen Similarity Index. The Sorensen similarity index (SI) [31] is a β diversity index that vary from 0 where the assemblages differ totally to 1 where they are identical (SI = $2c/(a + b)$), where c is the number of species shared by the two sites, and a and b are the total number of species at each site). The Sorensen similarity index was calculated for pairwise comparisons between different heterogeneity categories and also between each of them and the total biodiversity. This method was used to quantify the differences on bird communities between spatial heterogeneity gradient.

The nature and strength of the relationship between bird richness and environmental parameters on farmlands were examined using GLM [32], with the dependent variable (bird richness) modeled specifying Poisson errors. Independent predictive variables were expressed as arcsin square root in the case of proportions. In order to avoid multicolinearity of land-use and crop types, the parameters (regressors) showing the strongest correlation (>0.8) were manually eliminated. A stepwise backward procedure was carried out using all data pooled in order to select the best predictors using AIC criterion to select the lowest AIC values [33, 34]. In models, sites were treated as independent units, because the values of spatial autocorrelation between geographic distance and dissimilarity matrix of sites (obtained from landscapes variables) were low (Mantel test r = 0.13, n = 80 sites, not significant) [35]. Dissimilarity indices among sites were calculated by means of "vegdist" function of the vegan package [36], using five different landscape variables (altitude, urban, roads, hedgerows, and cultivated total). Internal validation of models was performed using a bootstrap resampling procedure (n bootstrap = 999). All tests were performed with R program (R Development Core Team 2011).

3. Results

Sixty-six bird species were recorded during the breeding period. For the analysis, a total of 1133 records were collected. The maximum potential bird richness for the studied farmland was fifty-five species (Tables 3 and 5) considering that were excluded not-breeding species, diurnal and nocturnal raptors.

The most common farmland types had high or medium spatial heterogeneity (over 73%), whereas farmlands with lower heterogeneity were only 27% (Table 3), mainly distributed near the coast, in lowland zones. Roads were present in all farmlands. The most common road type was paved (64%), while unpaved roads were 20%, and mixed (paved and unpaved) roads represented 16% of cases. Power lines were widespread in farmlands (90%), and no differences in this element were found between different farmland heterogeneity levels. The differences in crop diversity across the gradient of spatial heterogeneity were not significant (Table 4).

3.1. Effects of Spatial Heterogeneity on Bird Richness. The bird species richness on different farmland spatial heterogeneity categories was different, being higher in farmlands with greater heterogeneity (F: 37.8, df: 3, P value: 0.000, Figure 2). The most heterogeneous farmlands had a maximum richness value of fifty-one species, whereas in the most homogeneous farmlands the maximum richness value was twenty-five species.

The impact of FSH on biodiversity was quantified by means of similarity index pairwise between the FSH categories. The homogenization of farmlands was able to reduce the similarity index on bird richness up to a half compared to very heterogeneous farmlands. The total effect was quantified as a fall of circa 53% of diversity in comparison with the total amount of species and 40% between very heterogeneous and very homogeneous farmlands (Table 5). On the other hand, the difference between bird species in very heterogeneous areas and potential bird richness of farmlands was small, only 4%.

3.2. Parameters of Farmlands Useful to Explain Bird Distribution. Following the results of this work, the linear features as hedgerows were more important than point features as shrubs for the total bird richness (more dissimilarities from index were found between 1 and 3 than between 1 and 2 FSH categories) (Table 5).

The model that best described the relationships between bird species richness and environmental parameters in farmlands included three land-use variables (shrubs, uncultivated, and hedges) and one landscape variable (presence of power-lines). The ratio between explained and total deviance for the best model was 0.83. These four parameters were statistically significant for all the farmland typologies, and if compared to the initial model including all the variables, the model with only these four variables appeared to be the most suitable. Altitude had not effect on bird diversity in the study area. The results of backwise procedure have not selected altitude as variable related to the bird species richness (Tables 6 and 7).

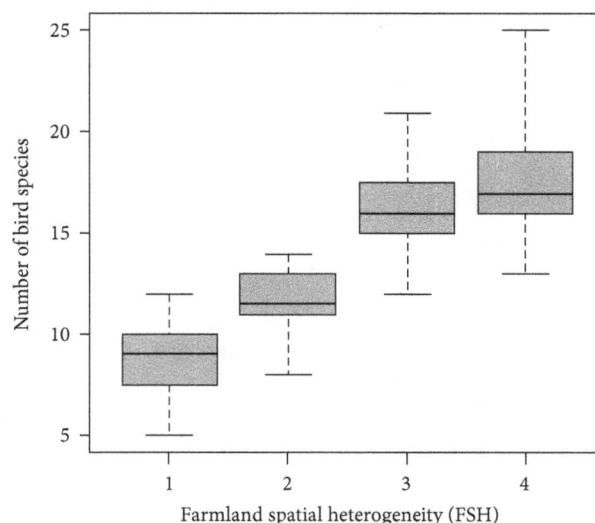

FIGURE 2: Relationship between bird species richness and gradient of spatial heterogeneity of farmlands in Central Italy.

3.3. Bird Species Frequencies on Farmland Type. The bird species most frequently observed in heterogeneous farmlands were *Sylvia atricapilla*, *Turdus merula*, and *Emberiza cirlus*. However, the species present only in this category were *Emberiza cirlus* and *Emberiza calandra*, followed by the forest birds *Oriolus oriolus*, *Troglodytes troglodytes*, and the raptor *Circus pygargus*, all considered vulnerable species in the Italian Red List. The only one endangered species recorded was *Jynx torquilla*.

The bird species most frequently observed in homogeneous farmlands were *Passer domesticus italiae*, *Alauda arvensis*, and *Motacilla flava*, all considered vulnerable species in the Italian Red List (Table 3).

Some species, such as *Serinus serinus*, *Carduelis chloris*, and *Emberiza hortulana* were indistinctly present in farmlands with different level of spatial heterogeneity.

4. Discussion

The use of environmental variables collected *in situ* and adjusted by interpretation of aerial photographs could be more accurate than only GIS analysis. When computing several indexes of fragmentation and spatial heterogeneity on landscapes (e.g., Fragstat and V-Late) the GIS tools working over land cover maps are limited by the spatial resolution of their maps. For this reason they may not be accurate enough in order to quantify small objects as shrubs and short hedges [11, 24, 37]. Although the link between biodiversity and farming practices or landscape structural characteristics has been underlined since the early nineties by Bennett et al. [38] and Beaufoy et al. [39], efforts in characterizing the high natural value of farmland have been carried out more recently [28, 40–42]. The HNV farmland concept has also been embedded in the Common Agricultural Policy (CAP) "to protect and enhance the EU's natural resources and landscapes in rural areas, the resources devoted to axis 2

TABLE 3: Total frequency (%), relative frequency, and concern status of bird species [29] in farmlands of Central Italy, classified into different spatial heterogeneity categories.

Species	FSH category				Total	Red List Italy
	4 ($n = 30$)	3 ($n = 28$)	2 ($n = 14$)	1 ($n = 8$)		
Sylvia atricapilla	90	96.3	84.6	62.5	86.3	LC
Serinus serinus	66.7	85.2	92.3	75	76.3	LC
Passer domesticus italiae	63.3	74.1	92.3	100	73.8	VU
Turdus merula	86.7	88.9	46.2	25	72.5	LC
Apus apus	70	77.8	84.6	50	71.3	LC
Luscinia megarhynchos	76.7	66.7	53.8	12.5	61.3	LC
Corvus cornix	76.7	77.8	23.1	25	61.3	LC
Hirundo rustica	56.7	63	53.8	100	61.3	NT
Emberiza cirlus	80	66.7	15.4	0	55	LC
Fringilla coelebs	76.7	55.6	30.8	12.5	53.8	LC
Sturnus vulgaris	53.3	51.9	76.9	25	52.5	LC
Streptopelia turtur	66.7	55.6	38.5	12.5	51.3	LC
Carduelis carduelis	46.7	51.9	61.5	12.5	46.3	NT
Streptopelia decaocto	40	51.9	46.2	25	42.5	LC
Carduelis chloris	33.3	40.7	38.5	50	37.5	NT
Oriolus oriolus	63.3	37	0	0	36.3	LC
Alauda arvensis	30	25.9	46.2	62.5	33.8	VU
Emberiza hortulana	36.7	29.6	53.8	12.5	33.8	DD
Emberiza calandra	50	29.6	23.1	0	32.5	LC
Cyanistes caeruleus	40	25.9	30.8	0	28.8	LC
Erithacus rubecula	30	37	7.7	0	25	LC
Parus major	26.7	33.3	15.4	12.5	25	LC
Garrulus glandarius	43.3	22.2	0	0	23.8	LC
Troglodytes troglodytes	30	33.3	0	0	22.5	LC
Sylvia communis	33.3	18.5	15.4	0	21.3	LC
Phoenicurus phoenicurus	20	22.2	7.7	12.5	17.5	LC
Pica pica	20	22.2	7.7	12.5	17.5	LC
Lanius collurio	30	11.1	0	12.5	16.3	VU
Phasianus colchicus	6.7	14.8	30.8	25	15	NA
Saxicola torquatus	23.3	11.1	7.7	0	13.8	VU
Coturnix coturnix	13.3	14.8	7.7	12.5	12.5	DD
Sylvia cantillans	23.3	7.4	0	0	11.3	LC
Cisticola juncidis	10	3.7	15.4	25	10	LC
Lullula arborea	10	7.4	15.4	0	8.8	LC
Passer montanus	0	11.1	15.4	12.5	7.5	VU
Corvus monedula	10	3.7	7.7	0	6.3	LC
Dendrocopos minor	13.3	0	7.7	0	6.3	LC
Picus viridis	10	7.4	0	0	6.3	LC
Columba palumbus	6.7	7.4	0	0	5	LC
Dendrocopos major	10	3.7	0	0	5	LC
Merops apiaster	6.7	3.7	7.7	0	5	LC
Regulus ignicapillus	6.7	3.7	0	0	3.8	LC
Aegithalos caudatus	6.7	0	0	0	2.5	LC
Alectoris rufa	6.7	0	0	0	2.5	DD
Circus pygargus	3.3	3.7	0	0	2.5	VU
Columba livia	3.3	3.7	0	0	2.5	DD
Motacilla alba	3.3	3.7	0	0	2.5	LC

TABLE 3: Continued.

Species	4 (n = 30)	3 (n = 28)	2 (n = 14)	1 (n = 8)	Total	Red List Italy
	FSH category					
Motacilla flava	0	0	0	25	2.5	VU
Periparus ater	6.7	0	0	0	2.5	LC
Phoenicurus ochruros	3.3	3.7	0	0	2.5	LC
Carduelis cannabina	0	3.7	0	0	1.3	NT
Emberiza citrinella	3.3	0	0	0	1.3	LC
Jynx torquilla	3.3	0	0	0	1.3	EN
Muscicapa striata	0	3.7	0	0	1.3	LC
Upupa epops	3.3	0	0	0	1.3	LC

TABLE 4: Descriptive elements of agricultural mosaic along the gradient of spatial heterogeneity of farmland in Central Italy.

Environmental parameter	1	2	3	4	Total	F	df	P value
	FSH category							
Altitude (m)	122.7 ± 77.2	183.3 ± 156.2	298.8 ± 168.2	382.5 ± 129.8	292.4 ± 168.8	10.1	3	0.000***
Terrain slope (mode)	Low	Low	Low	Low-medium	Low	—	—	—
Roads (mode)	Paved	Paved	Paved	Paved	Paved	—	—	—
Powerlines (mean)	3.1 ± 2.5	2.9 ± 2.0	2.5 ± 1.6	3.8 ± 2.8	3.1 ± 2.3	1.5	3	0.221
Crop diversity (mean)	0.632 ± 0.3	0.554 ± 0.3	0.533 ± 0.4	0.489 ± 0.4	0.530 ± 0.3	0.3	3	0.790

Significance codes: ***$P < 0.001$, **$P < 0.01$, *$P < 0.05$, '$P < 0.1$.

TABLE 5: Similarity index pairwise of the bird species richness among the farmlands according to their spatial heterogeneity gradient.

FSH category	1	2	3	4	Total
	Similarity index				
1	—	0.780	0.667	0.605	0.472
2	—	—	0.815	0.776	0.764
3	—	—	—	0.959	0.922
4	—	—	—	—	0.962
Number of bird species (total)	25	34	47	51	55

should contribute to three EU-level priority areas: biodiversity and the preservation and development of HNV farming and forestry systems and traditional agricultural landscapes water and climate change" (Community's Strategic Guidelines for rural development, 2007–2013, OJL55/20, 2006). The HNV is characterized by three main criteria: low intensity farming, presence of semi-natural vegetation (hedgerows, uncultivated, shrubs, etc.), and diversity of land cover mosaic.

The results showed how in Central Italy bird richness is related to spatial heterogeneity of farmlands, highlighting how the lack of several features, such hedgerows or scattered shrubs, causes the loss of more than 53% of the total potential bird species richness. These findings point out the important role of these few marginal components, typical of most heterogeneous mosaics, corresponding to HNV farmlands. The results also show that diversity of land-use rather than crop types seems to be important for bird species richness in farmlands. However, because these two variables are correlated, is hard determinate the real contribution of each

one. Shannon index of crop types decreases with farmland heterogeneity, because diversity of crop types corresponds to more extended and less complex farmlands from a structural point of view. The characteristics of farmlands seeming to support high numbers of bird species are hedgerows, scattered shrubs, uncultivated patches, and the presence of power lines. The first three elements could improve the land-use diversity, offering habitat availability to several species [7, 9, 43]. Other studies underlined how electricity wires may offer useful perches and singing posts, representing attractive structures to insectivorous and to those birds hunting from observation posts like the Lanidae [44, 45]. Moreover, a positive effect on bird detection may also be considered. However, the same structures could be dangerous and cause mortality by collision or electrocutions for several raptors species that were excluded in this study [46, 47]. In our study, the absence of all these features and structural items is associated with a deep decrease of biodiversity.

The comparison between similarities index provided a quantitative measure of the total influence of the spatial heterogeneity on bird community. The approach by mean of Sorensen index can constitute an easy and fast decisional tool, useful to compare and quantify how a gradient of urbanization can modify the biodiversity of ecosystems [48]. The quantification of the impact of landscape or land-use features on the biodiversity can constitute a useful tool for conservation planning, landscape restoring, and for maintaining HNV farmlands.

The results of this work can be useful to select the most important environmental parameters to maintain the biodiversity and to characterize the bird communities or the presence of threatened species on different farmland types too. The bird species frequencies on farmland types are

TABLE 6: Different candidate models, ranked according to the AIC values, used to select the best models to relate bird richness to landscape, land-use, and crop types in farmlands of Central Italy.

Variables included in the model	Total number of variables	Scale level	AIC
Alt + slo + pow + shr + for + unc + bad + gra + hed + tre + urb + cul + ofo + whe + alf + hay + sun + cri + sug + cor + gar + oat + bra + let + land-use div	25	Landscape, land-use, crop type	424.53
Alt + slo + pow + shr + for + unc + bad + gra + hed + tre + urb + cul	12	Landscape, land-use	429.73
Alt + pow + shr + unc + bad + gra + hed + urb	8	Landscape, land-use	422.51
Pow + shr + unc + hed	4	Landscape, land-use	417.31

TABLE 7: Results of Poisson regression for the best model relating the bird species richness to environmental parameters of farmlands in Central Italy. The table shows the most significant variables selected after a stepwise backward procedure using AIC criterion. A comparison of models with all possible variables, classified in the four different categories of spatial heterogeneity, showed that the significant parameters are the same for all the evaluated levels.

Environmental parameter	Estimate	SE	Z value	P-values
Farmlands ($n = 80$)				
Powerlines	0.02805	0.01242	2.257	0.0240*
Shrubs	1.23966	0.50681	2.446	0.0144*
Uncultivated	0.94663	0.53777	1.760	0.0784`
Hedgerows	1.51644	0.33585	4.515	$6.33e - 06$***
Intercept	2.21632	0.09018	24.576	$2e - 16$***

Significance codes: ***$P < 0.001$, **$P < 0.01$, *$P < 0.05$, `$P < 0.1$.

useful to develop any bioindicator approach (farmland bird index, monitoring of HNV farmlands, farmland focal species, etc.). For example Italian Sparrow, in decreasing throughout the whole Italian peninsula [49], was present mainly in homogeneous farmlands, almost twice times if compared to the only 11 records from more heterogeneous fields. The Ortolan Bunting that is considered in decline in Europe [50–54] did not show preferences between homogeneous or heterogeneous farmlands. The Western Yellow Wagtail was present only in farmlands with less spatial heterogeneity and completely absent in more heterogeneous fields. On the other hand, some species only occur in very heterogeneous farmlands and could be targeted as good indicators of biodiversity in farmlands, in the sense of the "focal species" approach [55]. Two typical farmland birds mainly present in more heterogeneous farmlands were the Cirl Bunting and the Corn Bunting (Table 3).

Finally, our results help filling the lack of data about farmland bird ecology in southern Europe [17] and suggest the necessity of coordinating the management actions established by conservations plans and the different farmlands involved, possibly through an agricultural policy on a regional scale, according to the guidelines of detailed EU rural policies. The importance found for several features such as linear hedgerows can suggest conservation policies as the creation and maintenance of hedges, field margins, retention of uncropped areas, or the use of intercropping cultures in the fields this way we can increase the spatial heterogeneity and help to enhancing, thus favoring the settlement of several concern bird species [9, 29, 56]. Furthermore, the importance of using species-specific strategies for maintaining and preserving the structure and the functionality of agroecosystems is highlighted, as the necessity of more local scale studies, from the applied point of view, considering that currently the farmland environment in Central-Eastern Europe is generally more extensive and more complex than in Western Europe, and this fact generates different biodiversity conditions in agricultural areas [57].

Acknowledgments

The author is very grateful to the following people: Dan Chamberlain, Jon Mc Kean, Marco Girardello, Raffaele Secchi, Maria Balsamo, Yanina Benedetti, and anonymous reviewers for their suggestions or valuable comments on earlier versions of the manuscript or in the field. He also thanks ION Proofreading for the critical revision of English version of the paper.

References

[1] D. E. Chamberlain, R. J. Fuller, R. G. H. Bunce, J. C. Duckworth, and M. Shrubb, "Changes in the abundance of farmland birds in relation to the timing of agricultural intensification in England and Wales," *Journal of Applied Ecology*, vol. 37, no. 5, pp. 771–788, 2000.

[2] P. F. Donald, R. E. Green, and M. F. Heath, "Agricultural intensification and the collapse of Europe's farmland bird populations," *Proceedings of the Royal Society B*, vol. 268, no. 1462, pp. 25–29, 2001.

[3] D. Kleijn, F. Kohler, A. Báldi et al., "On the relationship between farmland biodiversity and land-use intensity in Europe," *Proceedings of the Royal Society B*, vol. 276, no. 1658, pp. 903–909, 2009.

[4] A. C. Weibull, Ö. Östman, and Å. Granqvist, "Species richness in agroecosystems: the effect of landscape, habitat and farm management," *Biodiversity and Conservation*, vol. 12, no. 7, pp. 1335–1355, 2003.

[5] G. M. Siriwardena, H. Q. P. Crick, S. R. Baillie, and J. D. Wilson, "Agricultural land-use and the spatial distribution of granivorous lowland farmland birds," *Ecography*, vol. 23, no. 6, pp. 702–719, 2000.

[6] J. R. Krebs, J. D. Wilson, R. B. Bradbury, and G. M. Siriwardena, "The second silent spring?" *Nature*, vol. 400, no. 6745, pp. 611–612, 1999.

[7] H. J. W. Vermeulen, "Corridor function of a road verge for dispersal of stenotopic heathland ground beetles Carabidae," *Biological Conservation*, vol. 69, no. 3, pp. 339–349, 1994.

[8] J. Hoffmann and J. M. Greef, "Mosaic indicators—theoretical approach for the development of indicators for species diversity in agricultural landscapes," *Agriculture, Ecosystems and Environment*, vol. 98, no. 1-3, pp. 387–394, 2003.

[9] T. G. Benton, J. A. Vickery, and J. D. Wilson, "Farmland biodiversity: is habitat heterogeneity the key?" *Trends in Ecology and Evolution*, vol. 18, no. 4, pp. 182–188, 2003.

[10] F. Morelli, "Importance of road proximity for the nest site selection of the Red-backed shrike Lanius collurio in an agricultural environment in Central Italy," *Journal of Mediterranean Ecology*, pp. 21–29, 2011.

[11] F. Morelli, R. Santolini, and D. Sisti, "Breeding habitat of Red-backed Shrike Lanius collurio on farmland hilly areas of Central Italy: is functional heterogeneity an important key?" *Ethology Ecology & Evolution*, vol. 24, pp. 127–139, 2012.

[12] L. Norton, P. Johnson, A. Joys et al., "Consequences of organic and non-organic farming practices for field, farm and landscape complexity," *Agriculture, Ecosystems and Environment*, vol. 129, no. 1-3, pp. 221–227, 2009.

[13] N. A. Beecher, R. J. Johnson, J. R. Brandle, R. M. Case, and L. J. Young, "Agroecology of birds in organic and nonorganic farmland," *Conservation Biology*, vol. 16, no. 6, pp. 1620–1631, 2002.

[14] Y. Kisel, L. Mcinnes, N. H. Toomey, and C. D. L. Orme, "How diversification rates and diversity limits combine to create large-scale species-area relationships," *Philosophical Transactions of the Royal Society B*, vol. 366, no. 1577, pp. 2514–2525, 2011.

[15] D. J. Pain and M. W. Pienkowski, Eds., *Birds and Farming in Europe: The Common Agricultural Policy and Its Implications for Bird Conservation*, Academic Press, San Diego, Calif, USA, 1997.

[16] D. Kleijn and W. J. Sutherland, "How effective are European agri-environment schemes in conserving and promoting biodiversity?" *Journal of Applied Ecology*, vol. 40, no. 6, pp. 947–969, 2003.

[17] A. Báldi and P. Batáry, "Spatial heterogeneity and farmland birds: different perspectives in Western and Eastern Europe," *Ibis*, vol. 153, no. 4, pp. 875–876, 2011.

[18] P. Batáry, A. Báldi, and S. Erdős, "Grassland versus non-grassland bird abundance and diversity in managed grasslands: local, landscape and regional scale effects," *Biodiversity and Conservation*, vol. 16, no. 4, pp. 871–881, 2007.

[19] P. Batáry, J. Fischer, A. Báldi, T. O. Crist, and T. Tscharntke, "Does habitat heterogeneity increase farmland biodiversity?" *Frontiers in Ecology and the Environment*, vol. 9, no. 3, pp. 152–153, 2011.

[20] R. Barbault, *La Biodiversité: Introduction à la Biologie de la Conservation*, Collection les Fondamentaux, Hachette, Paris, France, 1997.

[21] C. J. Bibby, N. D. Burgess, and D. A. Hill, *Bird Census Techniques*, Academic Press, London, UK, 1997.

[22] R. Clarke, *Montagu's Harrier*, Arlequin Press, Chelmsford, UK, 1996.

[23] B. Arroyo, J. T. García, and V. Bretagnolle, "Conservation of the Montagu's harrier (Circus pygargus) in agricultural areas," *Animal Conservation*, vol. 5, no. 4, pp. 283–290, 2002.

[24] M. Brambilla, F. Casale, V. Bergero et al., "GIS-models work well, but are not enough: habitat preferences of Lanius collurio at multiple levels and conservation implications," *Biological Conservation*, vol. 142, no. 10, pp. 2033–2042, 2009.

[25] D. Baldock, G. Beaufoy, G. Bennett, and J. Clark, *Nature Conservation and New Directions in the Common Agricultural Policy*, Institute for European Environmental Policy, London, UK, 1993.

[26] E. M. Bignal and D. I. McCracken, "Low-intensity farming systems in the conservation of the countryside," *Journal of Applied Ecology*, vol. 33, no. 3, pp. 413–424, 1996.

[27] E. Andersen, D. Baldock, H. Bennett et al., "Developing a High Nature Value Farming area indicator," Report for the European Environment Agency, Copenhagen, Denmark, 2003, http://www.ieep.eu/assets/646/Developing_HNV_indicator.pdf.

[28] P. Pointereau, M. L. Paracchini, J.-M. Terres, F. Jiguet, Y. Bas, and K. Biala, "Identification of high nature value farmland in France through statistical information and farm practice surveys," Report EUR 22786 EN, Office for Official Publications of the European Communities, Brussels, Luxembourg, 2007.

[29] V. Peronace, J. G. Cecere, M. Gustin, and C. Rondinini, "Lista Rossa 2011 degli uccelli nidificanti in Italia," *Avocetta*, vol. 36, pp. 11–58, 2012.

[30] P. F. Donald and C. Forrest, "The effects of agricultural change on population size of corn buntings Miliaria calandra on individual farms," *Bird Study*, vol. 42, no. 3, pp. 205–215, 1995.

[31] R. H. G. Jongman, C. J. F. ter Braak, and O. F. R. Tongeren, *Data Analysis in Community and Landscape Ecology*, Cambridge University Press, Cambridge, UK, 1995.

[32] P. McCullagh and J. A. Nelder, *Generalized Linear Models*, Chapman and Hall, London, UK, 1989.

[33] H. Akaike, "A new look at the statistical model identification," *IEEE Transactions on Automatic Control*, vol. 19, no. 6, pp. 716–723, 1974.

[34] Anon, *S-PLUS 2000 Guide to Statistics*, vol. 1, MathSoft, Seattle, Wash, USA, 1999.

[35] M. G. Betts, A. W. Diamond, G. J. Forbes, M. A. Villard, and J. S. Gunn, "The importance of spatial autocorrelation, extent and resolution in predicting forest bird occurrence," *Ecological Modelling*, vol. 191, no. 2, pp. 197–224, 2006.

[36] J. Oksanen, F. G. Blanchet, R. Kindt et al., "vegan: Community Ecology Package. R package version 2.0-6," 2013, http://CRAN.R-project.org/package=vegan.

[37] F. Morelli, "Plasticity of habitat selection by red-backed shrikes Lanius collurio breeding in different landscapes," *The Wilson Journal of Ornithology*, vol. 124, pp. 52–57, 2012.

[38] A. F. Bennett, J. Q. Radford, and A. Haslem, "Properties of land mosaics: implications for nature conservation in agricultural environments," *Biological Conservation*, vol. 133, no. 2, pp. 250–264, 2006.

[39] G. Beaufoy, D. Baldock, and J. Clark, *The Nature of Farming: Low Intensity Farming Systems, in Nine European Countries*, Institute for European Environmental Policy, London, UK, 1994.

[40] EENRD/EC, "Guidance document. The application of the High Nature Value impact indicator. European Communities," 2009, http://ec.europa.eu/agriculture/rurdev/eval/hnv/guidance_en.pdf.

[41] M. L. Paracchini, J.-M. Terres, E. Petersen et al., "High nature value farmland in Europe. An estimate of the distribution patterns on the basis of land cover and biodiversity data. European Commission Joint Research Centre, Institute for

Environment and Sustainability," Report EUR 23480 EN, Office for Official Publications of the European Communities, Brussels, Luxembourg, 2008.

[42] European Environment Agency, "High nature value farmland: characteristics, trends and policy challenges," EEA Report No. 1/2004, Copenhagen, Denmark, 2004.

[43] H. J. W. Vermeulen and P. F. M. Opdam, "Effectiveness of roadside verges as dispersal corridors for small ground-dwelling animals: a simulation study," *Landscape and Urban Planning*, vol. 31, no. 1–3, pp. 233–248, 1995.

[44] A. Bechet, P. Isenmann, and R. Gaudin, "Nest predation, temporal and spatial breeding strategy in the Woodchat Shrike Lanius senator in Mediterranean France," *Acta Oecologica*, vol. 19, no. 1, pp. 81–87, 1998.

[45] N. Lefranc, *Les Pies-grieches d'Europa, d'Afrique du nord et du moyen-Orient*, Delachaux et Niestlé S.A., Lausanne, Paris, France, 1993.

[46] R. E. Harness and K. R. Wilson, "Electric-utility structures associated with raptor electrocutions in rural areas," *Wildlife Society Bulletin*, vol. 29, no. 2, pp. 612–623, 2001.

[47] K. Bevanger, "Biological and conservation aspects of bird mortality caused by electricity power lines: a review," *Biological Conservation*, vol. 86, no. 1, pp. 67–76, 1998.

[48] P. Clergeau, S. Croci, J. Jokimäki, M. L. Kaisanlahti-Jokimäki, and M. Dinetti, "Avifauna homogenisation by urbanisation: analysis at different European latitudes," *Biological Conservation*, vol. 127, no. 3, pp. 336–344, 2006.

[49] L. Fornasari, E. de Carli, L. Buvoli et al., "Secondo bollettino del progetto MITO2000: valutazioni metodologiche per il calcolo delle variazioni interannuali," *Avocetta*, vol. 28, pp. 59–76, 2004.

[50] S. Dale, "Causes of population decline of the Ortolan Bunting in Norway," in *Bunting Studies in Europe*, P. Tryjanowski, T. S. Osiejuk, and M. Kupczyk, Eds., pp. 33–41, Bogucki Wydawnictwo Naukowe, Poznan, Poland, 2001.

[51] P. F. Donald, F. J. Sanderson, I. J. Burfield, and F. P. J. van Bommel, "Further evidence of continent-wide impacts of agricultural intensification on European farmland birds, 1990–2000," *Agriculture, Ecosystems and Environment*, vol. 116, no. 3-4, pp. 189–196, 2006.

[52] F. Morelli, "Correlations between landscape features and crop type and the occurrence of the Ortolan Bunting Emberiza hortulana in farmlands of Central Italy," *Ornis Fennica*, vol. 89, pp. 264–272, 2012.

[53] I. Newton, "The recent declines of farmland bird populations in Britain: an appraisal of causal factors and conservation actions," *Ibis*, vol. 146, no. 4, pp. 579–600, 2004.

[54] G. M. Tucker and M. F. Heath, *Birds in Europe: Their Conservation Status*, BirdLife Conservation Series no. 3, BirdLife International, Cambridge, UK, 1994.

[55] R. J. Lambeck, "Focal species: a multi-species umbrella for nature conservation," *Conservation Biology*, vol. 11, no. 4, pp. 849–856, 1997.

[56] M. J. Whittingham and K. L. Evans, "The effects of habitat structure on predation risk of birds in agricultural landscapes," *Ibis*, vol. 146, no. 2, pp. 210–220, 2004.

[57] P. Tryjanowski, T. Hartel, A. Báldi et al., "Conservation of farmland birds faces different challenges in Western and Central-Eastern Europe," *Acta Ornithologica*, vol. 46, no. 1, pp. 1–12, 2011.

14

Spatial Distribution of Zooplankton Diversity across Temporary Pools in a Semiarid Intermittent River

Thaís X. Melo and Elvio S. F. Medeiros

Grupo Ecologia de Rios do Semiárido, Universidade Estadual da Paraíba-UEPB,
Rua Horácio Trajano de Oliveira, S/N, Cristo Redentor, 58020-540 João Pessoa, PB, Brazil

Correspondence should be addressed to Elvio S. F. Medeiros; elviomedeiros@uepb.edu.br

Academic Editor: José Manuel Guerra-García

This study describes the richness and density of zooplankton across temporary pools in an intermittent river of semiarid Brazil and evaluates the partitioning of diversity across different spatial scales during the wet and dry periods. Given the highly patchy nature of these pools it is hypothesized that the diversity is not homogeneously distributed across different spatial scales but concentrated at lower levels. The plankton fauna was composed of 37 species. Of these 28 were Rotifera, 5 were Cladocera, and 4 were Copepoda (nauplii of Copepoda were also recorded). We showed that the zooplankton presents a spatially segregated pattern of species composition across river reaches and that at low spatial scales (among pools or different habitats within pools) the diversity of species is likely to be affected by temporal changes in physical and chemical characteristics. As a consequence of the drying of pool habitats, the spatial heterogeneity within the study river reaches has the potential to increase β diversity during the dry season by creating patchier assemblages. This spatial segregation in community composition and the patterns of partition of the diversity across the spatial scales leads to a higher total diversity in intermittent streams, compared to less variable environments.

1. Introduction

The Brazilian semiarid region represents one of the major dry lands in South America, being characterized by a high hydric deficit and low thermal amplitude [1]. These factors interact with broad climatic patterns to create important variation between dry and wet periods [2]. This affects the hydrology of the region which is characterized by intermittent watercourses [3]. The natural variation associated with the alternation of dry and wet periods and the intermittency of flow in rivers and streams create a mosaic of temporary natural aquatic habitats, mostly strings of ephemeral pools in the dry river bed and more permanent larger ones [4]. Therefore, these systems are highly temporally variable and spatially heterogeneous [5, 6].

During the wet season, the mechanical force of the water flow changes the physical habitat and modifies water chemical characteristics and nutrient dynamics [5, 7], whereas during the dry season, discontinuation of flow and water volume reduction influence community structure and diversity by concentrating nutrients and affecting physical and chemical characteristics of the water [5, 8]. Furthermore, during this phase, the water retention in semipermanent and temporary pools represents refugia for the maintenance of aquatic species [9]. These small patchy ecosystems have been argued to contribute disproportionately to the regional assemblage of species given their high local diversity [10, 11]. This is specially the case for the diversity of the aquatic microinvertebrate fauna which is subject to dispersal limitations that create variation in community composition [12].

The role of dispersion/colonization of species in community composition is associated with the number of species that reach a pool or a suitable habitat and with the order in which these species reach a given habitat [13]. From then onwards species would be sorted out given other factors such as predation [14], species traits [15], and habitat structure [16].

In such pools, specific adaptations and strategies are important to cope with the variable and commonly extreme conditions [17], and such mechanisms may lead to the spatial and temporal segregation of the plankton fauna. Among the factors that influence species distribution and assemblage composition of zooplankton, physical and chemical characteristics have been argued to be the most important [18, 19]. Zooplankton assemblages respond rapidly to different water quality conditions most importantly so to temperature, conductivity, pH, and nutrient concentration.

Furthermore, studies have shown that the zooplankton community is sensitive to extreme variation in flow; thus species composition is changed and the succession of taxa is redirected after flow recession [20]. This results in different timing for the emergence of rotifers, cladocerans, and copepods from the inundated dry river beds [21]. However, the role of the hydrological disturbances in the spatial dynamics of zooplankton and the habitat variation in intermittent streams has received little attention. Studies indicate that water flow and its effects on habitat structure play a major role in producing and maintaining a mosaic of pools and microhabitats that can be used for colonization and refugia for aquatic organisms [22, 23]. This dynamic makes intermittent streams complex and heterogeneous systems, with organisms distributed in patches determined by environmental conditions affecting the fauna hierarchically [24, 25].

Therefore, a subdivision of diversity in hierarchical scales in stream ecosystems will result from these factors, which represent the interaction of physical and biological processes. It has been proposed that this hierarchization is represented by macrohabitats, representing distinct morphological zones [26]; mesohabitats, representing pools, runs, and riffles; and the microhabitats, representing stands of macrophytes or submerged vegetation and substrate types [24].

Studies in the aquatic systems of semiarid Brazil show a high diversity of species, including the plankton fauna, and demonstrate that the patterns of diversity are associated with the hydrological extremes and habitat types and/or structure (see [11, 18, 27]). Furthermore, zooplankton plays an important role in the nutrient dynamics and energy flow in dryland river systems [28]. Despite that, studies on patterns of distribution of zooplankton diversity and species composition in dry land intermittent streams are scarce and the effects of the habitat structure on plankton fauna are poorly known (see [11]).

In this study we describe the richness and density of zooplankton across temporary pools in an intermittent river of semiarid Brazil and evaluate the partitioning of diversity across different spatial scales during the wet and dry periods. Given the highly patchy nature of these pools it is hypothesized that the diversity is not homogeneously distributed across different spatial scales but concentrated at lower levels. We also aim to establish the degree of association between environmental variables and the composition of species in the zooplankton assemblages.

2. Materials and Methods

2.1. Study Area and Sampling Design. The present study was performed at the Paraíba River basin (Figure 1). This basin encompasses an area of 20072 km^2 and its altitudes range from 350 to 460 m a.s.l. [29]. Predominant vegetation in the basin is the Caatinga, an arboreal to shrubby open forest,

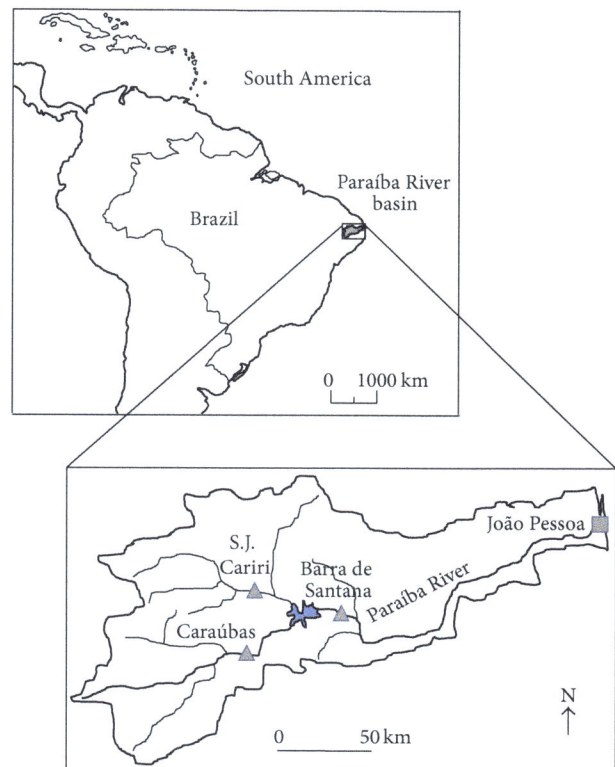

FIGURE 1: Study area showing the location of the Paraíba River basin and the study reaches.

characterized by the presence of xerophytic species [30]. This type of vegetation does not provide strong protection to the soil which increases water losses by evaporation, enhancing the temporal nature of river pools. The climate is classified as semiarid BS'h according to the classification of Köppen-Geiger modified by Peel et al. [31], with average temperature and precipitation of 26°C and 600 mm, respectively [32]. The wet season starts in January-February, with higher precipitation between April and June. The peak of the dry season lasts from September to January. The study area is classified as a being of extreme biological importance and identified as priority area for biodiversity conservation in the Caatinga biome [33].

In the study basin, zooplankton was collected from three river reaches of 100 to 500 m during the wet (June 2010) and dry (October 2010) seasons. These river reaches were located (1) at the Taperoá river, an affluent of the Paraiba River (location at the São João do Cariri municipality, 7°23'0"S; 36°34'24.4"W), (2) at the upper Paraiba River (municipality of Caraúbas, 7°43'29.7"S; 36°34'9.3"W), and (3) at the lower Paraíba River (municipality of Barra de Santana, 7°31'20.8"S; 36°1'29.8"W) (Figure 1). Henceforth, river reaches will be referred to as the municipality where they run through. At each river reach three habitat samples (10 m tows) were performed and represent different pools or distinct microhabitat types (macrophyte stands, shallow or deeper areas, and rocky outcrops) at the same pool when less than 3 pools were found. From each habitat type sample, three subsamples were taken to account for intrasample variability.

Zooplankton was collected quantitatively using a plankton net (opening diameter 30 cm, length 70 cm, and mesh size of 60 μm). The net was towed for a distance of 10 m on the surface of the water at dusk. To minimize variation in sampling efficiency across samples, velocity and length of tows were similar and the net was washed between each tow to prevent clogging (see [11]). The zooplankton collected was anesthetized with commercial sparkling water and preserved in 4% formalin. Sucrose was added to the preserved sample to prevent female cladocerans from losing eggs and to minimize carapace distortion [34]. In the laboratory, three sub-samples were taken from each habitat sample and all individuals were identified [35–37] and counted in a Sedgewick Rafter counting cell (1 mL), and the density was calculated. Only rotifers, cladocerans, and copepods were considered in the present study.

At each river reach, supportive environmental data was collected, comprising physical and chemical water variables, pool morphometry, substrate composition, and the physical structure of the habitat. The physical and chemical variables were estimated using portable meters for pH (tecnopon mpa-210), conductivity (μS/cm) (tecnopon MCS-150), dissolved oxygen (mg/L), and temperature (°C) (Lutron DO-5510). Transparency (cm) was measured with a Secchi disk and water velocity (m/s) was measured using the float method [38].

Stream reach morphology was evaluated by the average width (m) and depth (cm) taken from three transects randomly placed in the stream pool. Pool length was also measured and used to calculate pool area. The substrate composition and habitat structure were estimated in 9 to 12 survey points of 1 m^2 measured in the margins (see [5]). In each survey point, the proportion of the sediment composition (classified as mud, sand, gravel, and cobbles) and littoral and underwater structures (e.g., macrophytes, grass, submerged vegetation, overhanging vegetation, leaf litter, algae, and woody debris) were estimated visually.

2.2. Data Analysis. Zooplankton diversity was estimated using species richness (R), defined as the total number of species at any given sample, and density (D), defined as the number of individuals per cubic meter (ind m^{-3}). To test for changes in richness and density across samples a factorial ANOVA was carried out, where each sample measurement of the response variables (R and D) was classified according to three factors (river reach, habitat sample, and season), where river reach has three factor levels, habitat sample has three factor levels, and season has two levels (wet and dry). In the presence of interaction the next high-order terms were analyzed (reach $*$ season, reach $*$ habitat, and season $*$ habitat) and then the main effects. A $\log_{10}(x + 1)$ transformation was applied to the response variables to enhance normality of the data set and equality of variances [39]. *Post hoc* pairwise comparisons were performed using Tukey's HSD test ($\alpha = 0.05$).

Additive partitioning of diversity was used to decompose the total variation in zooplankton community composition (river reach) into its alpha and beta components. Alpha and beta diversity represent within-unit diversity (α) and between-unit diversity (β) on a given scale, respectively.

Zooplankton diversity was analyzed at three different spatial scales: (1) the sub-sample scale, representing individual counts in sub-samples of 1 mL for each sample ($n = 27$) (this is important to evaluate intrasample variability); (2) the local spatial scale representing the habitats at each reach ($n = 9$) and (3) the regional spatial scale representing the river reach ($n = 3$). Thus, the total diversity (γ) was partitioned into the diversity (expressed as percentage) within sub-samples ($\alpha 1$) and between subsamples ($\beta 1$), between reach habitats ($\beta 2$), representing different pools or pool habitats, and between river reaches ($\beta 3$).

To evaluate temporal (dry/wet season) variation, the data was analyzed separately for each season. The additive partition analyses were performed using PARTITION 3.0 [40], with individual-based randomizations (type I null model) [41]. The randomization process was repeated 10000 times to obtain null distributions of the alpha and beta diversity estimates at each hierarchical level. The probabilities obtained from randomization tests were interpreted as P values in the sense that a low P value ($P < 0.05$) indicates that the observed diversity is significantly higher than that expected under the null model and a high P value ($P > 0.95$) indicates that the observed diversity is significantly lower than that expected [42].

A 2-dimensional nonmetric multidimensional scaling (NMS) plot was obtained based on the Sorensen Bray-Curtis similarity between all samples [43] to evaluate composition of zooplankton across reaches, pool/habitats, and seasons. The significance of differences was tested using the multiresponse permutation procedure (MRPP) [44]. To all MRPP analyses, the value of A is presented as a measure of the degree of homogeneity between groups compared to random expectation. Data were transformed by the arcsine square root [39]. Indicator species analysis [45] was used to determine which species discriminated the different spatial and temporal scales. The significance of the discriminating power (IV) was determined by the Monte Carlo test (1000 permutations). This analysis assigns an indicator value for each species, based on the degree to which they discriminate among groups. These groups are determined *a priori* and, in the present study, represent river reaches at each season. The indicator value ranges from zero (no indication) to 100 (perfect indication) of a species for a group. Perfect indication means that the presence of a species points to that particular group [46]. To assess how well environmental constraints across river reaches and seasons were correlated with the assemblages composition, a canonical correspondence analysis (CCA) was performed. The data matrix was centered and normalized and the correlations tested by the Monte Carlo test with 999 runs. The environmental variables used in the CCA were temperature, dissolved oxygen, transparency, pH, and pool area. Density data were arcsine square root transformed and the environmental variables, except for pH, were $\log_{10}(x + 1)$ transformed [39, 47].

3. Results

Rainfall during the present study was highly variable, with most peaks and higher cumulative precipitation being recorded between April and June, whereas the dryer period was observed between July and November (Figure 2); as a consequence, surface water flow was observed only at the Cariri reach during the first sampling occasion (Table 1). Thus, river reach morphology varied regarding the presence and size of the temporary pools. During both wet and dry seasons, the Santana and Caraúbas reaches presented relatively large single pools of different sizes, whereas the Cariri reach presented two smaller pools of different sizes, during the dry season (Table 1).

River waters were neutral to slightly alkaline (range between 7.5 and 8.2) and well oxygenated (3.4 to 8.8 mg/L). Conductivity was below $5\,\mu S/cm$ at the Cariri and Santana reaches, whereas it remained above $900\,\mu S/cm$ at the Caraúbas reach. Water temperature amplitude was low, ranging between 26 and 30.9°C, and transparency relatively high, between 56 and 115 cm. Sediment was composed mostly of sand and gravel, whereas the aquatic habitat was diverse with higher proportional contributions of aquatic macrophytes, marginal grass, algae, woody debris and leaf litter, and lower contributions of vegetal cover and submerged vegetation (Table 1).

The plankton fauna was composed of 37 species. Of these 28 were Rotifera, 5 were Cladocera, and 4 were Copepoda (nauplii of Copepoda were also recorded). Overall densities were low and only 8 taxa showed densities greater than $5\,ind\,m^{-3}$. The ones with higher density were *Brachionus havanaensis* ($317.23\,ind\,m^{-3}$), *Brachionus caudatus* ($178.62\,ind\,m^{-3}$), and *Hexarthra mira* ($177.71\,ind\,m^{-3}$) (Table 2). The most frequently observed taxa were *Polyarthra dolicoptera* and Copepoda nauplii, occuring on all study reaches and in both sampling occasions. Cladocera was the taxa with the lowest overall density, with higher values presented by *Alonella granulata* and *Diaphanosoma spinulosum* ($0.28\,ind\,m^{-3}$ and $0.13\,ind\,m^{-3}$, resp.). Copepoda with higher densities were the naupliar stages ($153.07\,ind\,m^{-3}$) and *Mesocyclops longisetus* ($2.5\,ind\,m^{-3}$) (Table 2).

Factorial ANOVA showed significant interaction in richness and density across reaches, habitats, and seasons (three-way ANOVA, $F_R = 7.79$; d.f. = 4,36; $P < 0.001$; $F_D = 7.24$; d.f. = 4,36; $P < 0.001$). Thus, we interpreted the next high-order terms of the ANOVA, which were the two-way interactions between habitat * reach, habitat * season, and reach * season. These two-way interactions were significant for density (but not for richness) between habitat and river reach (three-way ANOVA, $F_D = 4.60$; d.f. = 4,36; $P = 0.004$) and habitat and season (three-way ANOVA, $F_D = 7.50$; d.f. = 2,36; $P < 0.002$). Analysis of simple main effects for these two-way interactions showed that habitat samples varied significantly for density only in the Cariri river reach (two-way ANOVA, $F_D = 11.6$; d.f. = 2,36; $P < 0.001$) and during the wet season (two-way ANOVA, $F_D = 12.3$; d.f. = 2,36; $P < 0.001$).

The two-way interaction remaining to be interpreted was the interaction between river reach and season, which was significant for both richness and density (three-way ANOVA, $F_R = 32.69$; d.f. = 2,36; $P < 0.001$; $F_D = 53.41$; d.f. = 2,36; $P < 0.001$). Multiple and pairwise comparisons showed that

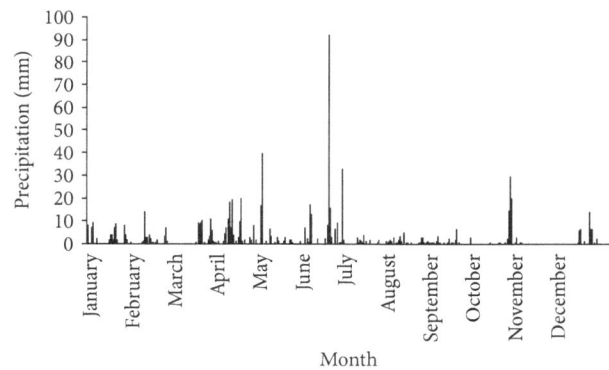

FIGURE 2: Precipitation at the São João do Cariri station from January to December 2010. Source: http://www.cptec.inpe.br/proclima/.

richness and density varied significantly across river reaches (Tukey tests <0.04) except for richness between Santana and Caraúbas during the dry season ($P = 0.547$) and between Cariri and Caraúbas during the wet season ($P = 0.385$) (Figure 3).

Partitioning of alpha and beta diversity components showed, in general, similar results for both wet and dry season collections (Figure 4). The partitioning of diversity showed that, for the sub-sample scale ($\alpha1$), diversity was lower than expected from the null model ($P > 0.999$) for both seasons, yet the mean number of taxa was similar between wet and dry seasons (6.78 and 7.37, resp.). The relative magnitude of the $\beta1$ component (among sub-samples) was generally low and similar between seasons. Nevertheless, $\beta1$ was higher than the expected from the null model ($P = 0.034$) in the wet season and lower in the dry season ($P = 0.992$). Values of the $\beta2$ component (among habitats or pools) were higher than expected from individual-based randomization during the dry season ($P = 0.002$) and equal to expectation during the wet season ($P = 0.101$). This component represented 8 and 19% (wet and dry season, resp.) of the total diversity of zooplankton. The $\beta3$ component (between river reaches) presented higher diversity than that expected from the null model ($P < 0.001$) and represented 67% of the diversity during the wet season and 52% during the dry season.

Nonmetric multidimensional scaling (NMS) (Figure 5) showed clear segregation in the zooplankton community across the three study reaches (MRPP, Cariri/Caraúbas, $A = 0.22$; $P = 0.003$; Cariri/Santana, $A = 0.34$; $P = 0.0004$; Caraúbas/Santana, $A = 0.29$; $P = 0.0008$) and between seasons within river reaches (MRPP, Cariri $A = 0.43$; $P = 0.023$; Santana, $A = 0.43$; $P = 0.022$; Caraúbas, $A = 0.43$; $P = 0.022$). Indicator species analysis (ISA) showed that the taxa that significantly contributed to the segregation between river reaches during the wet season were *Polyarthra vulgaris* (IV = 68.2; $P = 0.01$, Cariri reach), *Filinia longiseta* (IV = 69.5; $P = 0.02$, Caraúbas reach) and *Brachionus caudatus* (IV = 98.5; $P = 0.01$, Santana reach) and during the dry season were *Pompholyx sulcata* (IV = 100; $P = 0.01$) and *Rotaria* sp. (IV = 75.5; $P = 0.02$) for the Cariri reach, *Lecane leontina* (IV = 100; $P = 0.01$) and *Macrothrix* sp. (IV = 100; $P = 0.01$) for the Caraúbas reach, and *Brachionus havanaensis* (IV = 99.7;

TABLE 1: Environmental variables measured at the Paraiba River basin during the 2010 hydrological cycle.

River reach	Cariri		Caraúbas		Santana	
Season	Wet*	Dry**	Wet	Dry	Wet	Dry
Water quality						
pH	7.8	7.6	7.9	7.5	8.2	7.6
Dissolved oxygen (mg/L)	6.1	8.8	7.2	3.4	6.9	7.0
Conductivity (μS/cm)	2.2	3.9	915.3	1092.3	2.8	4.7
Temperature (°C)	27.1	30.9	26.0	27.6	26.1	27.4
Water transparency (cm)	58.3	56.0	75.0	115.0	69.0	63.0
Water velocity (m/s)	0.1	0	0	0	0	0
Morphometry						
Average depth (cm)	44.2	61.3/62.0	51.4	52.1	47.0	40.3
Average width (m)	19.0	23.9/17.8	9.3	10.3	65.1	63.7
Length (m)	n.a.	19.9/13.7	78.8	78.6	365.0	290.0
Total area (m^2)	n.a.	464.0	750.0	832.0	23400.0	17200.0
Reach altitude (m)	420.0	420.0	423.0	423.0	315.0	315.0
Substrate composition (%)						
Mud	0	0.9	0	1.0	0.7	3.0
Sand	78.3	90.4	98.8	99.5	89.5	83.8
Gravel	21.6	8.3	1.3	0	9.2	13.3
Cobbles	0	0.4	0	0	0.7	0
Habitat structure (%)						
Macrophytes	0.1	9.0	15.0	50.0	22.4	27.5
Grass	0	0	16.3	27.5	2.0	3.8
Submerged vegetation	3.8	0	2.0	0	0.2	0
Vegetal cover	5.0	6.3	0.3	0	0	1.3
Leaf litter	5.0	2.0	0.3	0	0	0
Algae	2.5	2.0	8.8	5.0	5.4	3.8
Woody debris	4.0	8.0	1.3	0	2.5	1.3

*Reach with surface water flow. **Two pools were surveyed.

P = 0.01), *Keratella valga* (IV = 98.9; P = 0.01), *Asplanchna sieboldi* (IV = 100; P = 0.007), and *Notodiaptomus* sp. (IV = 100; P = 0.007) for the Santana reach.

The first three axes of CCA explained 70.8% of the variation in zooplankton composition across river reaches, with a total variance (inertia) of 2.53. Most of the explained variations, based on the correlations between the environmental variables and the CCA axes, were explained by the first axis (27.2%), but the axes 2 and 3 were also important, explaining a substantial part of the variation in the data matrix (Table 3). The correlation between the zooplankton composition and the environmental variables was significant as shown by the Monte Carlo test for the eigenvalues and the species-environment correlations (P = 0.009) (Table 3, Figure 6). According to the interset correlations between the environmental variables and the CCA axes, the most important variables explaining the zooplankton composition were pH, dissolved oxygen, and water transparency (Table 3, Figure 6).

4. Discussion

Richness and composition of zooplankton observed in the present study are in accordance with other studies developed in the semiarid region of Brazil [11, 27, 48, 49] and elsewhere [50]. During the present study, Rotifera dominated in terms of richness and density. This has been reported for other systems [51] and is generally attributed to their high fecundity, parthenogenic reproduction, and high growth rates [15]. These characteristics, in association with the generalist feeding habit, make the Rotifera a typically r-strategist and opportunist group [52]. According to Medeiros et al. [11] these strategies are favored by the highly variable nature of the intermittent streams.

Among the Rotifera registered in the present study, Brachionidae and Lecanidae were the most representative in number of species (see also [18]). Brachionidae species are usually associated with the plankton whereas Lecanidae is mostly littoraneous, being associated with the benthos and periphyton or macrophyte stands [49]. The nonspecialized feeding of these taxa as well as the presence of a diverse array of littoral underwater structures and of aquatic macrophytes [16] associated with the low flows at the time of collection and the subsequent water retention is likely to have enhanced the richness of these taxa [53]. On the other hand, lower richness and density of Cladocera and Copepoda is explained by their more selective nature in relation to food and environmental changes. As reported by Walz and Welker [54] and Vieira et al. [18] these taxa tend to disappear or

Table 2: Density (ind m^{-3}) of zooplankton species at the Paraiba River basin during the 2010 hydrological cycle.

	Cariri		Caraúbas		Santana	
	Wet	Dry	Wet	Dry	Wet	Dry
Rotifera						
Asplanchnidae						
Asplanchna sieboldi	—	—	—	—	—	1.45
Brachionidae						
Brachionus angularis	—	12.34	—	0.08	31.45	104.57
B. havanaesis	—	0.01	—	—	317.22	—
B. calyciflorus	0.09	2.67	—	0.13	0.97	10.83
B. caudatus	0.03	—	0.03	—	178.57	—
B. plicatilis	—	0.08	—	—	1.41	0.44
B. quadridentatus	—	—	—	0.01	0.52	—
B. urceolaris	—	—	—	—	—	0.05
Keratella cochlearis	—	—	—	—	—	0.03
K. valga	0.01	—	—	—	31.24	—
Plationus patulus	—	—	0.28	0.25	—	—
Euchlanidae						
Euchlanis dilatata	—	—	—	0.03	—	—
Mytilinidae						
Mytilina ventralis	—	0.01	—	0.01	—	0.04
Hexarthridae						
Hexarthra mira	—	—	—	—	100.34	55.37
Synchaetidae						
Polyarthra dolicoptera	0.10	0.01	2.15	0.09	0.03	9.86
P. vulgaris	3.75	0.06	0.34	0.24	0.04	—
Lecanidae						
Lecane bulla	0.03	0.04	—	0.01	0.11	—
L. hamata	—	—	—	0.16	—	—
L. hastata	—	—	0.01	0.10	—	0.01
L. leontina	—	—	—	0.09	—	—
L. luna	—	—	—	0.03	—	—
L. lunaris	—	—	—	0.03	0.08	—
Colurellidae						
Lepadella sp.	—	—	—	0.01	—	—
Filiniidae						
Filinia longiseta	—	—	0.25	—	0.01	0.06
Trichocercidae						
Trichocerca sp.	—	0.01	0.04	—	0.01	—
Testudinellidae						
Pompholyx sulcata	—	3.15	—	—	—	—
Testudinella patina	—	—	—	—	0.14	—
Philodinidae						
Rotaria sp.	—	1.02	0.01	0.05	0.05	—
Cladocera						
Chydoridae						
Alonella granulata	—	—	—	0.05	0.23	—
Chydorus eurynotus	—	0.01	—	0.03	—	—
Daphniidae						
Ceriodaphnia cornuta	—	—	—	0.09	—	0.03
Diaphanosoma spinulosum	—	—	—	0.10	—	0.03
Macrothricidae						
Macrothrix sp.	—	—	—	0.05	—	—

TABLE 2: Continued.

	Cariri		Caraúbas		Santana	
	Wet	Dry	Wet	Dry	Wet	Dry
Copepoda						
Nauplii	4.50	1.92	6.59	4.66	76.26	59.13
Cyclopoida						
Mesocyclops longisetus	0.09	0.03	—	0.05	2.11	0.14
Paracyclops sp.	—	0.01	—	0.03	0.57	—
Thermocyclops sp.	—	—	—	—	0.09	—
Calanoida						
Notodiaptomus sp.	—	—	—	—	—	0.06
Total	8.59	21.38	9.70	6.35	741.43	242.09

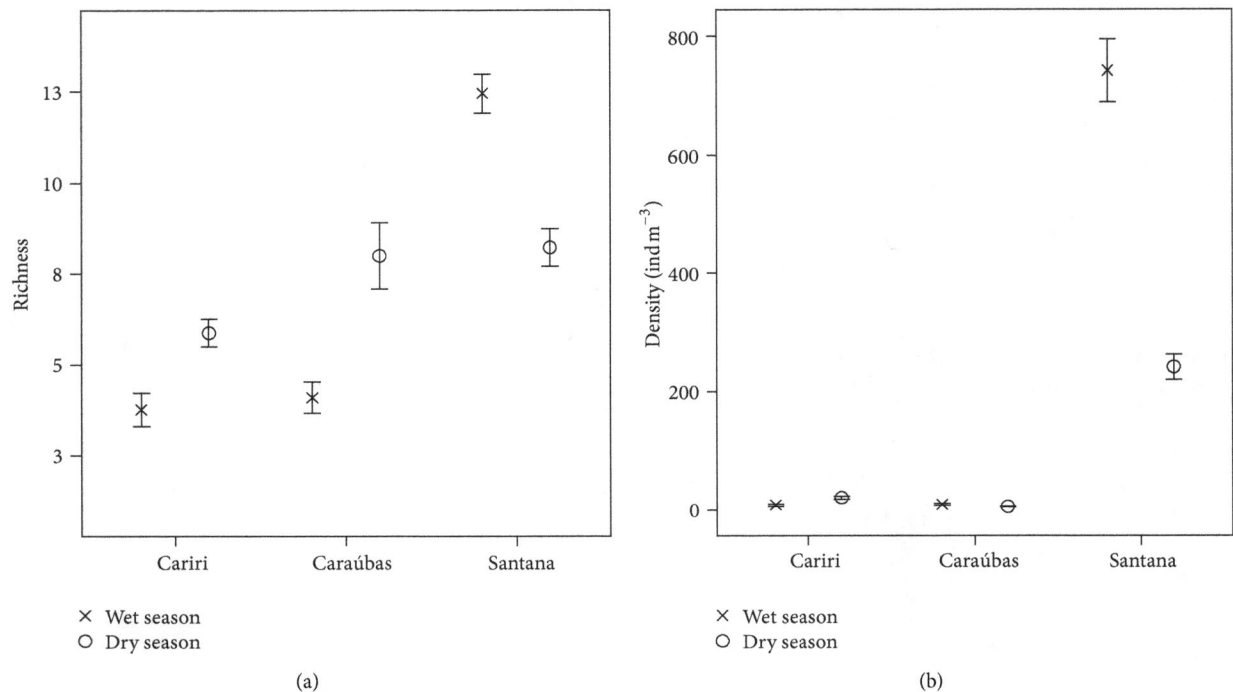

(a)

(b)

FIGURE 3: Average ± SE of richness and density of zooplankton at the Paraiba River basin during the 2010 hydrological cycle.

occur in very low abundance as the dry season progresses into harsh conditions and reappear when conditions become appropriate [55]. The higher densities of naupliar stages of Copepoda in the present study are the result of their ability to rapidly colonize temporary environments [56] associated with their high capacity to survive during recurrent events like the dry season in intermittent rivers [57].

During the present study, the river reaches were characterized by low to total absence of water flow, leading to the predominance of pools as the habitat available for the plankton fauna. The hydrological variation and dry periods have been reported to change the structure of the communities of zooplankton (see [11, 18]), where species with strategies to cope with the harsh and variable conditions during the dry season are selected among a larger possible array of species. Such environmental variability would lead to a spatial segregation, as the hydrological cycle progresses

into specific habitats and communities [58], and has been proposed to create a mosaic of local habitats at the pool scale and at the microhabitat within pool level in Brazilian semiarid rivers [5]. Thus, biological communities would be expected to assume different structures and composition along the river in accordance with the environmental heterogeneity. In the present study, analysis of variance supports that claim by showing significantly different richness across river reaches and habitats/pools and strong interaction with seasons. These interactions come, mostly, from the fact that richness was higher at the Santana reach during the wet period and density was higher at that reach during both seasons. Density and richness results for the other reaches were less conclusive, showing significant and nonsignificant variations across spatial and temporal scales, and only partially support the above argumentation. In particular, the significant variation observed in density across habitat samples in the Cariri reach

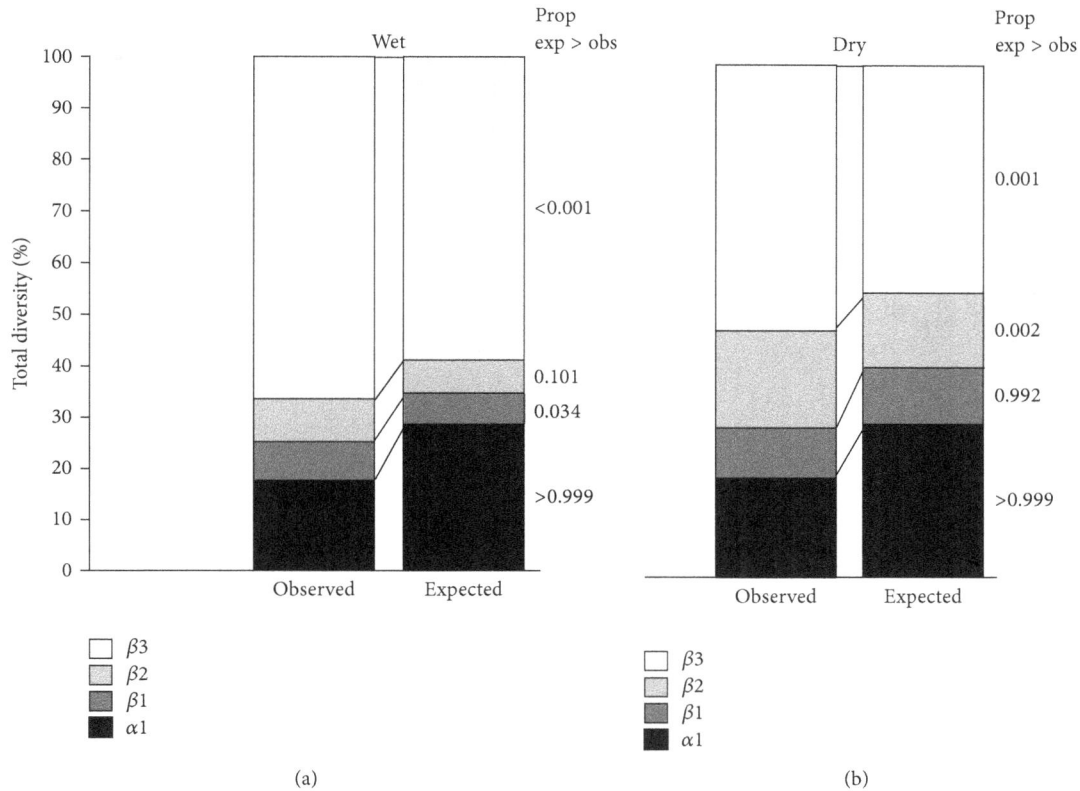

(a)   (b)

FIGURE 4: Observed and expected diversity, partitioned into one alpha and three beta components, expressed as percent of total species richness. Numbers indicate the proportion of randomized samples containing more species than the observed sample for each partition. $\alpha 1$: subsample species richness, $\beta 1$: beta diversity among sub-samples; $\beta 2$: beta diversity among pools or habitats within pool, and $\beta 3$: beta diversity among river reaches.

TABLE 3: Summary of axes for the canonical correspondence analysis of the zooplankton fauna and the environmental variables in the Paraiba River basin during the 2010 hydrological cycle.

	CCA axes		
	1	2	3
Eigenvalues	0.687	0.615	0.487
Monte Carlo test	0.009		
% variance explained	27.2	24.3	19.3
Pearson's correlations	1.000	1.000	1.000
Interset correlations			
Temperature	0.388	−0.237	0.633
D. O. (mg/L)	0.816	−0.405	0.385
Water transparency (cm)	0.794	0.479	−0.355
pH	0.862	0.404	−0.098
Area (m^2)	0.728	−0.306	−0.597
Species-environment correlations	1.000	1.000	1.000
Monte Carlo test	0.009		

The Santana reach was an important source of univariate (richness and density) variation across the study spatial and temporal scales, showing overall higher densities and the most abundant species. Furthermore, density and richness tended to be higher in the dry season at the Cariri and Caraúbas reaches, but not at the Santana reach. Higher density and richness during the dry season in intermittent streams has also been reported in other studies and was associated with increased concentration of nutrients at shrinking pools and higher water residence time [11, 15]. It is likely that the opposite relationship observed in the Santana reach is related with the greater human impact that this reach suffers in comparison with the others, since it receives organic inputs from local human communities and from an adjacent intermittent tributary that receives waste water from a nearby town. This probably resulted in the observed high abundance of species that are indicative of eutrophic environments such as the genus *Brachionus* [59] and specially *Brachionus angularis* and *B. havanaensis* [60].

Analysis of diversity partition provided additional support for the importance of local habitats/pools to the regional diversity, since the observed β was higher than the expected for both local habitats ($\beta 2$) and the river reaches ($\beta 3$). It also provides evidence that wet and dry periods play an important role in this partition, particularly at the lower

is likely the result of flooding during the wet season which decreased zooplankton numbers and enhanced intrareach variability.

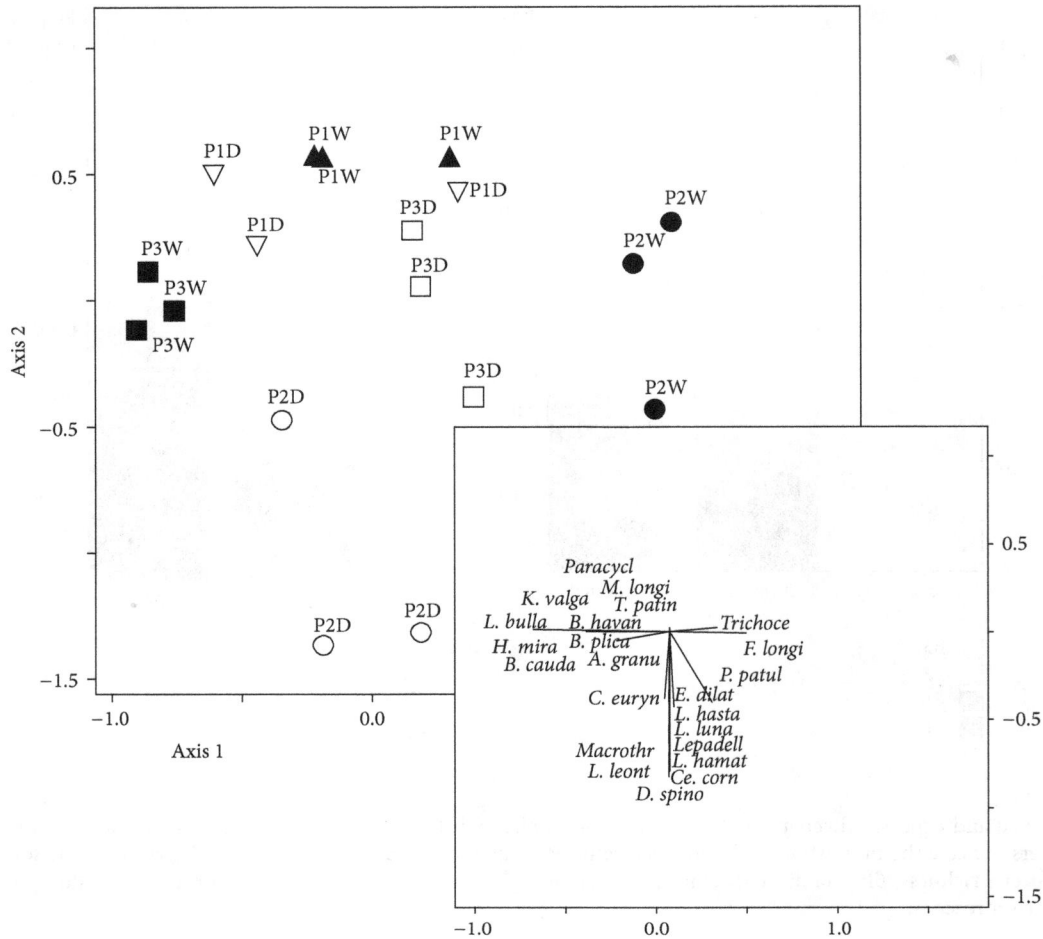

FIGURE 5: NMS results for zooplankton composition across the study reaches (P1 = Cariri, P2 = Caraúbas, and P3 = Barra de Santana) in the Paraíba River basin during the wet (W) and dry (D) seasons. Vectors (inset box) show taxa correlated ($R^2 > 0.2$) with samples in ordination space. The direction and length of vectors indicate strength of correlation.

scales of diversity ($\beta 2$), as shown by the higher observed and expected diversity during the dry period compared to the wet period. The importance of flooding on zooplankton composition has been recognized for intermittent aquatic environments in semiarid Brazil [18, 27]. In this study, we further support those views and highlight that the dry season or the absence of water flow enhances diversity by segregating the fauna composition among pools and therefore increasing β diversity at the pool/habitat scale (see also [11]). At larger spatial scales, in the present study at the river reach scale, the wet season has greater importance to β diversity. That is the result of the susceptibility of the zooplankton to water flow, which increases the potential for different assemblages at different reaches. In this context, the effects of varying magnitude, intensity, and frequency of flooding across river reaches remain untested.

At the larger spatial scale, ordination showed different compositions of the zooplankton fauna across river reaches and that, despite this spatial segregation, a few species of Brachionidae and Lecanidae dominated. The dominance of specific groups of zooplankton across segregated assemblages has been reported for Brazilian semiarid aquatic systems [11]. This has been linked to species' responses to the hydrological disturbances which can create conditions that select species from a common larger pool of species that is broken into more specific assemblages at local scales [61]. The $\alpha 1$ observed for both study seasons was low, and the $\beta 2$ diversity for the wet season represented a smaller fraction of the total diversity compared with the dry season. These results indicate not only higher intraspecific aggregation at the fine scales [62], but also that the habitat/pool diversity is more aggregated during the wet period. In theory this could be explained by the increased proportion of ecological interactions of conspecifics (intraspecific aggregation) in comparison with interspecific interactions, which reduces competitive exclusion and enhances diversity [63]. In the present study, this may be associated with physical and chemical changes in the water and most likely pool area and/or microhabitat diversity.

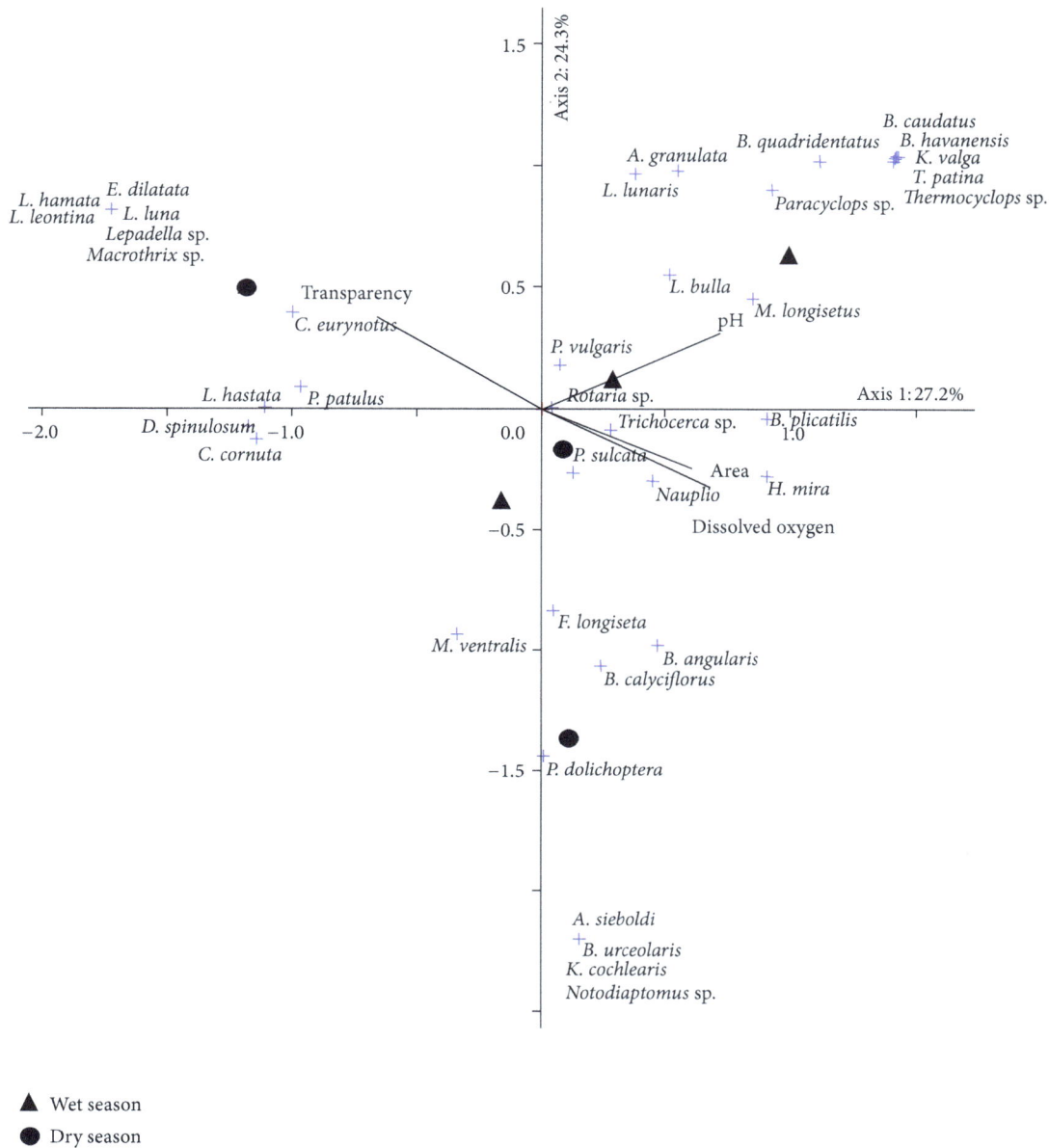

FIGURE 6: Biplot of CCA showing the composition of zooplankton in the sampling sites and seasons and the explanatory environmental variables defined by CCA.

Other physical and chemical factors, such as discharge, water temperature, pH, dissolved oxygen, and habitat complexity, have been shown to affect the zooplankton community [18].

Canonical correspondence analysis showed that a significant portion of the zooplankton distribution and composition was explained by the environmental variables. *Brachionus caudatus, B. havanaensis, B. quadridentatus, Keratella valga, Lecane bulla, L. lunaris, Polyarthra vulgaris, Testudinella patina, Rotaria* sp., *Alonella granulata, Mesocyclops longisetus, Paracyclops* sp. and *Thermocyclops* sp., were associated with habitats with higher pH, whereas *Plationus patulus, Lecane hastata, L. leontina, L. luna, Euchlanis dilatata, Lepadella* sp., *Ceriodaphnia cornuta, Diaphanosoma spinulosum, Chydorus eurynotus,* and *Macrothrix* sp., were more abundant

in habitats with greater water transparency. Pool area and dissolved oxygen explained substantial part in the distribution and abundance of *Trichocerca* sp., *Brachionus angularis, B. calyciflorus, B. plicatilis, B. urceolaris, Hexarthra mira, Pompholyx sulcata, Filinia longiseta, Polyarthra dolicoptera, Asplanchna sieboldi, Keratella cochlearis, Notodiaptomus* sp., and naupliar stages of Copepoda. The effects of pool area and the other environmental variables have been viewed as important to species' richness [64], even though the effects of these variables are difficult to be evaluated separately. Hessen et al. [65], for instance, pointed out that primary production and predation have greater effect on the zooplankton community than external factors such as area and geographic limits. Furthermore, physical and chemical variables in dryland rivers (e.g., conductivity and water transparency) have been reported to be associated with larger scale or geomorphologic processes [66] or with other variables operating at lower spatial scales (e.g., dissolved oxygen) [5].

5. Conclusions

This study showed that the zooplankton presents a spatially segregated pattern of species composition across river reaches and that at low spatial scales (among pools or different habitats within pools) the diversity of species is likely to be affected by temporal changes in physical and chemical characteristics. As a consequence of the drying of pool habitats, the spatial heterogeneity within the study river reaches has the potential to increase local β diversity during the dry season by creating patchier assemblages. This spatial segregation in community composition and the patterns of partition of the diversity across the spatial scales leads to a higher total diversity in intermittent streams, compared to less variable environments.

Acknowledgments

The authors are grateful to Professor. Maria Cristina Crispim (Universidade Federal da Paraíba) for confirming the identification of zooplankton. Thais Melo is grateful to the "Programa de Iniciação Científica UEPB/CNPq" for the scholarship granted (PIBIC/CNPq/UEPB 2010-2011). Elvio Medeiros holds a Brazilian Research Council (CNPq) Research Productivity Grant (312028/2012-1). This research was supported by funds from Edital 01/2008 PRPGP/UEPB - PROPESQ 007/2008.

References

[1] L. Maltchik and M. Florin, "Perspectives of hydrological disturbance as the driving force of Brazilian semiarid stream ecosystems," *Acta Limnologica Brasiliensia*, vol. 14, no. 3, pp. 35–41, 2002.

[2] W. T. Liu and R. I. Negrón Juárez, "ENSO drought onset prediction in northeast Brazil using NDVI," *International Journal of Remote Sensing*, vol. 22, no. 17, pp. 3483–3501, 2001.

[3] E. R. Steffan, "Hidrografia," in *Região Nordeste. Geografia Do Brasil*, IBGE, Ed., pp. 111–133, SERGRAF-IBGE, Rio de Janeiro, Brazil, 1977.

[4] L. Maltchik and E. S. F. Medeiros, "Conservation importance of semi-arid streams in north-eastern Brazil: implications of hydrological disturbance and species diversity," *Aquatic Conservation: Marine and Freshwater Ecosystems*, vol. 16, no. 7, pp. 665–677, 2006.

[5] E. S. F. Medeiros, M. J. Silva, and R. T. C. Ramos, "Application of catchment- and local-scale variables for aquatic habitat characterization and assessment in the Brazilian semi-arid region," *Neotropical Biology and Conservation*, vol. 3, no. 1, pp. 13–20, 2008.

[6] E. H. Stanley, S. G. Fisher, and N. B. Grimm, "Ecosystem expansion and contraction in streams: desert streams vary in both space and time and fluctuate dramatically in size," *BioScience*, vol. 47, no. 7, pp. 427–435, 1997.

[7] S. G. Fisher, N. B. Grimm, E. Martí, and R. Gómez, "Hierarchy, spatial configuration, and nutrient cycling in a desert stream," *Austral Ecology*, vol. 23, no. 1, pp. 41–52, 1998.

[8] M. C. Crispim and G. T. P. Freitas, "Seasonal effects on zooplanktonic community in a temporary lagoon of northeast Brazil," *Acta Limnologica Brasiliensia*, vol. 17, no. 4, pp. 385–393, 2005.

[9] E. S. F. Medeiros and L. Maltchik, "Diversity and stability of fishes (Teleostei) in a temporary river of the Brazilian semiarid region," *Iheringia*, no. 90, pp. 157–166, 2001.

[10] L. De Meester, S. Declerck, R. Stoks et al., "Ponds and pools as model systems in conservation biology, ecology and evolutionary biology," *Aquatic Conservation*, vol. 15, no. 6, pp. 715–725, 2005.

[11] E. S. F. Medeiros, M. P. Noia, L. C. Antunes, and T. X. Melo, "Zooplankton composition in aquatic systems of semi-arid Brazil: spatial variation and implications of water management," *Pan-American Journal of Aquatic Sciences*, vol. 6, pp. 290–302, 2011.

[12] D. G. Jenkins and A. L. Buikema Jr., "Do similar communities develop in similar sites? A test with zooplankton structure and function," *Ecological Monographs*, vol. 68, no. 3, pp. 421–443, 1998.

[13] C. E. Cáceres and D. A. Soluk, "Blowing in the wind: a field test of overland dispersal and colonization by aquatic invertebrates," *Oecologia*, vol. 131, no. 3, pp. 402–408, 2002.

[14] J. L. Brooks and S. I. Dodson, "Predation, body size, and composition of plankton," *Science*, vol. 150, no. 3692, pp. 28–35, 1965.

[15] R. Pourriot, C. Rougier, and A. Miquelis, "Origin and development of river zooplankton: example of the Marne," *Hydrobiologia*, vol. 345, no. 2-3, pp. 143–148, 1997.

[16] J. H. Stansfield, M. R. Perrow, L. D. Tench, A. J. D. Jowitt, and A. A. L. Taylor, "Submerged macrophytes as refuges for grazing Cladocera against fish predation: observations on seasonal changes in relation to macrophyte cover and predation pressure," *Hydrobiologia*, vol. 342-343, pp. 229–240, 1997.

[17] M. Seminara, D. Vagaggini, and F. G. Margaritora, "Differential responses of zooplankton assemblages to environmental variation in temporary and permanent ponds: zooplankton of temporary and permanent ponds," *Aquatic Ecology*, vol. 42, no. 1, pp. 129–140, 2008.

[18] A. C. B. Vieira, L. L. Ribeiro, D. P. N. Santos, and M. C. Crispim, "Correlation between the zooplanktonic community and environmental variables in a reservoir from the Northeastern semiarid," *Acta Limnologica Brasiliensia*, vol. 21, no. 3, pp. 349–358, 2009.

[19] W. Sousa, J. L. Attayde, E. D. S. Rocha, and E. M. Eskinazi-Sant'Anna, "The response of zooplankton assemblages to variations in the water quality of four man-made lakes in semi-arid northeastern Brazil," *Journal of Plankton Research*, vol. 30, no. 6, pp. 699–708, 2008.

[20] S. Tavernini, "Seasonal and inter-annual zooplankton dynamics in temporary pools with different hydroperiods," *Limnologica*, vol. 38, no. 1, pp. 63–75, 2008.

[21] K. M. Jenkins and A. J. Boulton, "Connectivity in a dryland river: short-term aquatic microinvertebrate recruitment following floodplain inundation," *Ecology*, vol. 84, no. 10, pp. 2708–2723, 2003.

[22] R. L. Farias, L. K. Carvalho, and E. S. F. Medeiros, "Distribution of Chironomidae in a semiarid intermittent river of Brazil," *Neotropical Entomology*, vol. 41, pp. 450–460, 2012.

[23] L. G. Rocha, E. S. F. Medeiros, and H. T. A. Andrade, "Influence of flow variability on macroinvertebrate assemblages in an intermittent stream of semi-arid Brazil," *Journal of Arid Environments*, vol. 85, pp. 33–40, 2012.

[24] C. A. Frissell, W. J. Liss, C. E. Warren, and M. D. Hurley, "A hierarchical framework for stream habitat classification: viewing streams in a watershed context," *Environmental Management*, vol. 10, no. 2, pp. 199–214, 1986.

[25] C. M. Pringle, C. M. Pringle, R. J. Naiman et al., "Patch dynamics in lotic systems: the stream as a mosaic," *Journal of the North American Benthological Society*, vol. 7, pp. 503–524, 1988.

[26] M. C. Thoms, P. J. Beyer, and K. H. Rogers, "Variability, complexity and diversity—the geomorphology of river ecosystems in dryland regions," in *Changeable, Changed, Changing: The Ecology of Desert Rivers*, R. T. Kingsford, Ed., p. 368, Cambridge University Press, Cambridge, UK, 2004.

[27] N. R. Simões, S. L. Sonoda, and S. M. M. S. Ribeiro, "Spatial and seasonal variation of microcrustaceans (Cladocera and Copepoda) in intermittent rivers in the Jequiezinho River Hydrographic Basin, in the Neotropical semiarid," *Acta Limnologica Brasiliensia*, vol. 20, no. 3, pp. 197–204, 2008.

[28] E. S. F. Medeiros and A. H. Arthington, "Allochthonous and autochthonous carbon sources for fish in floodplain lagoons of an Australian dryland river," *Environmental Biology of Fishes*, vol. 90, no. 1, pp. 1–17, 2011.

[29] AESA, "Agência Executiva de Gestão das Águas do Estado da Paraíba. Disponível em," Paraíba, Brasil, 2012, http://www.aesa.pb.gov.br/.

[30] I. R. Leal, M. Tabarelli, and J. M. C. Silva, Eds., *Ecologia e Conservação da Caatinga*, EDUFPE, Recife, Brazil, 2nd edition, 2005.

[31] M. C. Peel, B. L. Finlayson, and T. A. McMahon, "Updated world map of the Köppen-Geiger climate classification," *Hydrology and Earth System Sciences*, vol. 11, no. 5, pp. 1633–1644, 2007.

[32] SUDENE, *Dados Pluviométricos mensais do nordeste. Superintendência do Desenvolvimento do Nordeste*, Recife, Brazil, 1990.

[33] J. M. C. Silva, M. Tabarelli, M. Fonseca, and L. Lins, Eds., *Biodiversidade da Caatinga: áreas e ações prioritárias para a conservação*, Ministério do Meio Ambiente/Universidade Federal de Pernanbuco, Brasília, DF, Brazil, 2003.

[34] J. F. Haney and D. J. Hall, "Sugar-coated Daphnia: a preservation technique for Cladocerans," *Limnology and Oceanography*, vol. 18, no. 2, pp. 331–333, 1973.

[35] L. M. A. Elmoor-Loureiro, *Manual de identificação de cladóceros límnicos do Brasil*, Editora Universa, Brasilia, DF, Brazil, 1997.

[36] E. Suárez, J. W. Reid, T. M. Ilige et al., *Catálogo de los copépodos (Crustácea) continentales de la península de Yucatán, México*, ECOSUR/CONABIO, Mexico City, Mexico, 1996.

[37] H. Segers, "Annotated checklist of the rotifers (Phylum Rotifera), with notes on nomenclature, taxonomy and distribution," *Zootaxa*, no. 1564, pp. 1–104, 2007.

[38] P. S. Maitland, *Field Studies: Sampling in Freshwaters, in Biology of Fresh Waters*, Blackie, Glasgow, UK, 1990.

[39] R. R. Sokal and F. J. Rohlf, *Biometry: The Principles and Practice of Statistics in Biological Research*, W.H. Freeman, San Francisco, Calif, USA, 1969.

[40] J. A. Veech and T. O. Crist, "PARTITION: software for hierarchical partitioning of species diversity, version 3.0," 2009, http://www.users.muohio.edu/cristto/partition.htm.

[41] T. O. Crist, J. A. Veech, J. C. Gering, and K. S. Summerville, "Partitioning species diversity across landscapes and regions: a hierarchical analysis of α, β, and γ diversity," *American Naturalist*, vol. 162, no. 6, pp. 734–743, 2003.

[42] J. A. Veech and T. O. Crist, "PARTITION 3. 0 user's manual," unpublished document, 2009.

[43] B. McCune and J. B. Grace, *Analysis of Ecological Communities*, Gleneden Beach, Ore, USA, MjM Software Design, 2002.

[44] M. E. Biondini, C. D. Bonham, and E. F. Redente, "Secondary successional patterns in a sagebrush (Artemisia tridentata) community as they relate to soil disturbance and soil biological activity," *Vegetatio*, vol. 60, no. 1, pp. 25–36, 1985.

[45] M. Dufrêne and P. Legendre, "Species assemblages and indicator species: the need for a flexible asymmetrical approach," *Ecological Monographs*, vol. 67, no. 3, pp. 345–366, 1997.

[46] B. McCune and M. J. Mefford, *PC-ORD. Multivariate Analysis of Ecological Data. Version 4.27*, MjM Software Design, Gleneden Beach, Ore, USA, 1999.

[47] L. Maltchik, L. E. K. Lanés, C. Stenert, and E. S. F. Medeiros, "Species-area relationship and environmental predictors of fish communities in coastal freshwater wetlands of southern Brazil," *Environmental Biology of Fishes*, vol. 88, no. 1, pp. 25–35, 2010.

[48] M. C. Crispim and T. Watanabe, "Caracterização Limnológica das Bacias doadoras e receptoras de águas do Rio São Francisco:1—Zooplâncton," *Acta Limnologica Brasiliensia*, vol. 12, no. 2, pp. 93–103, 2000.

[49] V. L. S. Almeida, M. E. L. de Larrazábal, A. D. N. Moura, and M. de Melo Júnior, "Rotifera das zonas limnética e litorânea do reservatório de Tapacurá, Pernambuco, Brasil," *Iheringia*, vol. 96, no. 4, pp. 445–451, 2006.

[50] K. Fahd, L. Serrano, and J. Toja, "Crustacean and rotifer composition of temporary ponds in the Doñana National Park (SW Spain) during floods," *Hydrobiologia*, vol. 436, pp. 41–49, 2000.

[51] J. Green, "Diversity and dominance in planktonic rotifers," *Hydrobiologia*, vol. 255-256, no. 1, pp. 345–352, 1993.

[52] J. D. Allan, "Life history patterns in zooplankton," *American Naturalist*, vol. 110, no. 971, pp. 165–176, 1976.

[53] K. G. Porter, J. D. Orcutt Jr., and J. Gerritsen, "Functional response and fitness in a generalist filter feeder Daphnia magna (Cladocera: Crustacea)," *Ecology*, vol. 64, no. 4, pp. 735–742, 1983.

[54] N. Walz and M. Welker, "Plankton development in a rapidly flushed lake in the River Spree system (Neuendorfer See, Northeast Germany)," *Journal of Plankton Research*, vol. 20, no. 11, pp. 2071–2087, 1998.

[55] M. C. Crispim and T. Watanabe, "What can dry reservoir sediments in a semi-arid region in Brazil tell us about cladocera?" *Hydrobiologia*, vol. 442, no. 1–3, pp. 101–105, 2001.

[56] D. Frisch and A. J. Green, "Copepods come in first: rapid colonization of new temporary ponds," *Fundamental and Applied Limnology*, vol. 168, no. 4, pp. 289–297, 2007.

[57] G. A. Cole, "Contrasts among calanoid copepods from permanent and temporary ponds in Arizona," *American Midland Naturalist*, vol. 76, no. 2, pp. 351–368, 1966.

[58] F. Sheldon and K. F. Walker, "Spatial distribution of littoral invertebrates in the lower Murray-Darling River system, Australia," *Marine and Freshwater Research*, vol. 49, no. 2, pp. 171–182, 1998.

[59] K. E. Esteves and S. Sendacz, "Relações entre a biomassa do zooplâncton e o estado trófico de reservatórios do estado de São Paulo," *Acta Limnologica Brasiliensia*, vol. 2, no. 1, pp. 587–604, 1988.

[60] M. Serafim-Júnior, G. Perbiche-Neves, L. de Brito, A. R. Ghidin-
      i, and S. M. C. Casanova, "Variação espaço-temporal de Rotifera
      em um reservatório eutrofizado no sul do Brasil," *Iheringia*, vol.
      100, no. 3, pp. 233–241, 2010.

[61] P. Usseglio-Polatera, "Theoretical habitat templets, species
      traits, and species richness: aquatic insects in the Upper Rhone
      River and its floodplain," *Freshwater Biology*, vol. 31, pp. 417–437,
      1994.

[62] J. A. Veech, T. O. Crist, and K. S. Summerville, "Intraspecific
      aggregation decreases local species diversity of arthropods,"
      Ecology, vol. 84, no. 12, pp. 3376–3383, 2003.

[63] D. J. Murrell, D. W. Purves, and R. Law, "Uniting pattern and
      process in plant ecology," *Trends in Ecology and Evolution*, vol.
      16, no. 10, pp. 529–530, 2001.

[64] R. E. Ricklefs and I. J. Lovette, "The roles of island area per se
      and habitat diversity in the species-area relationships of four
      Lesser Antillean faunal groups," *Journal of Animal Ecology*, vol.
      68, no. 6, pp. 1142–1160, 1999.

[65] D. O. Hessen, B. A. Faafeng, V. H. Smith, V. Bakkestuen, and
      B. Walseng, "Extrinsic and intrinsic controls of zooplankton di-
      versity in lakes," *Ecology*, vol. 87, no. 2, pp. 433–443, 2006.

[66] J. C. Marshall, F. Sheldon, M. Thoms, and S. Choy, "The
      macroinvertebrate fauna of an Australian dryland river: spa-
      tial and temporal patterns and environmental relationships,"
      Marine and Freshwater Research, vol. 57, no. 1, pp. 61–74, 2006.

15

Extrapolative Estimation of Benthic Diatoms (Bacillariophyta) Species Diversity in Different Marine Habitats of the Crimea (Black Sea)

A. N. Petrov and E. L. Nevrova

Institute of Biology of the Southern Seas, National Academy of Sciences of Ukraine, Sevastopol 99011, Ukraine

Correspondence should be addressed to A. N. Petrov; alexpet-14@mail.ru

Academic Editor: Rafael Riosmena-Rodríguez

Benthic diatoms species richness was analyzed based on 93 samples collected at 8 areas of Crimea (Black Sea) on sandy/muddy bottoms within depth range 6–48 m. Totally 433 species were found. Expected species richness S_{exp} was estimated by application of Jack-knife -1 and -2, Chao-2, and Karakassis-S_∞ estimators. Magnitude of S_{exp}, resulted from S_∞, displayed the most similar values to the observed species number (S_{obs}). Overestimation of S_{obs} (10–13%) occurred for small number of samples (<12), and slight underestimation (3–5%) occurred when sample numbers exceeded 40–43. The other estimators gave large overestimated results (Chao—from 21 to 70% higher than S_{obs}, Jack-knife—23–58%). The relationship between number of samples (X) and number of observed species (Y) was calculated considering all 93 samples: $Y = 79.01 \ln(x) + 34.95$. Accordingly, not less than 10 samples are required for disclosing about 50% of the total species richness (433); to detect 80% (347 species) not less than 46 samples should be considered. Different configurations of S_∞ method were applied to optimize its performance. The most precise results can be achieved when the calculation of the S_{exp} is based on sequences of randomized samples with sampling lags of 10 to 15.

1. Introduction

Species richness is an essential attribute of a biological community and a widely used surrogate for the more complex concept of biological diversity. Quantitative change in species richness is an important characteristic underlying many biotic indices and integral assessment of community structure and condition in relation to habitat [1–4].

In the ecological study of benthic diatoms (Bacillariophyta), effective comparative assessment of the species structure of taxocene in various habitats, including protected marine areas, is a key problem that is important also to the establishment of conservation priorities. Therefore, the reliability and deviation of species richness measurements is one of the essential methodical tasks of diatomology that has been insufficiently studied as yet. The diatoms species richness may differ considerably even in adjacent sea bottom areas because of diverse environmental conditions and spatial microdistribution pattern of microalgae. In prognostic

assessment of species richness and diversity of taxocene in heterogeneous biotopes, it is methodologically important to determine the relationship between sampling effort and the number of species found in these samples. Hypothetically, the larger is the number of samples, the larger is the number of detected species. In practice, however, only a reasonable minimum of samples, usually on the researcher's request, is collected from a sampling site because of constraints inherent in sampling effort and further cameral treatment. The total number of benthic diatoms samples taken from a studying area is usually confined to 15–20 samples; however, often the species structure and diversity of a taxocene are assessed from only 3–5 samples [5–7]. Certainly, species composition of such microobjects as diatoms can rarely be completely determined for a sampling site even given a sufficient number of samples and their exhaustive taxonomical examination. In this case, prognostic algorithms (estimators) can provide a tool for estimation of expected species richness in different taxonomical groups of benthos [2, 8, 9]. It should be noted that application in our study the several widely used estimators for prognostic assessment of benthic diatoms species richness is one of the very few examples of such studies in diatomology [10, 11].

The objectives of the study were (1) to implement comparative prognostic estimation of the expected species richness of benthic diatoms for several near-shore sampling sites of the SW Crimea (Black Sea) and to evaluate the precision of each estimator used; (2) to derive and statistically estimate a generalized optimal ratio between an essential minimum of sampling effort and a maximum of relevant data on the taxocene species richness in the marine coasts of the Crimean peninsula.

2. Material and Methods

In prognostic estimation of the expected number of species to be found through examination of a certain number of samples (n), the records of benthic diatoms species composition from 93 samples were used. Material for the investigation was collected in 1996–2009 during the summer or autumn season from soft sediments (muddy sand) within the depth range 6–48 m at several sampling sites (or sampling areas) in Sevastopol, Balaklava, Karantinnaya, and Laspi bays and at the open coast Belbek (SW Crimea, Black Sea) (Figure 1).

Sediment samples were collected from soft-bottom substrate either using a Petersen grab-corer (at the most deep places) or by a diver using a hand-corer. Samples for diatom analysis (duplicate from the each station) were retrieved from the uppermost 2-3 cm layer of each sampled sediment bulk with the meiobenthic tube (surface area 15.9 cm^2). For better separation of epipelon and epipsammon, the sediment samples had been preliminary treated in an ultrasonic bath for 20 min; later samples were refined using the standard technique of cold burning in HCl and H_2SO_4 with the addition of $K_2Cr_2O_7$ [12]. Cleaned diatom valves were mounted using mountant of Eljashev for light microscopy and later examined for abundance and species richness. Diatom cells were quantified under the microscope (×400) in Goryayev chamber (7×10^{-3} cm^3), in three random replicates from each sample; later cells number (abundance) was recalculated per 1 cm^2 substrate surface as average value from three replicates.

The check-list of benthic diatoms was compiled for each of the sampling sites (see Supplementary Material available online at http://dx.doi.org/10.1155/2013/975459). The diatom species were counted and abundance values expressed per 1 cm^2 of seabed. The rated minimum abundance of a species in the samples was estimated 250 cell·cm^{-2}. Species not found in Goryayev chambers but registered only on permanent slides (i.e., rare or solitary species) were included in the rectangular matrix (species density versus samples number) as a conventional minimum value of 10 cells cm^{-2} for quantitative uni- and multivariate statistical analyses. Such values (not equal to 1) were used for the preliminary procedure of fourth-root transformation of the extensively ranged values of initial diatom abundance (250 to 3.56·10^6 cell·cm^{-2}) under further calculation of the similarity and diversity indices [13]. The

complete list of diatom species for each sample was identified to intraspecific level on permanent slides using the microscopes Zeiss Axiostar Plus and Nikon Eclipse E600 (×1000). The species were identified using the taxonomic atlases [14–20]. Afterwards, real species richness was compared with the expected estimates yielded by computation methods.

In the comparative prognostic estimation of expected species number we used commonly applied Chao and Jack-knife estimators [2, 9, 21, 22] and estimator S_∞ based on algorithms of regression analysis [23, 24].

The latter method implies that computation of a maximum expected number of species (S_{exp}) relies on the determination of a theoretical upper limit (asymptote) for the species-accumulation curve plotted from averages derived from many random permutations when two successive samples contain identical number of species using infinitely large number of samples. The expected number of species, that is, the asymptote magnitude, is calculated through solving the linear equation of the relationship between the ultimate species numbers accumulated in K samples ($S_{obs(k)}$) and in $k + 1$ samples ($S_{obs(k+1)}$) against parameters of equation $Y = X$ which is the bisecting line of 1st coordinate quarter. It was suggested to develop the estimator algorithm, as it has been done in our work, by taking into account the different sampling lag widths between the pair samples along their original sequence, that is, by constructing a series of regression equations $S_{obs(k)} = f(S_{obs(k+n)})$, where $k = 1, \ldots, n-1$ for sampling lags of different extension [25]. Such methodical amendments, though requiring far larger number of samples, enable more precise estimation of maximum expected species number [9].

The assessment of expected species richness by two other estimators, Chao-1 and Chao-2, involves relatively small number of samples [21, 24, 26]. Both the estimators are calculated by the formula $S_{total} = S_{obs} + (a^2/2b)$, where S_{total} is the total predicted species richness, S_{obs} is the number of species observed in the examined batch of samples, and a is the number of species represented by one individual (singleton species; Chao-1) or the number of species observed in only one sample (unique species; Chao-2). Coefficient b is the number of species represented by 2 individuals (Chao-1) or the number of species registered in only 2 samples (Chao-2). Since in our samples the admitted minimum of diatom cells was 10 cell·cm^{-2}, the curve of Chao-1 estimator overlaps the cumulative curve of detected species number; that is, $S_{total} = S_{obs}$; and therefore only estimator Chao-2 was used in the analysis.

The Jack-knife estimators rely on the record of expected number of rare species.

$S_{total} = S_{obs} + Q \cdot (m - 1/m)$, where Q is the number of species found only once in the studied samples and m is the total number of samples [2, 27]. This estimator performs effectively when relatively small number of samples is processed; it has been successfully applied in analysis of data sets pertaining to marine benthos [9, 25].

Two statistics, relative error (RE) and squared relative deviation (SRD), were used to evaluate the precision of the estimators relying on the deviation of the expected species number from the real number contained in a finite set of

Extrapolative Estimation of Benthic Diatoms (Bacillariophyta) Species Diversity in Different Marine
Habitats of the Crimea (Black Sea)

129

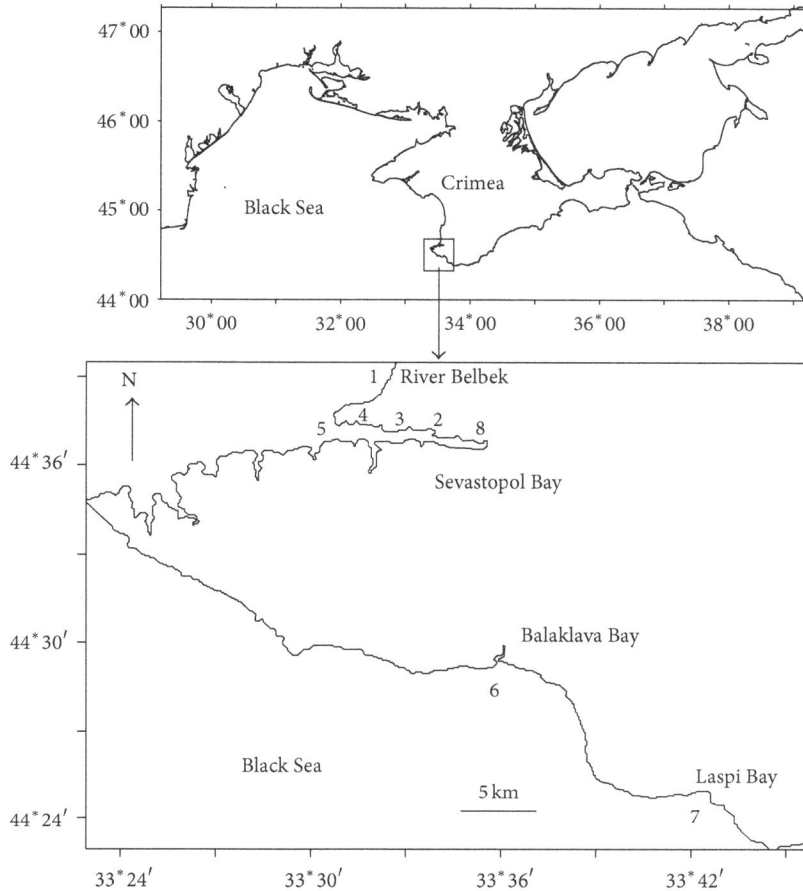

FIGURE 1: Schematic map of benthic diatoms sampling sites along the coast of SW Crimea. Designations of sampling sites: 1—Belbek; 2–4—parts of Sevastopol Bay (2—inner, 3—central, and 4—external); 5—Karantinnaya Bay; 6—Balaklava Bay; 7—Laspi Bay; 8—Inkerman.

samples, that is, over- and underestimation of real species richness: $RE = (S_{exp} - S_{obs})/S_{obs}$, where S_{exp} is the expected number of species determined using the estimator, and S_{obs} is the observed value of species richness computed from the upper limit of the species accumulation asymptote for the range of samples from 1 to n, multiple permutations taken into the account [28]. RE estimates relative difference between the value estimated and the true value of the species number under examination of different multitudes of samples. The square of RE (SRD) assesses the closeness of the estimator to the real number of species regardless of the deviation sign, that is, the measure of estimator inaccuracy [29].

In this study, we compared the performance of 4 estimation methods using real and simulated data sets. The reliability of each method was evaluated by calculating the bias and precision of its estimates against the known total diatom species richness. These two metrics allow an objective quantitative comparison of the performance of estimation methods. Bias measures whether an estimate consistently under- or overestimates the parameter. Precision measures

the overall closeness of simulated curve to the true number of species along the overall succession of samples:

$$\text{Bias} = \sum \left(\frac{(Ej - Aj)}{(Aj \cdot n)} \right), \qquad \text{Precision} = \sum \left(\frac{(Ej - Aj)^2}{(Aj^2 \cdot n)} \right),$$

(1)

where $j = 1$ to $j = n$, and n is the number of examined samples. E_j is the species richness as extrapolated by the respective estimation method and A_j is the asymptote of the species richness accumulation curve for j samples [22]. It is implied that a "good" estimator should have bias values close to zero and small precision values. Another measure of bias is the percentage of overestimates. If the estimator always overestimates A_j, it will have positive bias and 100% overestimates, and if it always underestimates A_j, it will have negative bias and 0% overestimates. An unbiased estimator returns zero bias and 50% overestimates [8].

Rarefaction method [30, 31] is an important diagnostic tool that consists in the plot of randomized richness against the sampling intensity used in comparing diatom species richness from different samples. The rarefaction (numerical species richness) index ($ES_{(n)}$) is based on different modes of species accumulation values in a large number of hypothetical subsamples with various numbers of diatom cells (10, 20,..., 500, etc.) having been repeatedly randomly selected from the whole sample, so that the variance among randomizations remains meaningful for large number of sampling units or individuals.

Multivariate analysis of diatom assemblage species structure was conducted using the PRIMER v5.2 software package [32]. Affinity of assemblage composition between sampling areas was estimated on ranked triangular similarity matrices based on the Bray-Curtis index on fourth-root transformed initial diatom abundance data. Results from nonmetric multidimensional scaling ($nMDS$) were used for a graphical representation of possible similarities between groups of samples (sampling sites) according to similarity of diatom taxocene species structure. Possible differences between sampling sites were tested for significance using analysis of similarity (ANOSIM). Smoothed species accumulation curves for each sampling sites were generated using 1000 random permutations. PRIMER's DIVERSE routine was used to calculate the number of individuals, number of species, values of Chao and Jack-knife estimators and rarefaction indices $ES(n)$ for each sample. Means of indices were then calculated for all data sets and various subsets of samples. Computation of data (S_{exp} values averaged over 1000 randomized runs) for Karakassis-S_∞ extrapolative model was performed also using DIVERSE routine with further calculations of regression equations using MS Excell.

3. Results and Discussion

The relationship between the observed species richness of benthic diatoms and the number of samples was estimated using the records from 93 samples taken in 8 near-shore seawater areas of the SW Crimea. Microscopic analysis revealed altogether of 433 species and intraspecific taxa (Annex 1), pooled in 96 genera, 51 families, 27 orders, and 3 classes of Bacillariophyta (Table 1). Species richness was highest for genera *Nitzschia* Hassall (53 species and intraspecific taxa), *Amphora* Ehrenberg (41), *Navicula* Bory (37), *Cocconeis* Ehrenberg (26), and *Diploneis* Ehrenberg ex Cleve (20). Aulacoseirales, Biddulphiales, Eunotiales, Paraliales, Rhabdonematales, Thalassionematales, and Toxariales were the most species-poor orders, where only one species recorded in each.

The list of species was compiled for each sampling sites (see Annex 1). These data have supplemented the created taxonomic base of the Black Sea diatom flora [33], based on the literature and own data ([11, 15, 16, 34, 35], etc.) According to this base, updated inventory of Black Sea benthic diatoms from 5 regions (Caucasian, Crimean, Bulgarian, Romanian coasts, and North-Western shelf) holds 1093 species and infraspecific taxa (ssp.), pooled in 942 species, 142 genera, 60 families, 32 orders and 3 classes, following the recent systems [17, 19, 20, 36]. The latest check-list of entire Crimean coast includes 886 sp. and ssp., belonging to 800 species, 130

TABLE 1: Representativeness of benthic diatoms (Bacillariophyta) at 8 investigated sampling sites in SW Crimea.

Class	Order	Family	Genus	Species	Intraspecific taxa
Coscinodiscophyceae	7	13	20	40	42
Fragilariophyceae	8	9	18	36	38
Bacillariophyceae	10	27	58	322	353
In total	25	49	96	398	433

genera, 55 families, and 29 orders [33]. However, previous studies of benthic diatoms diversity at Crimean shores were rather episodic and nonnumerous. Most of them had covered only spatially confined locations (e.g., one small bay [34]), only a few interseasonal samples in one point [37] or combined retrospective nonquantitative data on species richness throughout rather enlarged water area [15]. Therefore, their results have not provided a comprehensive data on Crimean diatoms diversity which could be considered as a quite exhaustive base on species wealth for evaluation of estimator's accuracy (ratio S_{obs}/S_{exp}) in our study. For comparison, in previous studies have performed in the Sevastopol region by various researchers, was found 93 sp. and ssp. of benthic diatoms [34], 136 sp. & ssp. [37], 161 sp. and ssp. [15].

Thus, the number of benthic diatoms species found in all our samples altogether (433) was much greater than in previous check-lists and accounted about 40% of revealed species richness for the Black Sea and almost of 50% of the total registered benthic diatoms diversity for the Crimean coast.

Sevastopol Bay (8.3 km^2) was divided into 3 parts: inner, central, and outer, with conspicuously different environmental parameters such as depth, grain-size composition, pH, Eh, O_2 concentration in near-bottom layers, and the industrial pollution level of bottom sediments with trace metals and organic pollutants such as PCBs, PAHs, and pesticides [38]. Such spatial division of the bay bottom area was based on the earlier obtained results on the assessment of key abiotic factors impact on the diatom taxocene structure in different part of Sevastopol Bay [13].

Results of $nMDS$ ordination (stress = 0.19) appeared to confirm the visual separation of Belbek samples from the others and the rather close interarrangement of the samples from Sevastopol and Balaklava bays and samples from Laspi and Karantinnaya bays (at 25% similarity level) (Figure 2). Visual differences in samples' interposition patterns on 2D plot were then statistically proved by the one-way ANOSIM test. The results indicated significant differences in the taxocene structure among almost all compared groups of samples (R_{global} = 0.672, $P < 0.001$), excepting pairs "Karantinnaya Bay versus Laspi Bay" and "Sevastopol central part versus Sevastopol outer part," where pairwise tests did not show significant differences ($R_{pairwise}$ = 0.356, $P < 0.2$).

Diverse environmental conditions and statistically significant differences in diatom taxocene structure between the compared sites are the prerequisites for comparative analysis of habitat-related differences in relationships between

Extrapolative Estimation of Benthic Diatoms (Bacillariophyta) Species Diversity in Different Marine
Habitats of the Crimea (Black Sea)

131

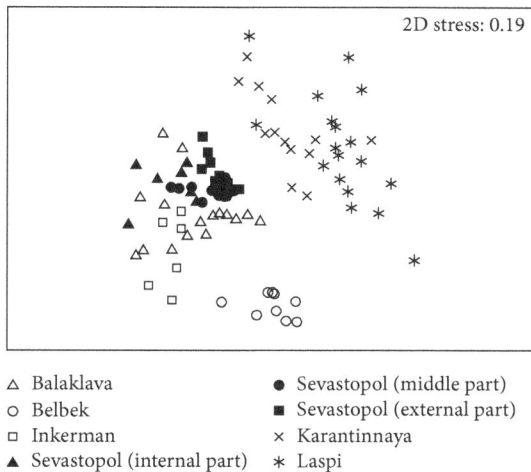

FIGURE 2: MDS ordination plot of all 93 samples (based on double square-root transformed abundance similarity matrice). Samples from eight different sampling sites are indicated on the plot by labels.

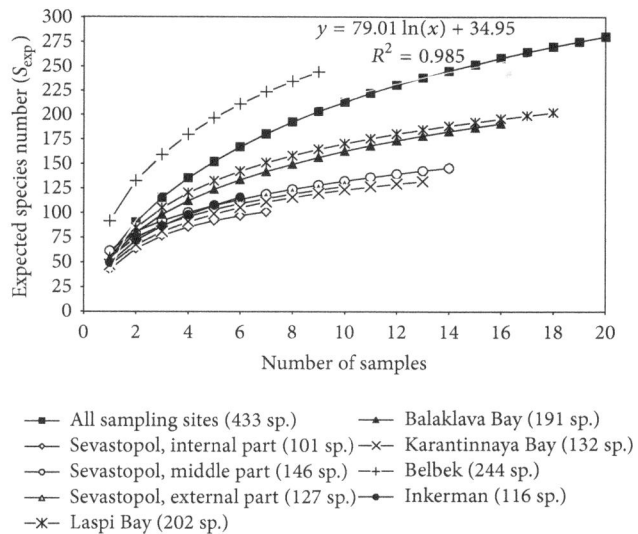

FIGURE 3: Cumulative randomized sequences of S_{exp} constructed for 8 sampling areas with different number of samples (6 to 18) as well as generalized species-accumulation curve (solid line) combining all 93 samples taken in SW Crimea.

number of samples and revealed species richness. On the other hand, such habitat-specific distinctions in species-accumulation pattern can take into account the variability of biotopes and, consequently, to improve the reliability of deductions under the most generalized model "sampling effort versus species richness" for the whole studied region (Crimea). Results based on this generalized region-specific curve can be applied for comparative interregional analysis of relationships between species richness, and sampling effort. Hence, the subsequent analysis of the diatom species-accumulation curves was performed both for each of 8 sampling sites and for the entire sequence of all 93 samples.

The number of samples within each of the sampling sites, the observed diatom species richness and the expected number of species assessed by different estimators are given in Table 2.

Application of estimators presumes that the prognostic estimation of species richness should overestimate the observed species number in the samples (S_{obs}) that conform to data in Table 2. The expected diatoms species richness (S_{exp}) estimated by the S_∞ method slightly overestimated (1–8%) the S_{obs} value for different sampling sites. The exception is Inkerman, where S_{exp} is about 18% as large as S_{obs} probably because only 6 samples were collected. Other estimators more considerably overestimated the S_{obs} values: Chao 21–70% and JN-1 24–36% and JN-2 33–58%, depending on sampling effort in each sampling site.

Habitat-dependent relationships of accumulation of new species (S_{exp}) with increasing sampling efforts were derived at each of the sampling sites (Figure 3). The average S_{exp} values were computed by 1000-fold randomized runs for different numbers of samples.

The most rapid rise of the S_{exp} with increasing number of samples (species accumulation curve) was observed in the Belbek area (open coast) where the total number of diatom species detected from 9 samples was 244, that is, 56% of the total list of diatom species registered at all 93 samples.

The accumulation curves corresponding to other sampling sites were more flat and, despite larger number of examined samples, showed a lesser number of observed species. Similar relationships were obtained between species accumulation and increasing number of samples derived for the sea bottom areas in Laspi Bay and Balaklava Bay, as well as for Sevastopol Bay, Karantinnaya Bay, and Inkerman. The latter set of 3 accumulation curves displays a similar mode though the number of taken samples and the number of diatom species found in each of the three sampling areas were different (see Table 2). The accumulation curves (S_{exp}) did not arrive at horizontal asymptote at any sampling area. Such results imply that the actual number of diatom species derived from the result of examination of the largest number of samples from the sampling areas is considerably lower than the expected species richness obtained by the estimators.

The cumulative curve integrating the results of all 93 samples from 8 sampling areas is also shown in Figure 3. Based on this randomized curve, the parameters of the generalized relationship between number of samples (X) and the number of observed species (Y) were calculated. This relationship is reliably described by log-equation $Y = 79.01 \ln(x) + 34.95$ with the correlation factor $R = 0.99$.

These results suggest that not less than 10 samples, that is, nearly 11% of the total studied number (93) should be considered for disclosing about 50% of the total species richness (433 species) of benthic diatoms which actually occur on sandy/muddy sediments at the near-shore marine areas of the SW Crimea. To detect 67% (or 290 species) and 80% (347 species) of the total species richness (on assumption of equal probability to reveal any diatom species in the sample), not less than 24 and 43% of the total number of considered samples, respectively, should be examined. Obviously, for other near shore water areas in which the number of collected

TABLE 2: Number of samples, observed (S_{obs}) and expected (S_{exp}) values of benthic diatom species richness at 8 investigated sampling sites in SW Crimea. S_{exp} values are calculated using 4 estimators.

Sampling site	Date of sampling	Number of samples	Number of observed species (S_{obs})	S_∞	Chao-2	Jack-knife-1	Jack-knife-2
Balaklava Bay	14.10.2006	16	191	200.8	259.4	250.1	281.5
Karantinnaya Bay	25.08.1996	13	132	136.5	159.5	163.4	175.9
Laspi Bay	27.06.1996	18	202	204.6	249.0	253.0	275.1
Belbek	05.11.2009	9	244	269.0	330.0	321.3	359.1
Inkerman	06.11.2009	6	116	138.2	197.3	158.5	184.0
Sevastopol Bay (inner part)	11.07.2001	7	101	107.2	122.0	125.9	134.7
Sevastopol Bay (central part)	11.07.2001	14	146	152.0	194.7	185.9	207.8
Sevastopol Bay (outer part)	11.07.2001	10	127	133.2	172.1	161.2	180.2
Altogether at all sites		93	433	412.6	541.7	531.9	585.3

samples and the total number of observed species could be different, the parameters of species-accumulation curves may differ, as well.

Some other researchers also reported similar percentage of observed species compared to the expected maximum depending on the size of sampling effort. For instance, the randomized estimation of zoobenthos species richness performed on the Norwegian shelf, and in the coastal sea of Hong Kong evidenced that analysis of 12 and 16% of the total number of samples (101) disclosed up to 50% of the species richness; to elicit 80% of the totality of 809 species in Norway and of 386 species in Hong Kong, 48 and 57% of the total number of samples, correspondingly, should be studied [39]. Results of methodically similar analysis applied to the macrozoobenthos samples closely taken in the Northern Sea has shown that examination of the first 7 samples in randomized row (10% of the total number of 70 samples) disclosed 50% of all species which dwell in the sampling area; eliciting that 80% of the species richness required not less than 26 samples, or 37% of the total sampling effort [25].

In the comparative estimation of diatom species richness, we also used the rarefaction index ES(n). This index entailed estimation of the expected number of species in the sequence of conditional subsets with different number of cells (10, 20,..., 500) randomly withdrawn from the totality of diatom abundance counted in the whole sample. The rarefaction curve is a math expectancy function of "species saturation" depending on abundance of the whole community (or taxocene of diatoms in our case). The ES(n) plot corresponding to the Belbek sampling site ascends highest comparatively with plots corresponding to other sampling areas (Figure 4).

In the Belbek area, the expected number of diatom species in the conditional subsets of 200, 300, and 500 cells was estimated to be 62.1, 71.9, and 84.5, correspondingly. This provides evidence about high species saturation in this taxocene, probably owing to species brought with the Belbek river inflow and to the slightly polluted level of the bottom deposits. Estimates of the expected species number in the various-sized subsets of cell abundance were lower (and nearly identical) for diatom taxocenes in Sevastopol Bay and in Balaklava Bay. The average expected number of species

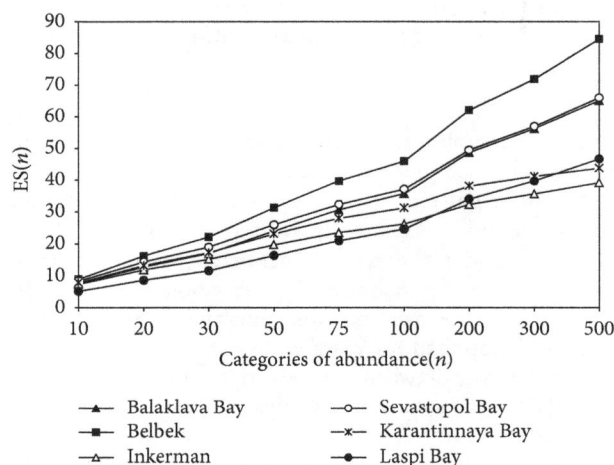

FIGURE 4: Relationships between expected number of diatoms species ES(n) and conventional abundance subsets, consisting of different numbers of cells (n) constructed for different sampling sites of SW Crimea.

in the subsets of 200, 300, and 500 cells was 49.2 ± 0.4, 56.6 ± 0.1, and 65.5 ± 0.2, respectively. The lowest species diversity subgroup includes the Laspi Bay, Karantinnaya Bay and Inkerman sampling sites in which the average expected number in the subsets of 200, 300, and 500 cells is 34.8 ± 2.3; 38.7 ± 1.9, and 43.2 ± 2.8 species, correspondingly (see Figure 4).

Hence, environmental differences, including level of pollution in a certain sampling area, can influence the relationship between new species accumulation and greater sampling effort and, eventually, the structural peculiarities of diatom taxocene.

As mentioned above, though the estimators applied in our study had generally overestimated the observed species number (S_{obs}), each of them is characterized by a different degree of deviation in S_{exp} values from S_{obs}. Such differences can be considered as accuracy measures of a certain estimator and designate its applicability for prognostic estimation of

Extrapolative Estimation of Benthic Diatoms (Bacillariophyta) Species Diversity in Different Marine
Habitats of the Crimea (Black Sea)

133

expected species richness, especially for such very abundant microobjects as benthic diatoms. Therefore, in further analysis we attempted to comparatively evaluate the accuracy of each of the applied estimators based on calculation of the expected diatom species number. Randomized accumulation curves, corresponding to 4 estimators and to simple species richness, are shown in Figure 5.

The species richness accumulation curve illustrates the relationship between S_{obs} against sampling effort. The curve monotonically increases, not converging to a horizontal asymptote at least as far as the extreme values in the entire series of 93 randomized samples. The Chao and Jack-knife estimators significantly overestimate the expected number of diatom species compared to the real species richness, especially given a small number of samples, for example, less than 10–12 samples. Beginning from this sampling effort level, the cumulative curves are plotted in parallel to the actual species accumulation curve, not approximating the horizontal asymptote along the whole randomized ascending sequence of samples number. The accumulation curve plotted for the Karakassis S_{∞} estimator is considerably closer to the curve of observed species number (S_{obs}); from the 1st sample rank to almost 40th rank in the sequence it slightly overestimates, and within the range from 40 to 93 sample rank S_{∞} underestimates the true species number for 3–5% (see Figure 5).

Earlier research applied to various groups of marine benthos found that all estimators (especially of the Chao family) often inaccurately estimate the observed species number when the sample number is small [2, 9]. When the number of samples increases, S_{exp} asymptote converges to the cumulative curve of observed species number (S_{obs}) irrespective of whether the curve S_{exp} over- or underestimates the true species richness [40]. Walther and Martin [22] also stated that not less than 30–40% of the total number of samples (about 100) in an ascending randomized series was required for adequately precise S_{exp} estimation. These authors denoted that all estimators considerably underestimated the real species number when less than 20–25% of the entire series was sampled. A reasonable precision level is when the estimator's asymptotic curve overestimates the real species richness value by not more than 20%.

Other authors [41] proved that the bootstrap, jack-knife-1 and -2 estimators can be applied to minimize underestimation of the expected species richness under comparing with actual species number in the samples. Given a small number of samples, for example, less than 25% of their total number in the randomized range, all these estimators similarly underestimate the species richness; however, the Jack-knife-2 estimator gives a lesser error. For a larger amount of samples, over 50% of the total number in the range, these estimators slightly overestimate the expected number of species and again the Jack-knife-2 estimator is more precise. Some other publications [24, 40] in which the Chao and Karakassis S_{∞} estimators were compared pointed out that both could slightly underestimate expected species richness given a large number of samples.

Such results presume that in the analysis of diatom species richness generally involving rather small number of

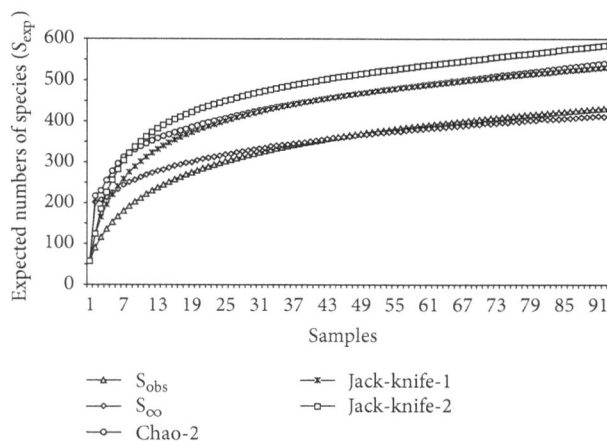

FIGURE 5: Species-accumulation curves (for randomized range of 93 samples), constructed for observed species richness (S_{obs}) and for expected diatom species number according to 4 estimators (S_{∞}, Chao-2, JN-1, and JN-2).

samples the estimator S_{∞} would estimate the expected species number (S_{exp}) close to the real level. In our study when 9–18 samples from each of the sampling sites were examined, the prognostic estimates S_{exp} were only 7–10% larger than the observed species number. In general, this degree of accuracy is acceptable with the average precision level of the estimator values for taxonomically different groups of benthos.

Considering these possible deviations, the statistical evaluation of precision of the 4 estimators was applied to various biotopes and different number of samples. It was found that S_{obs} index may vary considerably (high values of standard deviation, SD) using a relatively small number of samples (7–12). When the sample number increased to 15–18, the SD values consistently decreased, sometimes to zero, that is, narrowing the variability range of the expected species number occurred. Accuracy in the estimators was computed from the RE and SRD statistics for the overall randomized sequence of 93 samples (Figure 6).

Given a small number of samples (4–6), all the tested estimators highly overestimated the expected number of diatom species (see the peaks on the RE and SRD curves). When sample numbers increased to 15–20 or higher, the Chao and Jack-knife estimators gave lower relative error, that is, ratio ($S_{exp} - S_{obs})/S_{obs}$, and further convergence of the corresponding curves to the horizontal asymptote at level of 0.20–0.30 (RE) or 0.14–0.20 (SRD) was observed. As for the S_{∞} estimator, the RE and SRD relative errors plotted in relation to larger numbers of samples (n) showed a monotonic convergence to a zero asymptote. This estimator more or less adequately predicts the number of species in the taxocene starting from $n = 7$-8. Other authors [28] estimated species and generic richness of aquatic chironomids by 7 nonparametric estimators including the Chao and Jack-knife families and also concluded that all of these estimators may largely overestimate the expected number of species. Better accuracy could be attained only given a large number of the samples; the values of relative error (RE and SRD) were

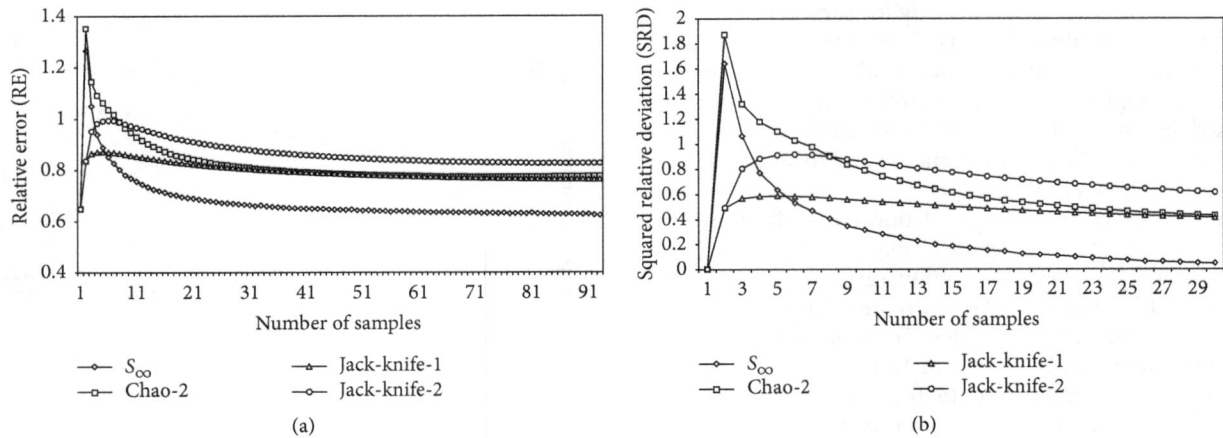

FIGURE 6: Evaluation of several estimators' inaccuracy based on relative error (RE) (a) and squared relative deviation, SRD (b) metrics. On the RE plot the data for all 93 samples are presented, while on the SRD plot only data for ranked sequence of the first 30 samples are shown (for clarity).

highest for the Chao-1 and receded from the Chao-2 to the Jack-knife-1 and to the Jack-knife-2 estimators.

A similar relationship between the accuracy of the S_{exp} averages relative to the increasing number of samples based upon standard deviation values has been yielded by the comparative estimation of benthic species richness in two areas of the Norwegian continental shelf [39]. The researchers concluded that the most precise estimates of S_{exp} (when SD estimates on the plots gradually declined to zero) could be obtained only with a sufficiently large number of samples (>20) taken from environmentally heterogeneous biotopes.

The results of the computed S_{exp}/S_{obs} ratio for the randomized sequences of samples in 4 sampling areas (Karantinnaya, Laspi, and central and outer parts of Sevastopol Bay) have shown that given a small number of samples all the estimators overestimated the S_{obs} value 1.3–1.8 times. With an increase in sampling effort, the S_{exp} gradually converges to the actual. The fact that estimators similarly evaluated S_{exp}/S_{obs} ratios for the Karantinnaya, Laspi the inner and outer parts of Sevastopol Bay can be explained by similarity of the habitats and, hence, similar species structure of the taxocenes in these pairs of sampling sites. Note, the points corresponding to samples of these areas on the MDS ordination plot also are arranged in dense patches according to the similarity of diatom species abundances (see Figure 2).

The results of estimators' reliability evaluation based on bias and precision metrics are represented in Table 3.

Both studied metrics have highest values considering the early 20% (sample ranks 1–19) of the overall ascending succession of 93 samples; here, all the estimators give considerably greater estimates of the actual species richness. Within the mid-range (sample ranks 20–58 or 21–60% of the entire series), the relevant estimates were substantially lower; the Karakassis-S_{∞} estimator most effectively approximated the true number of species. For the late samples in randomized row (ranks 59 to 93, or 61–100% of the total number of samples) all tested estimators predicted the expected species richness most precisely. Compared to the rest of

the estimators, the S_{∞} displays the minimum inaccuracy, that is, insignificantly underestimating (bias = −0.017) the real species number of benthic diatoms. In dealing with the overall range of 93 samples, the estimator S_{∞} had also displayed the best results, giving lowest average bias value. According to the application of the precision metrics, the Jack-knife-1 estimator was superior, estimating the expected species richness closest to the actually observed number of diatom species.

A methodically similar study to determine the optimum relationship between sampling effort and observed species richness by testing of 19 estimators was conducted by Walther and Martin [22]. Comparing the bias and the precision metrics in different segments of the randomized row of samples, they have shown that the majority of the estimators underestimate the real number of species in the first one-third (25–30%) of the total sample range. The precision receded from the Chao-2 to Chao-1 to Jack-knife-2 and to Jack-knife-1 estimators which were more accurate than the rest of the 15 tested estimators (the S_{∞} estimator was not used). Considering the latter segment of the sequence (50–100% of samples), the Chao-1 and Chao-2 estimators were also the most effective, giving the least bias and a slightly overestimated value of S_{exp} compared to the S_{obs}. Both Jack-knife estimators were also quite inaccurate for the late samples (50–100%). Thus, these authors proposed to consider both Chao estimators as the most fitting for species richness prognostication. The Jack-knife-1 and 2 estimators were inferior yet performing more precisely than the rest of tested estimators. Nevertheless, Walther and Moore [42] concluded that the nonparametric estimators of the Chao and Jack-knife families performed most reliably in the prediction of expected species richness.

Proceeding from all of the above stated facts, none of the studied estimators considered would provide a universal tool that would perform equally well in different groups of biota (although these conclusions were drawn from results of the investigation of organisms and habitats very different

Extrapolative Estimation of Benthic Diatoms (Bacillariophyta) Species Diversity in Different Marine
Habitats of the Crimea (Black Sea)

135

TABLE 3: Comparative assessment of reliability of 4 estimators based on bias and precision metrics calculated for different parts of a randomized ascending sequence of samples (1 to 93).

Randomized row of successively increased number of samples	S_∞	Chao-2	Jack-knife-1	Jack-knife-2
Bias				
Early 20% number of samples in a row	1.416	2.138	1.022	1.458
Further samples (21–60%)	0.032	0.355	0.340	0.505
Late samples (61–100%)	−0.017	0.126	0.121	0.181
All samples (1–100%) in a row	1.432	2.620	1.483	2.145
Precision				
Early 20% number of samples in a row	1.181	2.088	0.421	0.884
Further samples (21–60%)	0.002	0.115	0.104	0.231
Late samples (61–100%)	0.001	0.031	0.029	0.065
All samples (1–100%) in a row	1.184	2.235	0.554	1.181

from ours). Moreover, none of the estimators is fully precise to show near-zero convergence to the asymptote of actual species richness, especially given a rather small sampling effort. With respect to benthic diatoms, index S_∞ has the most precise predictions comparatively with the other estimators, though for large number of samples (more than 45–50 samples) it can slightly underestimate the true number of species.

Note that the precision of the expected number of species (S_{exp}) evaluated by the S_∞ estimator can depend also on the width of sampling lag between the pairs of samples in their ascending sequence [25]. The initially proposed estimation method was based on a regression of the species in $k + 1$ samples against the species contained in k samples. The main concept was that this index would provide the number of species expected when the difference in the cumulative number of species between two consecutive randomized samples (i.e., sampling lag = 1) would be zero. However, in that case there should also be a zero difference between higher sampling lags. It could be expected that increasing sampling lag would provide more precise results, since it could give a higher resolution in detecting trends in the increase of species richness.

In this part of our analysis, two additional samples taken in the coastal sea water southward of Balaklava Bay and northward of the mouth of Sevastopol Bay were included, thereby the total number of samples increased to 95 and the total list of observed diatoms to 471 species. Changes in the S_{exp} values determined from the linear regressions $S_{obs(k)} = f(S_{obs(k+n)})$, constructed for sequences of samples regarding various width of sampling lags are presented in Figure 7.

It was found that in developing the regressions of the species in $k + 1$ samples against k samples for variable sampling lags, increasing the lag from 1 to 15 within the entire range of 95 samples results in a greater expected number of species, with only minor underestimation of the real species richness. At the sampling lag of 15, the estimator gave the S_{exp} value of 463, closest to the S_{obs} number of 471 species. Further increase of the lag width brought about a conspicuous decline of and, hence, a greater underestimation of species richness

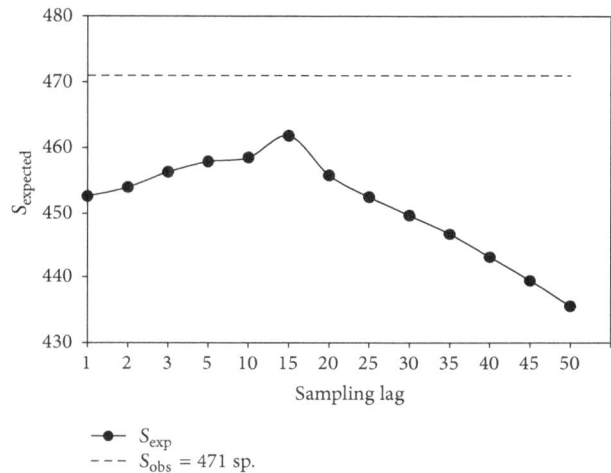

FIGURE 7: Changes in the expected diatom species number (S_{exp}) with respect to different sampling lag (lag = 1 for samples taken in initial sequence) between successive samples under construction of linear regressions $S_{obs}(n) = f(S_{obs}(n + 1))$ based on S_∞ estimator. The observed diatom species number ($S_{obs} = 471$) is also indicated on the plot by the dotted line.

in the taxocene (see Figure 7). Thus, most precise results under application of S_∞ estimator can be achieved when construction of the linear regression plots for calculation of the S_{exp} is based on sequences of randomized samples with sampling lags of 10 to 15.

The assessment of similarity from the pairwise comparison in the sequence of successively taken samples (considering the different width of sampling lags) elicited heterogeneity of the distribution pattern of the expected species richness and the diversity of the taxocene throughout certain sampling areas. So, in the triangular matrix of the intersample similarity of species richness, the first subdiagonal corresponds to samples taken in the initial sequence, displaying the similarity between the successive pairs of samples, that is, for the sampling lag 1; the second subdiagonal corresponds to the

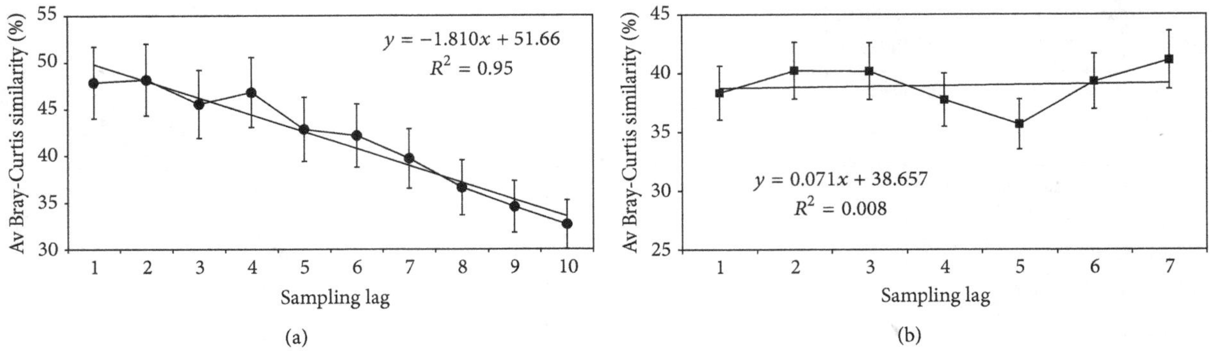

FIGURE 8: Trends in changes of average species similarity values (± SD) between pairs of samples in the sequences taken at different sampling lag ((a) Balaklava Bay and (b) Karantinnaya Bay).

sampling lag 2 when evaluating similarity between samples 1 and 3, 2, and 4, and so forth. Sampling lag 3 corresponds to the third subdiagonal in the matrice, displaying similarity values between pairs of samples 1 and 4, 2, and 5, and so forth. A possible trend of expected species richness distribution can be derived from comparison of the results averaged for each subdiagonal. A decreasing trend in the average similarity between all pairs of samples with increasing sampling lag assumes that strong patchiness or a "hidden" environmental gradient reduces the homogeneity of the data and along which the S_{exp} index changes throughout the biotope [25]. Accordingly, the absence of a negative trend in the quotient of similarity with increasing width of sampling lag presumes relatively homogeneous distribution of species richness.

This methodical approach has provided an insight into the distribution of the expected species richness of diatoms over the sampling sites in Balaklava and Karantinnaya bays (Figure 8).

At Balaklava Bay a pronounced negative linear trend ($R^2 = 0.95$) suggests evident heterogeneity in the S_{exp} distribution when comparing the inner and the outer parts of the bay. The uneven distribution can be attributed to the distinct integral gradient including factors such as depth, grain size of sediments, and the degree of anthropogenic impact along the bay water area towards the mouth [43]. On the contrary, in Karantinnaya Bay increases in the width of the sampling lag between pairwise comparing samples did not lead to a negative trend ($R^2 = 0.01$), the average similarity remained relatively constant for lags 1 to 7, ranging from 35 to 42%. These results imply the absence of an environmental gradient and, consequently, a relatively homogeneous pattern of S_{exp} distribution throughout different parts of this bay [44]. The results enable more precise estimation of the expected species richness of benthic diatoms along with reasonably minimal sampling efforts when studying other coastal habitats with similar bottom substrates and depth range.

4. Conclusion

The obtained results are one of the first attempts of prognostic estimation for benthic diatoms species richness in near shore habitats along the northern Black Sea coasts.

The results based on the randomized sequence of 93 samples taken at 8 sites of SW Crimea and applying 4 prognostic methods have shown that all estimators were dependent on the sampling effort in each data set. All applied indices overestimated the observed number (S_{obs}) of benthic diatom species, especially for a small number of samples (5-6). The expected species richness (S_{exp}) is most reasonably estimated by the Karakassis-S_∞ method. Lower-range overestimation of S_{obs} value (2–10%) occurred for small number of samples (<12), and slight underestimation (3–5%) occurred when sample numbers in a certain site exceeded 40–43. Level of S_{exp} averaged through all sampling sites overestimated S_{obs} value 6.8 ± 2.8%. Other considered estimators (Chao-2 and Jack-knife-1 and -2) more considerably overestimated S_{obs} level: 21–70% (on average 34.4 ± 8.0%), 24–36% (28.4 ± 2.2%), and 33–58% (42.5 ± 4.3%), respectively, depending on the number of samples in each sampling site. Thus, the estimator-S_∞ represents the best compromise choice for evaluation of expected diatoms species richness in the various habitats.

The empirical relationship between number of samples (X) and the number of observed species (Y) (considering all samples) is reliably described by log-equation $Y = 79.01 \ln(x) + 34.95$. Following this equation, nearly 10 samples are required for disclosing about 50% of the total species number of benthic diatoms (433 species) which were found on sandy/muddy sediments near Crimean coasts within the depth range 6–48 m. To detect about 80% species richness (347 species), not less than 46 samples should be considered (under the assumption of equal probability to reveal any diatom species in the sample). Results based on this generalized region-specific curve can be applied for evaluation of compromised ratio between minimum sampling effort and possibly higher number of revealed diatom species.

Acknowledgments

This research was partially supported by the USA Environmental Protection Agency, AED, through the mediation of the Science and Technology Center in Ukraine and is a contribution to the biodiversity assessment study of STCU's Project P-277a (2009-2010). The authors are grateful to Professor A. Witkowski and Prof. H. Lange-Bertalot (Szczecin University,

Extrapolative Estimation of Benthic Diatoms (Bacillariophyta) Species Diversity in Different Marine
Habitats of the Crimea (Black Sea)

137

Poland) for their valuable consultations during taxonomical identification of benthic diatoms. Thanks are also addressed to Dr. N. Revkov (IBSS, Sevastopol, Ukraine) for his friendly help during sampling surveys and to Dr. Kay T. Ho (US EPA, AED) for her kind assistance in improvement of the paper language. They also thank three anonymous reviewers for their constructive comments and criticism of the paper.

References

[1] H. L. Sanders, "Marine benthic diversity: a comparative study," *The American Nature*, vol. 102, no. 925, pp. 243–282, 1968.

[2] R. K. Colwell and J. A. Coddington, "Estimating terrestrial biodiversity through extrapolation," *Philosophical Transactions of the Royal Society of London B*, vol. 345, no. 1311, pp. 101–118, 1994.

[3] K. J. Gaston, "Species richness: measure and measurement," in *Biodiversity: A Biology of Numbers and Difference*, K. J. Gaston, Ed., pp. 77–113, Blackwell Science, Oxford, UK, 1996.

[4] J. S. Gray, "The measurement of marine species diversity, with an application to the benthic fauna of the Norwegian continental shelf," *Journal of Experimental Marine Biology and Ecology*, vol. 250, no. 1-2, pp. 23–49, 2000.

[5] T. Watanabe, K. Asai, and A. Houki, "Numerical water quality monitoring of organic pollution using diatom assemblages," in *Proceedings of the 9th International Diatom Symposium*, F. Round, Ed., pp. 123–141, Koeltz Scientific Books, Koenigstain, Germany, 1988.

[6] C. Izsak and A. R. G. Price, "Measuring β-diversity using a taxonomic similarity index, and its relation to spatial scale," *Marine Ecology Progress Series*, vol. 215, pp. 69–77, 2001.

[7] A. N. Petrov and E. L. Nevrova, "Database on Black Sea benthic diatoms (Bacillariophyta): its use for a comparative study of diversity pecularities under technogenic pollution impacts," in *Proceedings, Ocean Biodiversity Informatics: International Conference of Marine Biodiversity Data Management, Hamburg, Germany, November 2004*, vol. 202 of *IOC Workshop Report, VLIZ Special Publication no. 37*, pp. 153–165.

[8] B. A. Walther and S. Morand, "Comparative performance of species richness estimation methods," *Parasitology*, vol. 116, no. 4, pp. 395–405, 1998.

[9] A. Foggo, M. J. Attrill, M. T. Frost, and A. A. Rowden, "Estimating marine species richness: an evaluation of six extrapolative techniques," *Marine Ecology Progress Series*, vol. 248, pp. 15–26, 2003.

[10] R. J. Stevenson, "Epilithic and epipelic diatoms in the Sandusky River, with emphasis on species diversity and water pollution," *Hydrobiologia*, vol. 114, no. 3, pp. 161–175, 1984.

[11] E. L. Nevrova, N. K. Revkov, and A. N. Petrov, "Chapter 5. 3. Microphytobenthos," in *Modern Condition of Biological Diversity in Near-Shore Zone of Crimea (the Black Sea Sector)*, V. N. Eremeev and A. V. Gaevskaya, Eds., pp. 270–282, 288–302, 351–362, Ekosi-Gidrophyzika, Sevastopol, Ukraine, 2003.

[12] A. I. Proshkina-Lavrenko, *Diatomovye Vodorosli SSSR (Diatom Algae of USSR)*, Nauka, USSR, St Petersburg, Russia, 1974.

[13] A. N. Petrov, E. L. Nevrova, and L. V. Malakhova, "Multivariate analysis of benthic diatoms distribution across the multidimensional space of the environmental factors gradient in Sevastopol bay (the Black sea, Crimea)," *Marine Ecological Journal*, vol. 4, no. 3, pp. 65–77, 2005 (Russian).

[14] A. N. Kryshtofovich, *Diatom Analysis*, vol. 3, Gosgeolitizdat, USSR, Moscow, Russia, 1950.

[15] A. I. Proshkina-Lavrenko, *Benthic Diatom Algae of the Black Sea*, Academy of Sciences, USSR, Moscow, Russia, 1963.

[16] N. Guslyakov, O. Zakordonez, and V. Gerasim, *Atlas of Benthic Diatom Algae of the North-Western Part of the Black Sea and Adjacent Aquatories*, Naukova Dumka, USSR, Kiev, Ukraine, 1992.

[17] E. Fourtanier and J. P. Kociolek, "Catalogue of the diatom genera," *Diatom Research*, vol. 14, no. 1, pp. 1–190, 1999.

[18] E. Fourtanier and J. P. Kociolek, *Catalogue or Diatom Names*, California Academy of Sciences, San Francisco, Calif, USA, 2011, http://www.calacademy.org/research/diatoms/names/index.asp.

[19] A. Witkowski, H. Lange-Bertalot, and D. Metzeltin, "Diatom flora of marine coast," *Iconographia Diatomologica*, vol. 7, pp. 1–926, 2000.

[20] Z. Levkov, "Amphora sensu lato," *Diatoms of Europe*, vol. 5, pp. 1–916, 2009.

[21] A. Chao, "Estimating the population size for capture—recapture data with unequal catchability," *Biometrics*, vol. 43, no. 4, pp. 783–791, 1987.

[22] B. A. Walther and J. L. Martin, "Species richness estimation of bird communities: how to control for sampling effort?" *Ibis*, vol. 143, no. 3, pp. 413–419, 2001.

[23] I. Karakassis, "S_{∞}: a new method for calculating macrobenthic species richness," *Marine Ecology Progress Series*, vol. 120, pp. 299–303, 1995.

[24] K. I. Ugland and J. S. Gray, "Estimation of species richness: analysis of the methods developed by Chao and Karakassis," *Marine Ecology Progress Series*, vol. 284, pp. 1–8, 2004.

[25] H. Rumohr, I. Karakassis, and J. N. Jensen, "Estimating species richness, abundance and diversity with 70 macrobenthic replicates in the Western Baltic sea," *Marine Ecology Progress Series*, vol. 214, pp. 103–110, 2001.

[26] A. Chao, "Non-parametric estimation of the number of classes in a population," *Scandinavian Journal of Statistics*, vol. 11, pp. 265–270, 1984.

[27] J. F. Heltshe and N. E. Forester, "Estimating marine species richness using the jack-knife procedure," *Biometrics*, vol. 39, no. 1, pp. 1–11, 1983.

[28] D. Cogalniceanu, M. Tudorancea, E. Preda, and N. Galdean, "Evaluating diversity of Chironomid (Insecta: Diptera) communities in alpine lakes, Retezat National Park (Romania)," *Advanced Limnology*, vol. 62, pp. 191–213, 2009.

[29] U. Brose, N. D. Martinez, and R. J. Williams, "Estimating species richness: sensitivity to sample coverage and insensitivity to spatial patterns," *Ecology*, vol. 84, no. 9, pp. 2364–2377, 2003.

[30] S. H. Hurlbert, "The nonconcept of species diversity: a critique and alternative parameters," *Ecology*, vol. 52, no. 4, pp. 577–586, 1971.

[31] K. Soetaert and C. Heip, "Sample-size dependence of diversity indices and the determination of sufficient sample size in a high-diversity deep-sea environment," *Marine Ecology Progress Series*, vol. 59, pp. 305–307, 1990.

[32] K. R. Clarke and R. N. Gorley, *PRIMER V5: User Manual/Tutorial*, PRIMER-E, Plymouth, UK, 2001.

[33] E. Nevrova, "Benthic diatoms of the Black sea: inter-regional analysis of diversity and taxonomic structure," in *Proceedings of the 22nd International Diatom Symposium, Ghent, Belgium, 26–31 August 2012*, K. Sabbe, B. Van de Vijver, and W. Vyverman,

Eds., vol. 58 of *VLIZ Special Publication*, p. 35, Oral and Poster Programme, 2012.

[34] L. I. Ryabushko, *Microalgae of the Black Sea Benthos*, Ekosi-Gidrophyzika, Sevastopol, Ukraine, 2006.

[35] P. M. Tsarenko, S. P. Wasser, and E. Nevo, Eds., *Algae of Ukraine: Diversity, Nomenclature, Taxonomy, Ecology and Geography. 2. Bacillariophyta*, A.R.G. Gantner Verlag KG, Ruggell, Liechtenstein, 2009.

[36] F. E. Round, R. M. Crawford, and D. G. Mann, *The Diatoms: Biology and Morphology of the Genera*, Cambridge University press, Cambridge, UK, 1990.

[37] Z. S. Kucherova, *Diatoms and their role in fouling cenosis at the Black sea [Ph.D. thesis]*, IBSS, USSR, Sevastopol, Ukraine, 1973.

[38] R. M. Burgess, A. V. Terletskaya, M. V. Milyukin et al., "Concentration and distribution of hydrophobic organic contaminants and metals in the estuaries of Ukraine," *Marine Pollution Bulletin*, vol. 58, no. 8, pp. 1103–1115, 2009.

[39] K. I. Ugland, J. S. Gray, and K. E. Ellingsen, "The species-accumulation curve and estimation of species richness," *Journal of Animal Ecology*, vol. 72, no. 5, pp. 888–897, 2003.

[40] E. P. Smith and G. van Belle, "Nonparametric estimation of species richness," *Biometrics*, vol. 40, no. 1, pp. 119–129, 1984.

[41] J. J. Hellmann and G. W. Fowler, "Bias, precision and accuracy of four measures of species richness," *Ecological Applications*, vol. 9, no. 3, pp. 824–834, 1999.

[42] B. A. Walther and J. L. Moore, "The concepts of bias, precision and accuracy, and their use in testing the performance of species richness estimators, with a literature review of estimator performance," *Ecography*, vol. 28, no. 6, pp. 815–829, 2005.

[43] A. N. Petrov, E. L. Nevrova, A. V. Terletskaya, M. V. Milyukin, and V. Y. Demchenko, "Structure and taxonomic diversity of benthic diatom assemblage in a polluted marine environment (Balaklava bay, Black sea)," *Polish Botanical Journal*, vol. 55, no. 1, pp. 183–197, 2010.

[44] A. N. Petrov and E. L. Nevrova, "Comparative analysis of taxocene structures of benthic diatoms (Bacillariophyta) in regions with different level of technogenic pollution (the Black sea, Crimea)," *Marine Ecological Journal*, vol. 3, no. 2, pp. 72–83, 2004 (Russian).

16

Domestication, Conservation, and Livelihoods: A Case Study of *Piper peepuloides* Roxb.—An Important Nontimber Forest Product in South Meghalaya, Northeast India

H. Tynsong,[1] M. Dkhar,[2] and B. K. Tiwari[3]

[1] *Ministry of Environment and Forest, North Eastern Regional Office, Shillong 793021, Meghalaya, India*
[2] *Union Christian Collage Umiam, Ri Bhoi, Shillong 793122, Meghalaya, India*
[3] *Department of Environmental Studies, North-Eastern Hill University, Shillong 793022, Meghalaya, India*

Correspondence should be addressed to H. Tynsong; herotynsong@yahoo.com

Academic Editor: Alexandre Sebbenn

Wild pepper (*Piper peepuloides* Roxb., family: Piperaceae) is an evergreen climber which grows wild in tropical evergreen forests and subtropical evergreen forests of northeast India. This plant grows luxuriantly in the areas with high rainfall at lower elevations ranging from 100 to 800 m above mean sea level. In Meghalaya, to meet the market demand, farmers have domesticated it in arecanut agroforests and betel leaf agroforests. We found that the mean density of wild pepper in arecanut agroforest is 585 stem/ha and only 85 stem/ha in natural forest. In India, wild pepper is used in a variety of Ayurvedic medicines. Local people of Meghalaya uses powdered dry seeds mixed with honey and egg yolk for the treatment of severe cough. The study reveals that the average gross annual production of wild pepper is 7 quintals/ha, and final market price fetches Rs. 336,000/ha, out of which 42% of the money goes to the grower, 16% to local trader, 23% to dealer, 17% to retailer, 1.2% to wages of labourers, and 0.6% to transport.

1. Introduction

People in the rural areas across the world extract a wide variety of nontimber forest products (NTFPs), from nearby forests. NTFPs are important to rural households in terms of their contribution to health, food, energy, and other aspects of rural welfare [1]. In India, an estimated 50 million people living in and around forests rely upon NTFPs for their subsistence and cash income [2]. Bahuguna [3] and Mahapatra et al. [4] have studied at the contribution of NTFPs to cash income; however, such studies are very few, and our understanding of the subject remains inadequate. During recent years, forest managers have begun to consider the role of NTFPs in rural welfare, and in some cases, they have begun to manage forests in a way that promotes outputs other than timber [5, 6]. What is less understood and represented in policy is the contribution of these extracted products to a household's cash income [4]. When rural households use most of their agricultural output for subsistence consumption, cash from the sale of NTFPs can play an important role by allowing the households to use the same for vital cash-dependent transactions, namely, buying tools and paying for school [1]. It has been assumed that the extraction of NTFPs from natural forests could serve as the goals of biodiversity conservation and poverty alleviation [7, 8]. It is estimated that roughly 80% of the developing world including nearly 60 million indigenous people depend on wild fruits, seeds, poles for construction, and medicinal plants to meet subsistence and supplemental income needs [9]. NTFPs play a significant role in providing subsistence and cash income to local populations of the world [10–12]. India's National Forest Policy of 1988 and Joint Forest Management Notification of 1990 reflect the desire and need to ensure that rural people participate in the management of forests and capture benefits from those forests. Such people-oriented forest policy in India will be better implemented and have

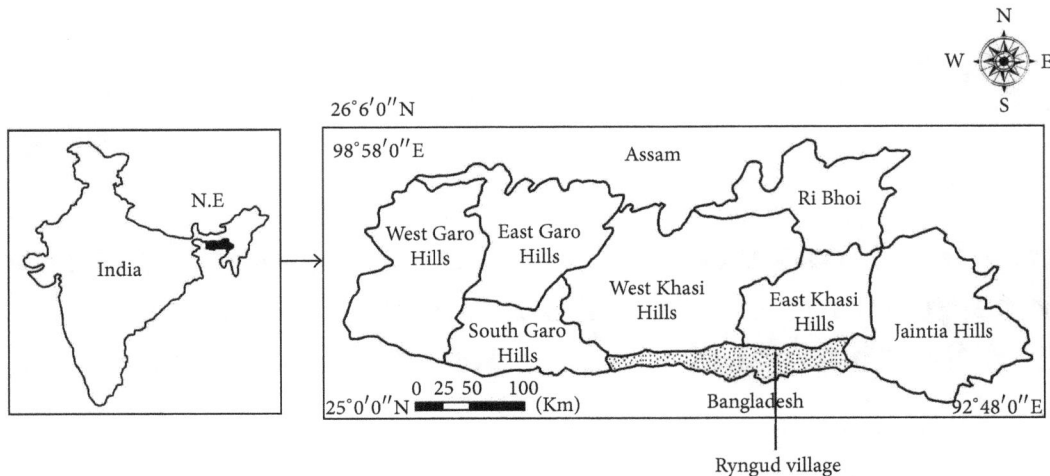

FIGURE 1: Location of the study area.

more impact if more researches on the analyses of NTFPs extraction quantities and values are undertaken across the diverse ecological, economic, and social settings in India. In Meghalaya, about 80% of the total populations are farmers, and a large section of them cultivates cash crops. Hence, the contribution of forest-based production to cash income of the people is potentially important for understanding the dependency of the rural poor on forests and forest products.

Piper peepuloides Roxb., commonly known as wild pepper, is an evergreen climber belonging to the family Piperaceae. It is found in the forests of Goalpara, North Cachar Hills, and on southern slope of Meghalaya including part of Jaintia Hills, Khasi Hills, and West Khasi Hills districts [13]. The area forms part of Eastern Himalayas, which is recognized for its exceptionally rich biodiversity [14]. In Meghalaya, it grows in areas falling within the altitude range of 100–800 m above mean sea level. Wild pepper grows wild in the forests and valleys along the streams. However, due to its high market demand, it has also been domesticated in arecanut agroforests as well as in betel leaf agroforests [13]. Seeds are used in a variety of Ayurvedic medicines, a traditional system of medicine known as Ayurveda which originated in India long back in Pre-Vedic period which is believed to be 5000 years BC [15]. Local people of Meghalaya use powdered dry seeds mixed with honey and egg yolk for the treatment of severe cough. There is very little research done on this plant except for the work of Singh [16] who reported that this plant can be used as spice and the research conducted by SFRI [17] on its distribution and medicinal use. In spite of its high market demand, there is little information with respect to habit, economic aspects, conservation, and harvesting and processing. This paper addresses the questions such as management, amount of seed production per plant and per unit area, harvesting, marketing, and its economic impact on the livelihood of the local people.

2. The Case Study

This paper is based on a case study conducted in the months of March–May 2008 at Ryngud village of south Meghalaya (25°13′ N, 91°46′ E) (Figure 1). The mean annual maximum and minimum temperatures are around 23°C and 13°C, respectively. The mean annual rainfall is 11,565 mm. The slope of the area is predominantly towards the south, and the angle of the slope varies between 10°–40°. The area has a large numbers of rivers and rivulets, which drain into the plains of Bangladesh. At times, narrow and deep river valleys separate one hill range from the other. The population density is sparse. Horticulture, forestry, and fisheries are the principal occupation of the people. Agriculture is limited to some small valleys, where mainly tuber crops are grown. Arecanut, orange, betel leaf, jack fruit, bayleaf, honey, and broom grass are the important produce of the region. The area is inhabited by *War Khasi* people, a tribal community having long tradition of forest conservation. People gather a variety of edibles from forests and water bodies including fish, frog, crustaceans, mollusks, bushmeats, tubers, and wild vegetables. The staple diet of the local inhabitants is rice, fish, and meat. People collect process and market a large variety of nontimber forest products (NTFPs) and medicinal and aromatic plants (MAPs) such as *Cinnamomum tamala*, *Phrynium capitatum*, bamboo, honey, mushrooms, nuts, tubers, edible worms, insects, and leafy vegetables from the forests [18].

The natural vegetation of the study area ranges from tropical evergreen to subtropical evergreen forests. The plant species in the forests are distributed in distinct vegetation layers. The important evergreen trees found in South Meghalaya include *Cinnamomum tamala*, *Daphniphyllum himalayense*, *Myrica esculenta*, *Sarcosperma griffithi*, and *Syzygium tetragonum*. The deciduous elements included *Betula alnoides*, *Cedrela toona*, *Engelharitia spicata*, and *Ficus roxburghii*. The shrub layer is thick and is predominantly

Domestication, Conservation, and Livelihoods: A Case Study of Piper peepuloides Roxb.—An Important Nontimber Forest Product in South Meghalaya, Northeast India

141

TABLE 1: Summary characteristics of wealth ranks by income and participatory method.

Parameter	Category		
	Rich	Middle	Poor
No. of households	20	29	73
Participatory Wealth Ranking Criteria	(i) Large plantations (5-6 plantations of approx.40 ha each) (ii) Hire labour (iii) Higher level of cash income (iv) Houses made of concrete cement (v) Good standard of living.	(i) Small plantations (2-3 plantations of approx. 40 ha each) (ii) Lower level of cash income (iii) Houses made of wood covered by tin (iv) Work as labourers for few months.	(i) Do not have plantation (ii) Generally labourers (iii) Very low level of cash income (iv) Houses made of bamboo covered by straw/tin.

composed of *Ardisia griffithii, Boehmeria malabarica, Goniothalamus sesquipedalis, Mahonia pycnophylla,* and *Wallichia densiflora.* The ground vegetation (herb) is dominated by *Borreria pilosa, Commelina benghalensis, Impatiens* spp., *Ophiorrhiza hispida, Sonerila khasiana,* and a large number of ferns. There are a good number of lianas and other climbers seen twining on the trees. The tree trunk and branches are covered with large number of mosses, epiphytic ferns, and different variety of orchids. The invasive weedy species like *Artemisia* spp., *Eupatorium* spp., and *Mikania micrantha* are also present in good number.

3. Methods

In this study, four complementary approaches were adopted, namely, (a) formal interview with the village headmen and secretary, (b) field observation, (c) interaction with the head of the selected households through questionnaires, and (d) phytosociological study in natural forests as well as arecanut agroforests [19]. The survey was administered to a random sample of 30 households. The total number of household at Ryngud village is 122. The production and marketing were studied by using household questionnaires and PRA methods as described by Mukherjee [22]. For understanding the economic value of the wild pepper, data were collected on costs of collection/production, harvesting, processing, value addition, transportation, storage, taxation, and benefit sharing. Marketing analysis was conducted by interviewing growers/local collectors, traders, dealers, and retailers [23]. The marketing channels were investigated using methods described by Raintree [24] and Karki [25]. To study the density and distribution of wild pepper, 40 quadrats (10 × 10 m size) were laid in the natural forests as well as 40 quadrats (10 × 10 m size) were laid in the arecanut agroforests. We selected an area of 1 ha in the arecanut agroforests to assess the average production (quantity) and analyze its monetary production in term of per unit area. Table 1 summarizes the characteristics of rich family, middle family, and poor family-based participatory wealth ranking criteria.

4. Results

4.1. Management. In Meghalaya, due to its high demand in market, this plant has been domesticated in arecanut agroforests and betel leaf agroforests by using the younger part of the stem tip. It is planted in the month of July and August at the base of trees, by digging the soil and without removing the shrubs and herbs growing in the vicinity of the trees. This is done to prevent wilting of the young stem especially during dry spells. After one month, cultivators do cleaning and lopping of tree branches to help the plant get enough sunlight. These management practices are particularly necessary when the plants are still young. When the plants become old (>3 years), it does not require much weeding and tree lopping. In colder places, fruits are smaller and not as healthy as those which are grown in warmer places. Wild pepper starts producing fruits from the 1st year of cultivation itself, and the production increases from 2nd year onward. No artificial fertilizer was needed for the cultivation of wild pepper. The plants obtain their nutrients from decomposing litter and weeds which people dump at the base of this wild pepper during weeding and tree lopping.

4.2. Density and Production. The mean density of wild pepper plant was found to be 585 (±34.43) stem/ha in arecanut agroforest and 85 (±7.09) stem/ha in natural forest. On an average, each plant bears 850–1000 fresh fruits which weighed about 2–4 kg. Based on the household data, it has been found that a total of 143 quintal/annum of dry fruits have been produced by Ryngud village with an average of 1.2 quintal/household. From 1 ha area of wild pepper cultivation, approximately 7 quintals of dry fruits were produced annually.

4.3. Harvesting and Processing. The harvesting of wild pepper starts from the month of March–December. The cultivators pluck the fruits manually by using their finger nail. Plucking usually starts from the ground level by sitting, and then the harvester moves upward by standing and also uses bamboo ladder for plucking fruits that grow at higher level. During harvesting, men, women, and children are all involved. After plucking, the fruits are sun dried for four to five days till they turn light black in color. During drying, care is taken so that it should not be over dried as it will lose its desired black color. Care should also be taken as it should not get wet while drying as it attracts fungus which deteriorates the fruit quality, reducing the profit margin of the cultivators. After

TABLE 2: Mean annual expenditure (ha^{-1}) incurred by the grower towards management, processing, and marketing of wild pepper.

Activity	Worker	Rate/day (Rs.)	No. of person	Total cost (Rs.)
Planting	Men	150	4	600
Weeding	Women	50	10	500
Harvesting	Men	100	2	200
	Women	50	4	200
	Boy	30	4	120
Drying	Men	100	2	200
	Women	50	2	100
	Boy	30	1	30
Sorting	Women	50	5	250
	Boy	30	5	150
Packing	Men	100	10	1000
Transport from village to road	Men	50 p/kg	—	350
	Women			
	Boy			
Tool	No.	Rate (Rs.) per unit	—	Total cost (Rs.)
Gunny bag	10	8	—	80
Knife (ka wait)	2	55	—	110
Bamboo ladder	2	80	—	160
Bamboo basket (ka ruh)	2	50	—	100
Grand total				4,150

Approximate currency conversion 1 USD = Rs. 50 (INR).

drying, the cultivators sort out and discard the fungal infested fruits. Wild pepper has been notified as a forest product, and therefore it attracts royalty and taxes. The traders had to pay the royalty to the District Council. The prevalent rate of royalty was Rs. 100/quintal at the time of study. Alleged illegal collections from the transporters at various check points are also prevalent. The monetary expenditure incurred by the growers towards management, processing, and marketing of wild pepper in 1 ha area is given in Table 2.

4.4. Marketing. The growers usually sell the produce to the local trader at the local market. A small number of growers sell it directly to the dealers in Shillong, which is the largest regional market in the state. The local traders then sell these products to the regional traders also known as dealers. From Shillong, it is transported to other metropolis, namely, Kolkata, Bangalore, Hyderabad, and Chennai via regional traders (wholesalers) based in Guwahati. The dealers sell a small quantity of produce to the retailers in Shillong. The marketing channel of wild pepper is given in Figure 2.

The price and cash flow of wild pepper in the year 2008 is given in Figure 3. The price is mostly controlled by regional traders (the business men who supply the goods outside the state). These regional traders control the market and are often determining the price at which the growers sell their produce. The growers, however, generally do not have any say in deciding the price. Market demand and price depend upon the quality of fruit and availability of goods. Proper storage was required to protect the product from the attack of fungi. A well-dried and uninfected product fetches best price.

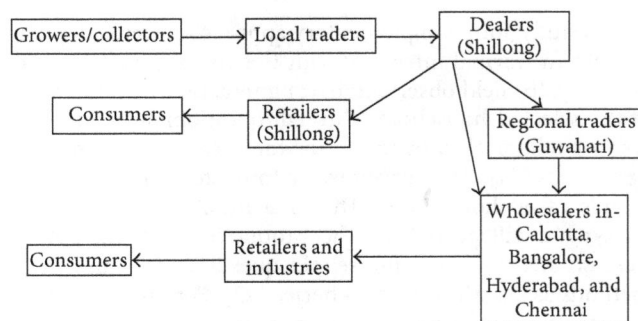

FIGURE 2: Marketing channel of the produce from the grower to the consumer.

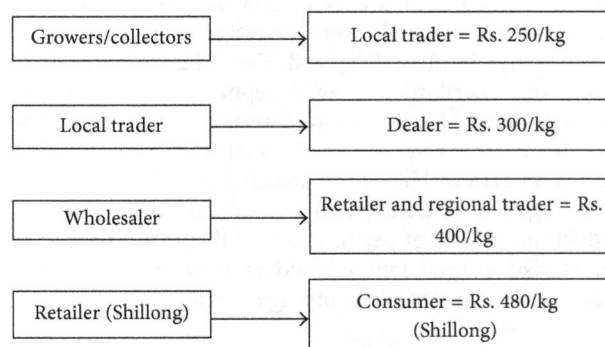

FIGURE 3: Pricing mechanism and average price of wild pepper at various stages of transaction (approximate currency conversion 1 USD = Rs. 50 (INR).

Domestication, Conservation, and Livelihoods: A Case Study of Piper peepuloides Roxb.—An Important Nontimber
Forest Product in South Meghalaya, Northeast India

143

4.5. Economic Impact. The average gross production of 7 quintals of wild pepper from 1 ha area fetches Rs. 1,75,000/annum (@ Rs. 250/kg), out of which Rs. 4,150 was spent by the growers in management, harvesting, processing, and marketing. Therefore, a grower in Ryngud village annually earns Rs. 1,70,850 ha^{-1} from the selling of wild pepper. With an average production of 1.2 quintal/household, each household earned an average income amounting to Rs. 30,000/annum from wild pepper. Based on our household survey, the average annual income at Ryngud village for those households engaged in wild pepper cultivation was found to be Rs. 71,724. Thus, 42% of mean annual income of the farmers engaged in wild pepper cultivation and trade came from wild pepper only. However, this value does not reflect the amount earned by every household in the village or area as only the rich and middle class families (40%) of the village have wild pepper cultivation. The final market price of 7 quintals of wild pepper produced in 1 ha goes to the grower 42%, the rest was shared by local trader 16%, dealer 23%, retailer 17%, wages of labourers 1.2%, and 0.6% in transport and taxes. The wild pepper market pathway and cost from grower to consumer at the various stages of the movement and processing of the product within the state of Meghalaya is given in Figure 4.

5. Discussion

Wild pepper in natural habitats is not the dominant species and thus restricted in its distribution. We recorded that the mean density of wild pepper in natural forest is 85 stem/ha, while in arecanut agroforest, it is 585 stem/ha. Low in its density; results in limited production per unit area that often fail to fulfill market demand. Domestication of wild pepper in arecanut agroforest ensures greater densities and greater production. From the conservation point of view, domestication limits human interference on wild population, thus minimizing disturbance of biodiversity. Murali et al. [26] are of the opinion that NTFP extraction in natural populations even at moderate levels may change floristic composition and erode species diversity and affect population structure. The reason leading to its domestication in Meghalaya is attributed mainly to high market demand, less production in natural forests, and high profit from cultivation.

In term of cash income, the study revealed that 100% of growers were from rich and middle class families. This is true because in the state of Meghalaya, rich and middle class families have land for cultivations, which the poor families do not have [27]. Therefore, this product can be called as forest product for the rich and middle class families. This is in contrast to *Phrynium capitatum*, another important NTFPs of Meghalaya in which mostly poor families are involved in collection [19]. The finding of the present study highlighted that cultivation, processing, and marketing of wild pepper are economically viable cropping practices in Meghalaya due to its demand. The value chain has developed and the growers are getting good return to their investments. The trade benefits almost all sections of the society as the landowners benefit from the cultivation, the landless benefit by working

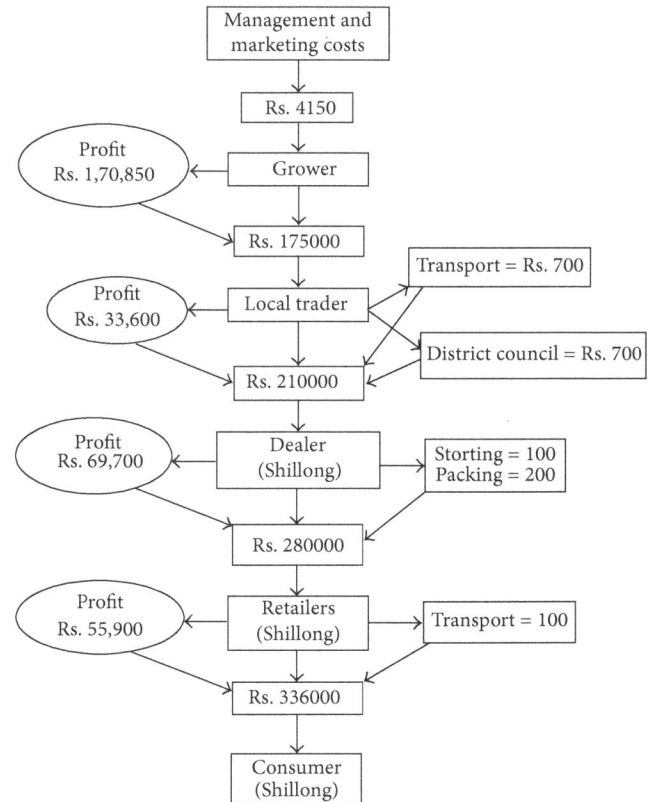

FIGURE 4: Wild pepper market pathway from the grower to the consumer depicting price appreciation at various stages of movement of the product in Meghalaya. For each hectare area of wild pepper, the grower earns a profit of Rs. 1,70,850 and sells to the local trader for Rs. 1,75,000 who in turn sells it to the dealer who earns a profit of Rs. 69,700 and sell it to the retailer who earns a profit of Rs. 55,900, and the consumer finally have to pay Rs. 3,36,000.

as daily wage labour, and the traders and the transporters earn their livelihood by marketing the produce. The farmers are getting good returns because of low investment and quick and good return as the produce from the plant may be harvested after one year of plantation. In this study a cost-benefit analysis revealed an output/input ratio of 42.16 which is higher than 6.84 obtained from broom grass [16], 3.1 obtained from bayleaf [13], and 1.8 obtained from broom grass [28]. A final value of 42% accrued to the grower from wild pepper was much better return when compared with the forest produce growers of Solika in Karnataka who received only 4% of the final value of *Phyllanthus emblica* fruits [20], *War* community of south Meghalaya who benefited only 23% from Bayleaf (*Cinnamomum tamala*) [13] and 28% from *Phrynium* leaf [19], and local people of Darjeeling, Sikkim who received 35% from the cultivation of broom grass [21] (Table 3).

6. Conclusion

The study revealed that domestication and marketing of wild pepper fetche high percent profit to the grower.

TABLE 3: Comparison of benefits accrued by collectors from different forest products.

NTFPs	States	Net benefit of collector (%)	Reference
Piper peepuloides (wild pepper)	Meghalaya	42	This study
Phrynium capitatum (packing leaf)	Meghalaya	28	Tynsong and Tiwari [19]
Cinnamomum tamala (bayleaf)	Meghalaya	23	Tynsong [13]
Phyllanthus emblica (amla)	Karnataka	4	Shankar et al. [20]
Thysanolaena maxima (broom grass)	Sikkim	35	Shankar et al. [21]

The expenditure incurred on the management of wild pepper in agroforest is minimal. It also highlights that domestication of nontimber forest products such as wild pepper could become an effective instrument for rural development and less impact on natural habitat, because its cultivation needs minimum input and labour and generates very attractive economic return.

Wild pepper limited production in the state of Meghalaya may be attributed to the plant limited adaptability, as it requires specific local habitat of slopy nature, high rainfall, and altitudes ranging from 100 to 800 m above mean sea level. This is the main reason why mass cultivation of this plant is not possible in the state of Meghlaya and also in the entire northeastern region of India. Our findings also highlighted that cultivation, processing, and marketing of wild pepper are economically viable cropping practices only in those area where local habitats are favorable for its growth. The value chain has developed, and the growers are receiving a good return for their investments. The trade benefits almost all sections of society, as the growers benefit from the cultivation, the landless benefit by working as daily wage labour, and the traders and the transporters earn their livelihood by marketing the produce.

Acknowledgments

The authors are thankful to the Head of Department of Environmental Studies, North-Eastern Hill University, Shillong, for providing necessary laboratories facilities and are also deeply indebted to Dr. (Mrs.) S. J. Phukan, Deputy Director, BSI, Eastern Circle, Shillong, for allowing them to consult the herbaria and deputing her staff to help in the identification of plant specimen. Financial support received from UGC-RGNF is gratefully acknowledged.

References

[1] W. Cavendish, "Empirical regularities in the poverty-environment relationship of rural households: evidence from zimbabwe," *World Development*, vol. 28, no. 11, pp. 1979–2003, 2000.

[2] D. N. Tewari, *Tropical Forestry in India*, vol. 387, International Book Distributors, Mumbai, India, 1992.

[3] V. K. Bahuguna, "Forests in the economy of the rural poor: an estimation of the dependency level," *Ambio*, vol. 29, no. 3, pp. 126–129, 2000.

[4] A. K. Mahapatra, H. J. Albers, and E. J. Z. Robinson, "The impact of NTFP sales on rural households' cash income in India's dry deciduous forest," *Environmental Management*, vol. 35, no. 3, pp. 258–265, 2005.

[5] A. B. Anderson, *Alternative To Deforestation: Steps Towards Sustainable Use of the Amazon Rain Forests*, Columbia University Press, New York, NY, USA, 1990.

[6] J. Clay, "A rainforest emporium," *Garden*, vol. 14, pp. 2–7, 1990.

[7] M. Ros-Tenen, W. Dijkman, and E. Lammerts van Bueren, *Commercial and Sustainable Extraction of Non-Timber Forest Products: Towards a Policy and Management Oriented Research Strategy*, The Tropenbos Foundation, Wageningen, The Netherlands, 1995.

[8] M. Ruiz Peres, *Current Issues in Non-Timber Forest Product Research*, CIFOR, Bogor, Indonesia, 1996.

[9] FAO, Global Forest Resources Assessment, http://www.fao.org/forestry/site/28699/en/, 2005.

[10] B. Belcher, M. Ruíz-Pérez, and R. Achdiawan, "Global patterns and trends in the use and management of commercial NTFPs: implications for livelihoods and conservation," *World Development*, vol. 33, no. 9, pp. 1435–1452, 2005.

[11] B. Belcher and K. Schreckenberg, "Commercialisation of non-timber forest products: a reality check," *Development Policy Review*, vol. 25, no. 3, pp. 355–377, 2007.

[12] S. Wunder, "Poverty alleviation and tropical forests—what scope for synergies?" *World Development*, vol. 29, no. 11, pp. 1817–1833, 2001.

[13] H. Tynsong, *Plant diversity and NTFP management in community forests of War area Meghalaya [Ph.D. thesis]*, North-Eastern Hill University, ShillongIndia, India, 2009.

[14] N. Myers, R. A. Mittermeler, C. G. Mittermeler, G. A. B. Da Fonseca, and J. Kent, "Biodiversity hotspots for conservation priorities," *Nature*, vol. 403, no. 6772, pp. 853–858, 2000.

[15] B. K. Tiwari, H. Tynsong, and S. Rani, "Medicinal and aromatic plants: medicinal plants and human health," in *Encyclopedia of Forest Sciences*, J. J. Burley Evans and J. A. Youngquist, Eds., pp. 515–523, Elsevier, Oxford, UK, 2004.

[16] A. K. Singh, "Probable agricultural biodiversity heritage sites in India: VI. The northeastern hills of Nagaland, Manipur, Mizoram, and Tripura," *Asian Agri-History*, vol. 14, no. 3, pp. 217–243, 2010.

[17] SFRI, State Forest Research Institute Department of Environment and Forests Government of Arunachal Pradesh, Itanagar-791 111, Bulletin 2001, 1994-1000.

[18] H. Tynsong and B. K. Tiwari, "Diversity of plant species in arecanut agroforests of south Meghalaya, north-east India," *Journal of Forestry Research*, vol. 21, no. 3, pp. 281–286, 2010.

[19] H. Tynsong and B. K. Tiwari, "Contribution of Phrynium capitatum Willd. leaf a non-timber forest product to the livelihoods of rural poor of South Meghalaya, North-East India," *Indian Journal of Natural Products and Resources*, vol. 2, no. 2, pp. 229–235, 2011.

[20] U. Shankar, K. S. Murali, R. Uma Shaanker, K. N. Ganeshaiah, and K. S. Bawa, "Extraction of non-timber forest products in the

forests of Biligiri Rangan Hills, India. 3. Productivity, extraction and prospects of sustainable harvest of Amla Phyllanthus Emblica, (Euphorbiaceae)," *Economic Botany*, vol. 50, no. 3, pp. 270–279, 1996.

[21] U. Shankar, S. D. Lama, and K. S. Bawa, "Ecology and economics of domestication of non-timber forest products: an illustration of broomgrass in Darjeeling Himalaya," *Journal of Tropical Forest Science*, vol. 13, no. 1, pp. 171–191, 2001.

[22] N. Mukherjee, *Participatory Methods and Rural Knowledge. Participatory Rural Appraisal Methodology and Applications*, Concept Publishing Company, New Delhi, India, 1993.

[23] B. K. Tiwari, "Forest biodiversity management and livelihood enhancing practices of War Khasi of Meghalaya, India," in *Himalayan Medicinal and Aromatic Plants, Balancing Use and Conservation*, Y. Thomas, M. Karki, K. Gurung, and D. Parajuli, Eds., pp. 240–255, His majesty Government of Nepal Ministry of Forests and Soil Conservation, 2005.

[24] J. Raintree, "Developing and marketing of non-timber forest products: methods used in protected areas in Vietnam," in *Shifting Cultivation Towards Sustainability and Resource Conservation in Asia*, IFAD, IDRC, CIIFAD, ICRAF, and IIRR, Eds., pp. 269–271, International Institute of Rural Reconstruction Y.C.James Yen Centre, Biga, Silang Cavite, Philippines, 2001.

[25] M. Karki, "Medicinal plants for sustainable management of uplands in south and south-east Asia," in *Shifting Cultivation Towards Sustainability and Resource Conservation in Asia*, IFAD, IDRC, CIIFAD, ICRAF, and IIRR, Eds., pp. 225–231, International Institute of Rural Reconstruction Y.C.James Yen Centre Biga, Silang Cavite, Philippines, 2001.

[26] K. S. Murali, U. Shankar, R. Uma Shaanker, K. N. Ganeshaiah, and K. S. Bawa, "Extraction of non-timber forest products in the forests of Biligiri Rangan Hills, India. 2. Impact of NTFP extraction on regeneration, population structure, and species composition," *Economic Botany*, vol. 50, no. 3, pp. 252–269, 1996.

[27] H. Tynsong and B. K. Tiwari, "Contribution of NTFPs to the cash income of *War Khasi* community of South Meghalaya, North-East India," *Forestry Studies in China*, vol. 14, no. 1, pp. 47–54, 2012.

[28] A. K. Gangwar and P. S. Ramakrishnan, "Ethnobiological notes on some tribes of arunachal pradesh, northeastern India," *Economic Botany*, vol. 44, no. 1, pp. 94–105, 1990.

17

Holistic Management: Misinformation on the Science of Grazed Ecosystems

John Carter,[1] Allison Jones,[2] Mary O'Brien,[3] Jonathan Ratner,[4] and George Wuerthner[5]

[1] Kiesha's Preserve, Paris, ID 83261, USA
[2] Wild Utah Project, Salt Lake City, UT 84101, USA
[3] Grand Canyon Trust, Flagstaff, AZ 86001, USA
[4] Western Watersheds Project, Pinedale, WY 82941, USA
[5] Foundation for Deep Ecology, Bend, OR 97708, USA

Correspondence should be addressed to John Carter; johncarter@hughes.net

Academic Editor: Lutz Eckstein

Over 3 billion hectares of lands worldwide are grazed by livestock, with a majority suffering degradation in ecological condition. Losses in plant productivity, biodiversity of plant and animal communities, and carbon storage are occurring as a result of livestock grazing. Holistic management (HM) has been proposed as a means of restoring degraded deserts and grasslands and reversing climate change. The fundamental approach of this system is based on frequently rotating livestock herds to mimic native ungulates reacting to predators in order to break up biological soil crusts and trample plants and soils to promote restoration. This review could find no peer-reviewed studies that show that this management approach is superior to conventional grazing systems in outcomes. Any claims of success due to HM are likely due to the management aspects of goal setting, monitoring, and adapting to meet goals, not the ecological principles embodied in HM. Ecologically, the application of HM principles of trampling and intensive foraging are as detrimental to plants, soils, water storage, and plant productivity as are conventional grazing systems. Contrary to claims made that HM will reverse climate change, the scientific evidence is that global greenhouse gas emissions are vastly larger than the capacity of worldwide grasslands and deserts to store the carbon emitted each year.

1. Introduction

Lands grazed by livestock include 3.4 billion ha worldwide with 73% estimated to be suffering soil degradation [1]. The solution presented during Allan Savory's February 2013 TED Talk was to use holistic management (HM) to reverse desertification and climate change [2]. He reported that we are creating "too much bare ground" (1:30 in video) in the arid areas of the world and, as a consequence, rainfall runs off or evaporates, soils are damaged, and carbon is released back to the atmosphere. Grasslands, even in high rainfall areas, may contain large areas of bare ground with a crust of algae, leading to increased runoff and evaporation. Desertification is caused by livestock, "overgrazing the plants, leaving the soil bare, and giving off methane" (4:20 in video).

HM is also called holistic resource management, time controlled grazing, Savory grazing method, or short-duration grazing. It is designed to mimic the behavior of grazing animals that are regulated by their predators to gather in large groups. As Savory puts it [2], "What we had failed to understand was that these seasonal humidity environments of the world, the soil and the vegetation developed with very large numbers of grazing animals, and that these grazing animals developed with ferocious pack-hunting predators. Now, the main defense against pack-hunting predators is to get into herds, and the larger the herd, the safer the individuals. Now, large herds dung and urinate all over their own food, and they have to keep moving, and it was that movement that prevented the overgrazing of plants, while the periodic trampling ensured good cover of the soil, as we see where a herd has passed." (9:28 in video). In view of the large amount of attention received from the TED talk it is important to examine the validity of Savory's claims.

Savory's writings lack specifics that could be used for implementation of HM or for scientific testing. Details regarding setting of stocking rates, allowable use by livestock, amount of rest needed for recovery, or ecological criteria to be met for biodiversity, sustainability, wildlife, and watershed protection are absent [3–7]. These publications by Savory and his colleagues show that HM is based on the following assumptions: (1) plant communities and soils of the arid, semiarid, and grassland systems of the world evolved in the presence of large herds of animals regulated by their predators; (2) grasses in these areas will become decadent and die out if not grazed by these large herds or their modern day equivalent, livestock; (3) rest from grazing by these large herds of livestock will result in grassland deterioration; (4) large herds are needed to break up decadent plant material and soil crusts and trample dung, urine, seeds, and plant material into the soil, promoting plant growth; and (5) high intensity grazing of these lands by livestock will reverse desertification and climate change by increasing production and cover of the soil, thereby storing more carbon.

We address these five assumptions of HM with a focus on western North American arid and semiarid ecosystems, principally in the desert, steppe, grassland, and open conifer woodland biomes as described by [8]. We use the broad term, grassland, to be inclusive of these types.

2. Are Western North American Ecosystems Adapted to Herds of Large Hooved Animals?

Not all of today's grasslands, arid, and semiarid systems evolved with herds of large, hooved animals. The Great Plains of North America and subtropical grasslands in Africa that receive moisture during the long, warm, and moist growing season historically supported millions of herbivores [9–11]. Lands west of the Continental Divide of the USA, including the Great Basin, Sonoran, Mojave, and Colorado Plateau deserts, along with the Palouse Prairie grasslands of eastern Washington, western Montana, and northern Idaho, did not evolve with significant grazing pressure from bison (*Bison bison*) [9, 12, 13]. Though bison were abundant east of the Rockies on the Great Plains, they only occurred in limited numbers across western Wyoming, northeastern Utah, and southeastern Idaho [12, 14]. These low numbers and patchy occurrence would not have played the same ecological role as in the plains. Historically, pronghorn antelope (*Antilocapra americana*) were more widespread than bison west of the Rockies, but these animals are smaller and lighter than bison and are not ecologically comparable [9]. Evidence for this general lack of large herds of grazing animals west of the Rockies also includes the lack of native dung beetles in the region. Whereas 34 native species of dung beetle (g. *Onthophagus*) are found east of the Rockies on the plains where bison were numerous, none are found west of the Rockies [15].

The supposition that current western North American plant communities are adapted to livestock grazing because the region supported a diverse herbivore fauna during the Pleistocene epoch ignores that the plant communities have changed in the intervening time [16]. There was rapid evolutionary change following the Pleistocene glaciations in North America with "the establishment of open xeric grasslands in the west central part of the USA . . . less than 10,000 years ago" [17]. Many of the grasses of the Pleistocene have disappeared from the plains and western USA and the fauna of the Pleistocene was altered by the arrival of bison from Eurasia. These were destructive to long-leaved bunchgrasses found west of the Rockies, while the rhizomatous grasses found east of the Rockies in the prairies were more resistant to their grazing pressure. These rhizomatous grasses are the types found in the prairies of the central and western USA in conjunction with fossil remains of bison. In summary, the western USA of the Pleistocene is not the western USA of today. The climate was much wetter and cooler and the vegetation more mesic in the Pleistocene than today [8, 18]. The drier periods following the Pleistocene as temperatures warmed have altered soil conditions and fire cycles and contributed to the changing flora [19, 20].

Grasslands cover large areas that could support forests but are maintained by grasses outcompeting woody species for belowground resources and providing fuel for fires that limit encroachment of woody species [21]. Climate change and increasing CO_2 concentrations have been implicated in the post industrial expansion of woody species and invasive species into grasslands; however, studies from paired locations in grasslands have shown that under similar climatic and CO_2 environments, herbivory by domestic livestock has caused the shifts in woody species and increased invasive species [20, 22, 23]. Elevated CO_2 and climate warming appear to contribute to, but do not explain, the shrub encroachment in these semiarid areas which is due to intensive grazing by domestic livestock [24].

Conclusion. Western US ecosystems outside the prairies in which bison occurred are not adapted to the impact of large herds of livestock. Recent changes to these grassland ecosystems result from herbivory by domestic livestock which has altered fire cycles and promoted invasive species at the expense of native vegetation.

3. Do Grasses Senesce and Die If Not Grazed by Livestock?

A major premise of HM is that grass species depend on large grazing ungulates in some way, and thus grasses become moribund and die if not grazed, leading to deterioration or eventual loss of the entire grass community [6]. The dead or dormant residual leaves and stalks that remain attached to ungrazed grasses at the end of each growing season can be deceptive. The plants are still alive and healthy, with living buds at the plant base. Bunchgrass canopies collect snow and funnel rainwater to the plant base and soil and increase infiltration [25]. The plants and dead leaves in contact with the soil reduce overland flow and erosion [26]. Plant canopies moderate temperature and protect the growing points from temperature extremes [27]. The standing dead litter provides cover and food for wildlife species including large and small mammals, ground-living birds, and insects. The loss of these leaves and stems through heavy grazing by livestock which occurs under HM destroys these natural attributes.

Grasses with attached dead leaves are more productive than grasses from which the dead leaves have been removed. Loss of these dead tissues to grazers increases thermal damage to the growing shoots and reduces the vigor of the entire plant [28]. Dead leaves and flowering stalks on ungrazed grasses inhibit livestock grazing, allowing those grasses to grow larger than their neighbors [29]. Grazing and trampling by domestic livestock damage plants in natural plant communities [30–32], reduce forage production as stocking rates increase [33], and can lead to simplification of plant communities, establishment of woody vegetation in grasslands, and regression to earlier successional stages [20] or conversion to invasive dominated communities [23] and altered fire cycles [20]. In contrast to the assertion that grasses will die if not grazed by livestock, bunchgrasses in arid environments are more likely to die if they are heavily grazed by domestic animals [34, 35].

Conclusion. Grasses, particularly bunchgrasses, have structure that protects growing points from damage, harvests water, and protects the soil at the plant base. Removal of the standing plant material exposes the growing points, leading to loss or replacement by grazing tolerant species, including invasives.

4. Does Rest Cause Grassland Deterioration?

Another principle of HM is that grasslands and their soils deteriorate from overrest, a term that implies insufficient grazing by livestock. However, grasslands that have never been grazed by livestock have been found to support high cover of grasses and forbs. Relict sites throughout the western USA, such as on mesa tops, steep gorges, cliff sides, and even highway rights of way, which are inaccessible to livestock or most ungulates, can retain thriving bunchgrass communities [36–38]. For example, herbaceous growth was vigorous on never-grazed Jordan Valley kipukas in southeast Oregon [37] and on a once-grazed butte called The Island in south central Oregon [36]. Published comparisons of grazed and ungrazed lands in the western USA have found that rested sites have larger and more dense grasses, fewer weedy forbs and shrubs, higher biodiversity, higher productivity, less bare ground, and better water infiltration than nearby grazed sites. These reports include 139 sites in south Dakota [39], as well as sites that had been rested for 18 years in Montana [40], 30 years in Nevada [41], 20–40 years in British Columbia [42], 45 years in Idaho [43], and 50 years in the Sonoran Desert of Arizona [44]. None of the above studies demonstrated that long periods of rest damaged native grasslands. A list and description of such sites can be found in [45].

The HM misinterpretation of the natural history of grazed and ungrazed grasslands is apparent in Savory's description of the Appleton-Whittell Research Ranch in southeastern Arizona [6]. This ranch has been protected from livestock grazing since 1968 with the grasses on the ranch described by Savory as becoming "moribund" (page 211), with "bare spots opening up" (page 211). In contrast to those claims, plant species richness on the ranch increased from 22 species in 1969 to 49 species in 1984, while plant cover increased from 29% in 1968 to 85% in 1984 [46]. Total grass cover on the ranch was significantly higher on ungrazed sites when compared to grazed sites ($P < 0.01$) [47]. These well designed studies produced quantitative data showing that the HM view of the ranch is not the case.

Conclusion. Contrary to the assumption that grasses will senesce and die if not grazed by livestock, studies of numerous relict sites, long-term rested sites, and paired grazed and ungrazed sites have demonstrated that native plant communities, particularly bunchgrasses, are sustained by rest from livestock grazing.

5. Is Hoof Action Necessary for Grassland Health?

A key premise of HM is that livestock can be made to emulate native ungulate responses to predators by moving them frequently in large numbers and tight groups. This promotes very close cropping which is said to benefit grasses and other forage, as well as hoof action that breaks up soil crusts, increases infiltration, plants seeds, and incorporates plant material, manure, and urine into the soil [6]. Other than bison in the plains states, the evidence indicates a low frequency of large hooved mammals in the western USA during pre-Columbian times [48], so the opportunity for hoof action to sustain grasslands and deserts appears limited at best. In contrast to HM claims, elk (Cervus canadensis), mule deer (Odocoileus hemionus), and other ungulates may avoid areas where predators have an advantage in capturing them [49]. Avoidance is not the same as a panicked flight or tight groupings of animals promoting hoof action. Rather, the major response is greater vigilance and sometimes avoidance of risky areas. While the presence of wolves (Canis lupus) affects elk behavior by reducing browsing on willows and aspen [50], snow depth and other ecological needs appear to outweigh the effect of wolves leading to grazing and browsing in areas of higher risk [51, 52]. We found no documentation of native animal responses to predators generating hoof action or herd effect in tight groupings in the western USA.

Soils in arid and semiarid grasslands often have significant areas covered by biological crusts [53–55]. These are made up of bacteria, cyanobacteria, algae, mosses, and lichens and are essential to the health of these grasslands. Biological crusts stabilize soils, increase soil organic matter and nutrient content, absorb dew during dry periods, and fix nitrogen [53, 56–60]. Crusts enhance soil stability and reduce water runoff by producing more microcatchments on soil surfaces. They increase water absorbing organic matter, improve nutrient flow, germination and establishment for some plants, while dark crusts may stimulate plant growth by producing warmer soil temperatures and water uptake in cold deserts [61]. Some crusts are hydrophobic, shedding water [60]. Biological soil crusts are fragile, highly susceptible to trampling [61–63], and are slow to recover from trampling impacts [64]. Loss of these crusts results in increased erosion and reduced soil fertility. The loss of crusts in the bunchgrass communities of the western USA may be largely responsible for the widespread establishment of cheatgrass and other exotic annuals [23, 58, 65]. The rapid spread of introduced weeds throughout the arid western USA is estimated at over 2000 hectares per day [66], largely due to livestock disturbance.

The HM assumption that increasing hoof action will increase infiltration has been disproven. Livestock grazing can compact soil, reduce infiltration, and increase runoff, erosion, and sediment yield [67–71]. Major increases in erosion and runoff occur under normal stocking when comparing grazed to ungrazed sites [68, 71–74]. Extensive literature reviews report the negative impacts of livestock grazing on soil stability and erosion [75–77]. For example, a study of wet and dry meadows in Oregon found the infiltration rate in ungrazed dry meadows was 13 times greater and 2.3 times greater in ungrazed wet meadows, compared to similar grazed meadows [78].

Hoof action is not needed to increase soil fertility and decomposition of litter. It is well-established that soil protozoa, arthropods, earthworms, microscopic bacteria, and fungi decompose plant and animal residues in all environments [79, 80]. Even the driest environments contain 100 million to one billion decomposing bacteria and tens to hundreds of meters of fungal hyphae per gram of soil [81]. Brady and Weil [80] discuss the importance of mammals in the decay process, mentioning burrowing mammals, but not large grazers such as cattle and bison. Removal of plant biomass and lowered production resulting from livestock grazing can reduce fertility and organic content of the soil [70, 82–84].

Conclusion. We found no evidence that hoof action as described by Savory occurs in the arid and semiarid grasslands of the western USA which lacked large herds of ungulates such as bison that occurred in the prairies of the USA or the savannahs of Africa. No benefits of hoof action were found. To the contrary, hoof action by livestock has been documented to destroy biological crusts, a key component in soil protection and nutrient cycling, thereby increasing erosion rates and reducing fertility, while, increasing soil compaction and reducing water infiltration.

6. Can Grazing Livestock Increase Carbon Storage and Reverse Climate Change?

Among the most recent HM claims is that livestock grazing will lead to sequestration of large amounts of carbon, thus potentially reversing climate change [2]. However, any increased carbon storage through livestock grazing must be weighed against the contribution of livestock metabolism to greenhouse gas emissions due to rumen bacteria methane

emissions, manure, and fossil fuel use across the production chain [85, 86]. Nitrous oxide, 300 times more potent than methane in trapping greenhouse gases [87], is also produced and released with livestock production. The livestock industry's contribution to greenhouse gases also includes CO_2 released by conversion of forests to grasslands for the purpose of grazing [86].

Worldwide, livestock production accounts for about 37 percent of global anthropogenic methane emissions and 65 percent of anthropogenic nitrous oxide emissions with as much as 18% of current global greenhouse gas emissions (CO_2 equivalent) generated from the livestock industry [85]. It is estimated that livestock production, byproducts, and other externalities account for 29.5 billion metric tons of CO_2 per year or 51 percent of annual worldwide greenhouse gas emissions from agriculture [88]. Lower amounts of greenhouse gas emissions due to livestock may be estimated by using narrower definitions of livestock-related emissions that include feed based emissions only and exclude externalities [89].

Some suggest that grass-fed beef is a superior alternative to beef produced in confined animal feeding operations [90]. However, grass provides less caloric energy per pound of feed than grain and, as a consequence, a grass-fed cow's rumen bacteria must work longer breaking down and digesting grass in order to extract the same energy content found in grain, while the bacteria in its rumen are emitting methane [89]. Comparisons of pasture-finished and feedlot-finished beef in the USA found that pasture-finished beef produced 30% more greenhouse gas emissions on a live weight basis [91].

It is estimated that three times as much carbon resides in soil organic matter as in the atmosphere [92], while grasslands and shrublands have been estimated to store 30 percent of the world's soil carbon with additional amounts stored in the associated vegetation [93]. Long term intensive agriculture can significantly deplete soil organic carbon [94] and past livestock grazing in the United States has led to such losses [95, 96]. Livestock grazing was also found to significantly reduce carbon storage on Australian grazed lands while destocking currently grazed shrublands resulted in net carbon storage [97]. Livestock-grazed sites in Canyonlands National Park, Utah, had 20% less plant cover and 100% less soil carbon and nitrogen than areas grazed only by native herbivores [98]. Declines in soil carbon and nitrogen were found in grazed areas compared to ungrazed areas in sage steppe habitats in northeastern Utah [84]. As grazing intensity increased, mycorrhizal fungi at the litter/soil interface were destroyed by trampling, while ground cover, plant litter, and soil organic carbon and nitrogen decreased [84]. A review by Beschta et al. [20] determined that livestock grazing and trampling in the western USA led to a reduction in the ability of vegetation and soils to sequester carbon and also led to losses in stored carbon.

Conclusion. Livestock are a major source of greenhouse gas emissions. Livestock removal of plant biomass and altering of soil properties by trampling and erosion causes loss of carbon storage and nutrients as evidenced by studies in grazed and ungrazed areas.

7. What Is the Evidence That Holistic Management Does Not Produce the Claimed Effects?

HM is a management system that includes setting goals, monitoring, and adapting in order to continually move towards the goals established by the producer [6]. This more goal-oriented and adaptive management aspect of the HM system, its promise of environmental benefits, and increased production make it attractive to many ranchers [99]. However, researchers who have studied HM in South Africa and Zimbabwe, where Savory originated his theories, have rejected many of HM's underlying assumptions and found that HM approaches result in reduced water infiltration into the soil, increased erosion, reduced forage production, reduced soil organic matter and nitrogen, reduced mineral cycling, and increased soil bulk density [82, 100, 101].

In a recent evaluation of HM by Briske et al. [102], three of its principle claims were addressed: (1) all nonforested lands are degraded; (2) these lands can store all fossil fuel carbon in the atmosphere; and (3) intensive grazing is necessary to prevent the degradation. The authors pointed out that there are well managed lands that are not degraded; deserts are a consequence of climate and soils as well as improper management; and degradation is largely a function of growing populations of humans and livestock, land fragmentation, and other societal issues. As to the claim that these nonforested lands could store all the carbon emissions that humans produce, the researchers show that the potential carbon sequestration of these lands is only about one to two billion metric tons per year (mtpy), a small fraction of global carbon emissions of 50 billion mtpy. They further point out that these lands would have to produce much larger vegetation biomass than they are capable of producing in order to sequester human-caused carbon emissions and that much of the carbon is released back to the atmosphere through respiration as CO_2. They note that grass cover increases dramatically with rest and intensive grazing delays this recovery; many desert grassland soils are sandy, so hoof action does not increase infiltration; and biological crusts stabilize these soils and protect them from wind erosion and carbon loss.

A review of short-duration grazing studies in the western USA by Holechek et al. [83] included locations in the more arid western states as well as prairie types. The researchers found that this grazing system, which is equated with HM, resulted in decreased infiltration, increased erosion, and reduced soil organic matter and nitrogen. Forage production and range condition were similar under short-duration and continuous grazing with the same stocking rates. Under short-duration grazing, standing crop of forage declined as stocking rates increased, while bare ground and vegetation composition were a function of stocking rate as opposed to grazing system. Grazing distribution was not improved over continuous grazing and the claims for hoof action and improved range condition under increased stocking rates and densities were not realized [83]. Another review of grazing systems by Briske et al. [103], including HM, versus continuous grazing concluded that plant and animal production were equal or greater in continuous grazing than in rotational grazing systems.

Even though the ecosystems of the Great Plains states evolved with the pressure of bison, Holechek et al. [83] and Briske et al. [103] found that HM did not differ from traditional, season-long grazing for most dependent variables compared. Studies commonly held up as supporting HM [104–108] used HM paddocks that were grazed with light to moderate grazing, not the heavy grazing that Savory recommends. Further, long-term range studies have shown that it is reductions in stocking rate that lead to increased forage production and improvements in range condition, not grazing system [33, 109, 110]. While HM advocates allowing recovery to take place following grazing, recovery can take many years to decades even under total rest from livestock, but it does occur [43, 111]. Native, western USA bunchgrass species such as bluebunch wheatgrass (*Pseudoroegneria spicata*) and Idaho fescue (*Festuca idahoensis*) are sensitive to defoliation and can require long periods (years) of rest following each period of grazing in order to restore their vigor and productivity [34, 35].

Conclusion. Studies in Africa and the western USA, including the prairies which evolved in the presence of bison, show that HM, like conventional grazing systems, does not compensate for overstocking of livestock. As in conventional grazing systems, livestock managed under HM reduce water infiltration into the soil, increase soil erosion, reduce forage production, reduce range condition, reduce soil organic matter and nutrients, and increase soil bulk density. Application of HM cannot sequester much, let alone all the greenhouse gas emissions from human activities because the sequestration capacity of grazed lands is much less than annual greenhouse gas emissions.

8. What about Riparian Areas and Biodiversity?

In the western USA, riparian areas are rare and valuable ecological systems supporting a disproportionate number of species and providing many ecosystem services [112]. How does HM, with its emphasis on high stocking rates and trampling, affect these systems? Soil compaction from livestock is a common and widespread problem in grazed riparian areas, reducing infiltration rates and water storage and increasing surface runoff and soil erosion during storm events [78, 112]. Soil compaction from livestock increases with increased numbers, as in an HM application [70]. Livestock grazing in riparian areas reduces willow and herbaceous production and canopy cover of shrubs and grasses compared to ungrazed controls [113]. The most effective way to restore damaged riparian areas is to remove livestock [110, 112].

We found very little information about total number of plant, animal, or invertebrate species present when HM is compared to other grazing methods or nongrazed areas, and, further, what proportion of total plant species or total cover of plant species was native or nonnative. Moreover, we did not see other biodiversity considerations addressed in any of the published studies investigating HM. Rotational grazing systems do not improve range condition and plant production over conventional grazing systems [83, 103], while stocking rate is considered the most important variable affecting vegetation production and range condition [33]. Range condition is determined based on the current plant community composition and production as compared to the potential natural community [114]. The relative composition of Increasers, those plants with tolerance for grazing, Decreasers, those plants with low tolerance for grazing, and Invasives, those plants occupying a site that are grazing tolerant and nonnative, forms the basis of the determination of condition. Higher range condition ratings reflect greater similarity to the native plant community for a site [115]. This basic concept reflects the biodiversity of the native plant community, which necessarily declines as range condition declines. Application of HM with its large herd size and density of use, like other grazing systems with high stocking rates, must necessarily decrease native plant diversity and productivity. This affects the animal communities accordingly as habitat structure and production are altered.

A review, by Fleischner [76], of the effects of livestock grazing on plant and animal communities in the western USA found that livestock grazing reduced species richness and abundance of plants, small mammals, birds, reptiles, insects, and fish compared to conditions following removal of livestock. A quantitative review by Jones [77] of published studies of ecosystem attributes in North American arid ecosystems affected by livestock grazing, compared to ungrazed conditions, found decreases in rodent species richness and diversity and vegetation diversity in the grazed areas. Livestock grazing-induced simplified plant communities in western USA arid and semiarid lands have negative effects on pollinators, birds, small mammals, amphibians, wild ungulates, and other native wildlife [20]. Riparian songbird abundance increases as riparian systems recover after livestock exclusion [116, 117], while overall biodiversity increases under long term rest from livestock grazing [46, 47, 118]. Invasives such as cheatgrass (*Bromus tectorum*) are favored and increase in abundance in the presence of livestock grazing [23, 65] and are inversely related to abundance of native perennial grasses [43].

Conclusion. HM does not address riparian areas and biodiversity with its focus on livestock production, although operators could choose these as goals. We have seen no studies of HM impacts on riparian areas and biodiversity, although livestock grazing impacts on riparian areas and biodiversity have been well documented. Livestock degrade riparian areas by removal of streamside vegetation, reduction of cover and food for fish and wildlife, and soil compaction, erosion, and sedimentation. These impacts lead to loss of native fish and wildlife populations. Studies in areas from which livestock have been removed demonstrate increases in diversity and abundance of birds, mammals, insects, and fish.

9. Is Scientific Evidence Important?

Effectiveness studies of HM have been undertaken by ranchers and farmers who were selected because of their

commitment to HM [119]. In other words, such studies were neither experimental nor were the participants randomly selected. Livestock producers who may have had negative experiences with HM were not included in the studies. Nearly all of the support and confirmation for HM come from articles developed at the Savory Institute or testimonials by practitioners. Most of the published literature that attempts to rigorously test HM in any scientific fashion does not support its principal assumptions.

Holechek et al. [83] stated that "No grazing approach, including that of Savory, will overcome the adverse effects of drought and/or chronic heavy stocking on forage production." These researchers were also critical of government agencies for adopting these unproven theories rather than basing management on "scientifically proven range management practices and principles" [83] (page 25). Briske et al. [103] stated,

> the rangeland profession has become mired in confusion, misinterpretation, and uncertainty with respect to the evaluation of grazing systems and the development of grazing recommendations and policy decisions. We contend that this has occurred because recommendations have traditionally been based on perception, personal experience, and anecdotal interpretations of management practices, rather than evidence-based assessments of ecosystem responses. [103] (page 11).

Briske et al. [102] state, "Mr. Savory's attempts to divide science and management perspectives and his aggressive promotion of a narrowly focused and widely challenged grazing method only serve to weaken global efforts to promote rangeland restoration and C sequestration." [102] (page 74).

Conclusion. Studies supporting HM have generally come from the Savory Institute or anecdotal accounts of HM practitioners. Leading range scientists have refuted the system and indicated that its adoption by land management agencies is based on these anecdotes and unproven principles rather than scientific evidence. When addressing the application of HM or any other grazing systems, practitioners, including agencies managing public lands, private livestock operations, and scientists, should (1) consider inclusion of watershed-scale ungrazed reference areas of suitable size to encompass the plant and soil communities found in the grazed area, (2) define ecological (plant, soil, and animal community) and production (livestock) criteria on which to base quantitative comparisons, (3) use sufficient replication in studies, (4) and include adequate quality control of methods. Economic analysis of grazing systems should compare all expenditures with income, including externalized costs such as soil loss, water pollution, reduction of water infiltration, and carbon emissions and capture.

10. Management Implications

This review shows that the underlying assumptions of HM regarding the evolutionary adaptation of western North

American landscapes to large herds of hooved animals only applies to prairie grasslands and that most arid and semiarid areas of western North America are not adapted to their impacts. The premise that rest results in degradation of grassland ecosystems by allowing biological crusts to persist and grasses to senesce and die has been disproven by a large body of research. Reliance on hoof action to promote recovery by trampling seeds and organic matter into the soil and breaking up soil crusts needs to be considered in the context of increased soil compaction, lower infiltration rates, and the destruction of biological crusts that normally provide long-term stability to soil surfaces, enhance water retention, and promote nutrient cycling. The use of HM in an attempt to capture atmospheric greenhouse gases and incorporate them into soils and plant communities, thereby reducing climate change effects, is demonstrably impossible because the nonforested grazed lands of the world do not have the capacity to sequester this amount of emissions. Even in the prairie regions of the United States, which are evolutionarily adapted to large herbivores such as bison, research indicates that not only does HM not produce results superior to conventional season-long grazing, but also that stocking rate, rest, and livestock exclusion represent the best mechanisms for restoring grassland productivity, ecological condition, and sustainability. Various studies indicate livestock grazing reduces biodiversity of native species and degrades riparian areas, with nearly all studies finding livestock exclusion to be the most effective, reliable means to restore degraded riparian areas. Claims of the benefits of HM or other grazing systems should be validated by quantitative, scientifically valid studies.

Conflict of Interests

The authors declare that there is no conflict of interests regarding the publication of this paper.

Funding

Research was funded by the Foundation for Deep Ecology, Grand Canyon Trust, Kiesha's Preserve, Western Watersheds Project and Wild Utah Project.

Acknowledgment

Joy Belsky (1944–2001), Range Ecologist, who made available a draft analysis of holistic management before her death, provided us with much of the material presented here.

References

[1] E. Gabathuler, H. Liniger, C. Hauert, and M. Giger, *Benefits of Sustainable Land Management*, World Overview of Conservation Approaches and Technologies, Center for Development and Environment, University of Bern, Bern, Switzerland, 2009.

[2] A. Savory, "How to fight desertification and reverse climate change," 2013, http://www.ted.com/talks/allan_savory_how_to_green_the_world_s_deserts_and_reverse_climate_change.html.

[3] A. Savory and S. D. Parsons, "The Savory grazing method," *Rangelands*, vol. 2, pp. 234–237, 1980.

[4] A. Savory, "The Savory grazing method or holistic resource management," *Rangelands*, vol. 5, pp. 155–159, 1983.

[5] A. Savory, *Holistic Resource Management*, Island Press, Washington, DC, USA, 1988.

[6] A. Savory and J. Butterfield, *Holistic Management: A New Framework for Decision Making*, Island Press, Washington, DC, USA, 1999.

[7] C. J. Hadley, "The wild life of Allan Savory," *Rangelands*, vol. 22, pp. 6–10, 2000.

[8] R. S. Thompson and K. H. Anderson, "Biomes of western North America at 18,000, 6000 and 0 ^{14}C yr BP reconstructed from pollen and packrat midden data," *Journal of Biogeography*, vol. 27, no. 3, pp. 555–584, 2000.

[9] R. N. Mack and J. N. Thompson, "Evolution in steppe with few large hooved mammals," *American Naturalist*, vol. 119, no. 6, pp. 757–773, 1982.

[10] A. R. E. Sinclair and M. Norton-Griffiths, *Serengeti: Dynamics of an Ecosystem*, University of Chicago Press, 1979.

[11] A. R. E. Sinclair, S. A. R. Mduma, J. G. C. Hopcraft, J. M. Fryxell, R. Hilborn, and S. Thirgood, "Long-term ecosystem dynamics in the serengeti: lessons for conservation," *Conservation Biology*, vol. 21, no. 3, pp. 580–590, 2007.

[12] R. Daubenmire, "The western limits of the range of the American bison," *Ecology*, vol. 66, no. 2, pp. 622–624, 1985.

[13] G. Wuerthner, "Are cows just domestic bison? Behavioral and habitat use differences between cattle and bison," in *Proceedings of an International Symposium on Bison Ecology and Management in North America*, L. Irby, L. Knight, and J. Knight, Eds., pp. 374–383, Bozeman, Mont, USA, June 1998.

[14] F. G. Roe, *The North American Buffalo: A Critical Study of the Species in Its Wild State*, University of Toronto Press, Toronto, Calif, USA, 1951.

[15] J. F. Howden, "Some possible effects of the Pleistocene on the distributions of North American Scarabaeidae (Coleoptera)," *Canadian Entomologist*, vol. 98, no. 11, pp. 1177–1190, 1966.

[16] J. W. Burkhardt, *Herbivory in the Intermountain West*, vol. 58 of *Station Bulletin*, University of Idaho Forest, Wildlife and Range Experiment Station, Moscow, Idaho, USA, 1996.

[17] J. M. J. de Wet, "Grasses and the culture history of man," *Annals Missouri Botanical Garden*, vol. 68, no. 1, pp. 87–104, 1981.

[18] H. Wanner, J. Beer, J. Bütikofer et al., "Mid- to Late Holocene climate change: an overview," *Quaternary Science Reviews*, vol. 27, no. 19-20, pp. 1791–1828, 2008.

[19] D. K. Grayson, "Mammalian responses to middle Holocene climatic change in the Great Basin of the western United States," *Journal of Biogeography*, vol. 27, no. 1, pp. 181–192, 2000.

[20] R. L. Beschta, D. L. Donahue, A. DellaSala et al., "Adapting to climate change on western public lands: addressing the ecological effects of domestic, wild, and feral ungulates," *Environmental Management*, vol. 51, no. 2, pp. 474–491, 2012.

[21] W. J. Bond, "What limits trees in C4 grasslands and savannas?" *Annual Review of Ecology, Evolution, and Systematics*, vol. 39, pp. 641–659, 2008.

[22] S. Archer, D. S. Schimel, and E. A. Holland, "Mechanisms of shrubland expansion: land use, climate or CO_2?" *Climatic Change*, vol. 29, no. 1, pp. 91–99, 1995.

[23] M. D. Reisner, J. B. Grace, D. A. Pyke, and P. S. Doescher, "Conditions favoring *Bromus tectorum* dominance of endangered

sagebrush steppe ecosystems," *Journal of Applied Ecology*, vol. 50, no. 4, pp. 1039–1049, 2013.

[24] O. W. van Auken, "Shrub invasions of North American semiarid grasslands," *Annual Review of Ecology and Systematics*, vol. 31, pp. 197–215, 2000.

[25] G. W. Gee, P. A. Beedlow, and R. L. Skaggs, "Water balance," in *Shrub-Steppe Balance and Change in a Semi-Arid Terrestrial Ecosystem*, W. H. Rickard, L. E. Rogers, B. E. Vaughan, and S. F. Liebetrau, Eds., pp. 61–81, Elsevier, New York, NY, USA, 1988.

[26] W. H. Wischmeier and D. D. Smith, *Predicting Rainfall Erosion Losses: A Guide to Conservation Planning*, vol. 537 of *Agriculture Handbook*, US Department of Agriculture, Washington, DC, USA, 1978.

[27] W. T. Hinds and W. H. Rickard, "Soil temperatures near a desert steppe shrub," *Northwest Science*, vol. 42, pp. 5–8, 1968.

[28] R. H. Sauer, "Effect of removal of standing dead material on growth of Agropyron spicatum," *Journal of Range Management*, vol. 31, no. 2, pp. 121–122, 1978.

[29] D. Ganskopp, R. Angell, and J. Rose, "Response of cattle to cured reproductive stems in a caespitose grass," *Journal of Range Management*, vol. 45, no. 4, pp. 401–404, 1992.

[30] L. Ellison, "Influence of grazing on plant succession of Rangelands," *The Botanical Review*, vol. 26, no. 1, pp. 1–78, 1960.

[31] A. J. Belsky, "Does herbivory benefit plants? A review of the evidence," *American Naturalist*, vol. 127, no. 6, pp. 870–892, 1986.

[32] E. L. Painter and A. J. Belsky, "Application of herbivore optimization theory to rangelands of the western United States," *Ecological Appplications*, vol. 3, no. 1, pp. 2–9, 1993.

[33] J. L. Holechek, H. Gomez, F. Molinar, and D. Galt, "Grazing studies: what we've learned," *Rangelands*, vol. 21, no. 2, pp. 12–16, 1999.

[34] W. F. Mueggler, "Rate and pattern of vigor recovery in Idaho fescue and bluebunch wheatgrass," *Journal of Range Management*, vol. 28, no. 3, pp. 198–204, 1975.

[35] L. D. Anderson, *Bluebunch Wheatgrass Defoliation, Effects and Recovery—A Review*, vol. 91-2 of *BLM Technical Bulletin*, Bureau of Land Management, Idaho State Office, Boise, Idaho, USA, 1991.

[36] R. S. Driscoll, "A relict area in the central Oregon juniper zone," *Ecology*, vol. 45, no. 2, pp. 345–353, 1964.

[37] R. R. Kindschy, "Pristine vegetation of the Jordan Crater kipukas: 1978–1991," in *Proceedings-Ecology and Management of Annual Rangelands*, S. B. Monsen and S. G. Kitchen, Eds., INT-GTR-313, pp. 85–88, US Department of Agriculture, Forest Service, Boise, Idaho, USA, May 1992.

[38] N. Ambos, G. Robertson, and J. Douglas, "Dutchwoman butte: a relict grassland in central Arizona," *Rangelands*, vol. 22, no. 2, pp. 3–8, 2000.

[39] D. F. Costello and G. T. Turner, "Vegetation changes following exclusion of livestock from grazed ranges," *Journal of Forestry*, vol. 39, pp. 310–315, 1941.

[40] A. B. Evanko and R. A. Peterson, "Comparisons of protected and grazed mountain rangelands in southwestern Montana," *Ecology*, vol. 36, no. 1, pp. 71–82, 1955.

[41] J. H. Robertson, "Changes on a sagebrush-grass range in Nevada ungrazed for 30 years," *Journal of Range Management*, vol. 24, no. 5, pp. 397–400, 1971.

[42] A. McLean and E. W. Tisdale, "Recovery rate of depleted range sites under protection from grazing," *Journal of Range Management*, vol. 25, no. 3, pp. 178–184, 1972.

[43] J. E. Anderson and R. S. Inouye, "Landscape-scale changes in plant species abundance and biodiversity of a sagebrush steppe over 45 years," *Ecological Monographs*, vol. 71, no. 4, pp. 531–556, 2001.

[44] J. Blydenstein, C. R. Hungerford, G. I. Day, and R. R. Humphrey, "Effect of domestic livestock exclusion on vegetation in the Sonoran Desert," *Ecology*, vol. 38, no. 3, pp. 522–526, 1957.

[45] G. Wuerthner and M. Matteson, "A guide to livestock-free landscapes," in *Welfare Ranching: The Subsidized Destruction of the American West*, G. Wuerthner and M. Mattson, Eds., pp. 327–329, Island Press, Washington, DC, USA, 2004.

[46] W. W. Brady, M. R. Stromberg, E. F. Aldon, C. D. Bonham, and S. H. Henry, "Response of a semidesert grassland to 16 years of rest from grazing," *Journal of Range Management*, vol. 42, no. 4, pp. 284–288, 1989.

[47] C. E. Bock, J. H. Bock, W. R. Penney, and V. M. Hawthorne, "Responses of birds, rodents, and vegetation to livestock exclosure in a semidesert grassland site," *Journal of Range Management*, vol. 37, no. 3, pp. 239–242, 1984.

[48] W. J. Ripple and B. van Valkenburgh, "Linking top-down forces to the pleistocene megafaunal extinctions," *BioScience*, vol. 60, no. 7, pp. 516–526, 2010.

[49] J. W. Laundré, L. Hernández, and W. J. Ripple, "The landscape of fear: ecological implications of being afraid," *The Open Ecology Journal*, vol. 3, no. 2, pp. 1–7, 2010.

[50] C. Eisenberg, S. T. Seager, and D. E. Hibbs, "Wolf, elk, and aspen food web relationships: context and complexity," *Forest Ecology and Management*, vol. 299, pp. 70–80, 2013.

[51] S. Creel and D. Christianson, "Wolf presence and increased willow consumption by Yellowstone elk: implications for trophic cascades," *Ecology*, vol. 90, no. 9, pp. 2454–2466, 2009.

[52] J. Winnie Jr. and S. Creel, "Sex-specific behavioural responses of elk to spatial and temporal variation in the threat of wolf predation," *Animal Behaviour*, vol. 73, no. 1, pp. 215–225, 2007.

[53] J. Belnap, D. Eldridge, J. H. Kaltenecker, S. Leonard, R. Rosentreter, and J. Williams, *Biological Soil Crusts Ecology and Management*, TR-1730-2, US Department of Interior, Bureau of Land Management, Denver, Colo, USA, 2001.

[54] N. E. West, "Western intermountain sagebrush steppe," in *Temperate Deserts and Semi-Deserts*, N. E. West, Ed., pp. 351–373, Elsevier Scientific Publishing Company, Amsterdam, The Netherlands, 1983.

[55] J. Belnap and O. L. Lange, Eds., *Biological Soil Crusts: Structure, Function, and Management*, Springer, New York, NY, USA, 2003.

[56] O. L. Lange, E. D. Schulze, L. Kappen, U. Buschbom, and M. Evenari, "Adaptations of desert lichens to drought and extreme temperatures," in *Environmental Physiology of Desert Ecosystems*, N. F. Hadley, Ed., pp. 27–30, Dowden, Hutchinson and Ross, Stroudsberg, Pa, USA, 1975.

[57] J. A. R. Ladyman and E. Muldavin, *Terrestrial Cryptogams of Pinyon-Juniper Woodlands in the Southwestern US: A Review*, RM-GTR-280, US Department of Agriculture, Forest Service, Fort Collins, Colo, USA, 1996.

[58] A. J. Belsky and J. L. Gelbard, *Livestock Grazing and Weed Invasions in the Arid West*, Oregon Natural Desert Association, Bend, Ore, USA, 2000.

[59] G. Wuerthner, "The soil's living surface: biological crusts," in *Welfare Ranching: The Subsidized Destruction of the American West*, G. Wuerthner and M. Mattson, Eds., pp. 199–204, Island Press, Washington, DC, USA, 2004.

[60] L. Deines, R. Rosentreter, D. J. Eldridge, and M. D. Serpe, "Germination and seedling establishment of two annual grasses on lichen-dominated biological soil crusts," *Plant and Soil*, vol. 295, no. 1-2, pp. 23–35, 2007.

[61] J. Belnap, "Potential role of cryptobiotic soil crust in semi-arid rangelands," in *Proceedings-Ecology and Management of Annual Rangelands*, S. B. Monsen and S. G. Kitchen, Eds., INT-GTR-313, pp. 179–185, US Department of Agriculture, Forest Service, Boise, Idaho, USA, May 1992.

[62] E. F. Kleiner and K. T. Harper, "Environment and community organization in grasslands of Canyonlands National Park," *Ecology*, vol. 53, no. 2, pp. 299–309, 1972.

[63] M. L. Floyd, T. L. Fleischner, D. Hanna, and P. Whitefield, "Effects of historic livestock grazing on vegetation at chaco culture National Historic Park, New Mexico," *Conservation Biology*, vol. 17, no. 6, pp. 1703–1711, 2003.

[64] J. M. Ponzetti and B. P. McCune, "Biotic soil crusts of Oregon's shrub steppe: community composition in relation to soil chemistry, climate, and livestock activity," *Bryologist*, vol. 104, no. 2, pp. 212–225, 2001.

[65] R. N. Mack, "Invasion of *Bromus tectorum* L. into western North America: an ecological chronicle," *Agro-Ecosystems*, vol. 7, no. 2, pp. 145–165, 1981.

[66] US Bureau of Land Management, "Partners against weeds: an action plan for the Bureau of land management," Tech. Rep. BLM/MT/ST-96/003+1020, US Bureau of Land Management, Billings, Mont, USA, 1996.

[67] L. Ellison, "Influence of grazing on plant succession of Rangelands," *The Botanical Review*, vol. 26, no. 1, pp. 1–78, 1960.

[68] G. C. Lusby, "Effects of grazing on runoff and sediment yield from desert rangeland at Badger Wash in western Colorado, 1953–1973," US Geological Survey Water Supply Paper 1532-1, 1979.

[69] S. D. Warren, M. B. Nevill, W. H. Blackburn, and N. E. Garza, "Soil response to trampling under intensive rotation grazing," *Soil Science Society of America Journal*, vol. 50, no. 5, pp. 1336–1341, 1986.

[70] S. W. Trimble and A. C. Mendel, "The cow as a geomorphic agent—a critical review," *Geomorphology*, vol. 13, no. 1–4, pp. 233–253, 1995.

[71] G. P. Asner, A. J. Elmore, L. P. Olander, R. E. Martin, and T. Harris, "Grazing systems, ecosystem responses, and global change," *Annual Review of Environment and Resources*, vol. 29, pp. 261–299, 2004.

[72] W. P. Cottam and F. R. Evans, "A comparative study of the vegetation of grazed and ungrazed canyons of the Wasatch Range, Utah," *Ecology*, vol. 26, no. 2, pp. 171–181, 1945.

[73] J. L. Gardner, "The effects of thirty years of protection from grazing in desert grassland," *Ecology*, vol. 31, no. 1, pp. 44–50, 1950.

[74] J. B. Kauffman, W. C. Krueger, and M. Vavra, "Effects of late season cattle grazing on riparian plant communities," *Journal of Range Management*, vol. 36, no. 6, pp. 685–691, 1983.

[75] G. F. Gifford and R. H. Hawkins, "Hydrologic impact of grazing on infiltration: a critical review," *Water Resources Research*, vol. 14, no. 2, pp. 305–313, 1978.

[76] T. L. Fleischner, "Ecological costs of livestock grazing in western North America," *Conservation Biology*, vol. 8, no. 3, pp. 629–644, 1994.

[77] A. Jones, "Effects of cattle grazing on North American arid ecosystems: a quantitative review," *Western North American Naturalist*, vol. 60, no. 2, pp. 155–164, 2000.

[78] J. B. Kauffman, A. S. Thorpe, and E. N. J. Brookshire, "Livestock exclusion and belowground ecosystem responses in riparian meadows of eastern Oregon," *Ecological Applications*, vol. 14, no. 6, pp. 1671–1679, 2004.

[79] R. E. Ingham, J. A. Trofymow, E. R. Ingham, and D. C. Coleman, "Interactions of bacteria, fungi, and their nematode grazers: effects on nutrient cycling and plant growth," *Ecological Monographs*, vol. 55, no. 1, pp. 119–140, 1985.

[80] N. C. Brady and R. R. Weil, *The Nature and Properties of Soils*, Prentice-Hall, Upper Saddle River, NJ, USA, 12th edition, 1999.

[81] E. R. Ingham, *Soil Biology Primer*, US Department of Agriculture, Natural Resources Conservation Service, Soil Quality Institute, 1999.

[82] J. Skovlin, "Southern Africa's experience with intensive short duration grazing," *Rangelands*, vol. 9, pp. 162–167, 1987.

[83] J. L. Holechek, H. Gomes, F. Molinar, D. Galt, and R. Valdez, "Short-duration grazing: the facts in 1999," *Rangelands*, vol. 22, no. 1, pp. 18–22, 2000.

[84] J. Carter, B. Chard, and J. Chard, "Moderating livestock grazing effects on plant productivity, carbon and nitrogen storage," in *Proceedings of the 17th Wildland Shrub Symposium*, T. A. Monaco et al., Ed., pp. 191–205, Logan, Utah, USA, May 2010.

[85] H. Steinfeld, P. Gerber, T. Wassentaar, V. Castel, M. Rosales, and C. de Haan, *Livestock's Long Shadow*, Food and Agriculture Organization of the United Nations, Rome, Italy, 2006.

[86] R. Goodland and J. Anhang, "Livestock and climate change," *World Watch*, vol. 22, no. 6, pp. 10–19, 2009.

[87] Environmental Protection Agency, "Climate change overview of nitrous oxide," 2013, http://epa.gov/climatechange/ghgemissions/gases/n2o.html.

[88] L. Reynolds, "Agriculture and livestock remain major sources of greenhouse gas emissions," 2013, http://www.worldwatch.org/agriculture-and-livestock-remain-major-sources-greenhouse-gas-emissions-1.

[89] K. A. Johnson and D. E. Johnson, "Methane emissions from cattle," *Journal of Animal Science*, vol. 73, no. 8, pp. 2483–2492, 1995.

[90] L. Abend, "How cows (grass-fed only) could save the planet," Time Magazine, 2010.

[91] N. Pelletier, R. Pirog, and R. Rasmussen, "Comparative life cycle environmental impacts of three beef production strategies in the Upper Midwestern United States," *Agricultural Systems*, vol. 103, no. 6, pp. 380–389, 2010.

[92] R. R. Allmaras, H. H. Schomberg, J. Douglas C.L., and T. H. Dao, "Soil organic carbon sequestration potential of adopting conservation tillage in U.S. croplands," *Journal of Soil and Water Conservation*, vol. 55, no. 3, pp. 365–373, 2000.

[93] J. Grace, J. S. José, P. Meir, H. S. Miranda, and R. A. Montes, "Productivity and carbon fluxes of tropical savannas," *Journal of Biogeography*, vol. 33, no. 3, pp. 387–400, 2006.

[94] D. K. Benbi and J. S. Brar, "A 25-year record of carbon sequestration and soil properties in intensive agriculture," *Agronomy for Sustainable Development*, vol. 29, no. 2, pp. 257–265, 2009.

[95] R. F. Follett, J. M. Kimble, and R. Lal, Eds., *The Potential of US Grazing Lands to Sequester Carbon and Mitigate the Greenhouse Effect*, Lewis Publishers, Boca Raton, Fla, USA, 2001.

[96] C. Neely, S. Bunning, and A. Wilkes, "Review of evidence on drylands pastoral systems and climate change: implications and opportunities for mitigation and adaptation," Land and Water Discussion Paper 8, Food and Agriculture Organization of the United Nations, Rome, Italy, 2009.

[97] S. Daryanto, D. J. Eldridge, and H. L. Throop, "Managing semi-arid woodlands for carbon storage: grazing and shrub effects on above and belowground carbon," *Agriculture, Ecosystems and Environment*, vol. 169, pp. 1–11, 2013.

[98] D. P. Fernandez, J. C. Neff, and R. L. Reynolds, "Biogeochemical and ecological impacts of livestock grazing in semi-arid southeastern Utah, USA," *Journal of Arid Environments*, vol. 72, no. 5, pp. 777–791, 2008.

[99] D. D. Briske, N. F. Sayre, L. Huntsinger, M. Fernandez-Gimenez, B. Budd, and J. D. Derner, "Origin, persistence, and resolution of the rotational grazing debate: integrating human dimensions into rangeland research," *Rangeland Ecology and Management*, vol. 64, no. 4, pp. 325–334, 2011.

[100] D. M. Gammon, "An appraisal of short duration grazing as a method of veld management," *Zimbabwe Agriculture Journal*, vol. 81, pp. 59–64, 1984.

[101] PJ. O'Reagain and J. R. Turner, "An evaluation of the empirical basis for grazing management recommendations for Rangeland in southern Africa," *Journal of the Grassland Society of Southern Africa*, vol. 9, no. 1, pp. 38–49, 1992.

[102] D. D. Briske, B. T. Bestelmeyer, J. R. Brown, S. D. Fuhlendorf, and H. W. Polley, "The Savory method cannot green deserts or reverse climate change," *Rangelands*, vol. 35, no. 5, pp. 72–74, 2013.

[103] D. D. Briske, J. D. Derner, J. R. Brown et al., "Rotational grazing on Rangelands: reconciliation of perception and experimental evidence," *Rangeland Ecology and Management*, vol. 61, no. 1, pp. 3–17, 2008.

[104] J. T. Manley, G. E. Schuman, J. D. Reeder, and R. H. Hart, "Rangeland soil carbon and nitrogen responses to grazing," *Journal of Soil and Water Conservation*, vol. 50, no. 3, pp. 294–298, 1995.

[105] J. M. Earl and C. E. Jones, "The need for a new approach to grazing management—is cell grazing the answer?" *The Rangeland Journal*, vol. 18, no. 2, pp. 327–350, 1996.

[106] G. Sanjari, H. Ghadiri, C. A. A. Ciesiolka, and B. Yu, "Comparing the effects of continuous and time-controlled grazing systems on soil characteristics in southeast Queensland," *Australian Journal of Soil Research*, vol. 46, no. 4, pp. 348–358, 2008.

[107] W. R. Teague, S. L. Dowhower, S. A. Baker, N. Haile, P. B. DeLaune, and D. M. Conover, "Grazing management impacts on vegetation, soil biota and soil chemical, physical and hydrological properties in tall grass prairie," *Agriculture, Ecosystems and Environment*, vol. 141, no. 3-4, pp. 310–322, 2011.

[108] K. T. Weber and B. S. Gokhale, "Effect of grazing on soil-water content in semiarid rangelands of southeast Idaho," *Journal of Arid Environments*, vol. 75, no. 5, pp. 464–470, 2011.

[109] H. W. van Poolen and J. R. Lacey, "Herbage response to grazing systems and stocking intensities," *Journal of Range Management*, vol. 32, no. 4, pp. 250–253, 1979.

[110] W. P. Clary and B. F. Webster, "Managing grazing of riparian areas in the Intermountain Region," Tech. Rep. GTR-INT-263, US Department of Agriculture, Forest Service, Ogden, Utah, USA, 1989.

[111] H. K. Orr, "Recovery from soil compaction on bluegrass range in the Black Hills," *Transactions of the American Society of Agricultural and Biological Engineers*, vol. 18, no. 6, pp. 1076–1081, 1975.

[112] A. J. Belsky, A. Matzke, and S. Uselman, "Survey of livestock influences on stream and riparian ecosystems in the western United States," *Journal of Soil and Water Conservation*, vol. 54, no. 1, pp. 419–431, 1999.

[113] T. Tucker Schulz and W. C. Leininger, "Differences in riparian vegetation structure between grazed areas and exclosures," *Journal of Range Management*, vol. 43, no. 4, pp. 295–299, 1990.

[114] E. J. Dyksterhuis, "Condition and management of rangeland based on quantitative ecology," *Journal of Range Management*, vol. 2, no. 3, pp. 104–115, 1949.

[115] E. F. Habich, "Ecological site inventory, technical reference 1734-7," Tech. Rep. BLM/ST/ST-01/003+1734, Bureau of Land Management, Denver, Colo, USA, 2001.

[116] D. S. Dobkin, A. C. Rich, and W. H. Pyle, "Habitat and avifaunal recovery from livestock grazing in riparian meadow system of the northwestern Great Basin," *Conservation Biology*, vol. 12, no. 1, pp. 209–221, 1998.

[117] S. L. Earnst, J. A. Ballard, and D. S. Dobkin, "Riparian songbird abundance a decade after cattle removal on Hart Mountain and Sheldon National Wildlife Refuges," in *Proceedings of the 3rd International Partners in Flight Conference*, C. J. Ralph and T. Rich, Eds., General Technical Report PSW-GTR-191, pp. 550–558, US Department of Agriculture, Forest Service, Albany, Calif, USA, 2005.

[118] C. E. Bock and J. H. Bock, "Cover of perennial grasses in southeastern Arizona in relation to livestock grazing," *Conservation Biology*, vol. 7, no. 2, pp. 371–377, 1993.

[119] D. H. Stinner, B. R. Stinner, and E. Martsolf, "Biodiversity as an organizing principle in agroecosystem management: case studies of holistic resource management practitioners in the USA," *Agriculture, Ecosystems and Environment*, vol. 63, no. 2-3, pp. 199–213, 1997.

18

Characterization of Annur and Bedakam Ecotypes of Coconut from Kerala State, India, Using Microsatellite Markers

M. K. Rajesh,[1] K. Samsudeen,[1] P. Rejusha,[1] C. Manjula,[1,2] Shafeeq Rahman,[1] and Anitha Karun[1]

[1] *Division of Crop Improvement, Central Plantation Crops Research Institute, Kasaragod Kerala, 671124, India*
[2] *Nehru Arts and Science College, Kanhangad Kerala, 671314, India*

Correspondence should be addressed to M. K. Rajesh; mkraju.cpcri@gmail.com

Academic Editor: Arianna Azzellino

The coconut palm is versatile in its adaptability to a wide range of soil and climatic conditions. A long history of its cultivation has resulted in development of many ecotypes, which are adapted to various agro-eco factors prevalent in a particular region. These ecotypes usually are known by the location where they are grown. It is important to explore such adaptation in the coconut population for better utilization of these ecotypes in coconut breeding programs. The aim of the present study was to identify the genetic diversity of the Bedakam and Annur ecotypes of coconut and compare these ecotypes with predominant West Coast Tall (WCT) populations, from which they are presumed to have been derived, using microsatellite markers. All the 17 microsatellite markers used in the study revealed 100% polymorphism. The clustering analysis showed that Annur and Bedakam ecotypes were two separate and distinct populations compared to WCT. It was also evident from the clustering that Annur ecotype was closer to WCT than Bedakam ecotype.

1. Introduction

Coconut (*Cocos nucifera* L.), a monotypic species, is one of the major perennial oil crops of the tropics. The palm forms the basis in many developing countries for food products as well as serving industrial purposes [1]. It is often referred to as "*Kalpavriksa*," the tree which provides all the necessities of life. The coconut palm is versatile in its adaptability to a wide range of soil and climatic conditions. In India, it grows well in coastal alluvium of both the West and East coasts, river alluvium of the deltaic regions, and the literate and red loam soil of the inland areas. It is estimated that about 70% of the cropped area is under sandy loam soil. Sandy loam soil with good cation-exchange capacity and soil water level at about 4 m depth is considered as the best soil type for coconut [2]. Coconut cultivars are generally classified into tall and dwarf types. The tall types are primarily outcrossing, while the dwarf types are predominantly self-pollinated [3]. In a cultivar, certain members have been reported to differ from one another in a single or a constellation of characters [4].

Coconut is a perennial crop with indeterminate flowering and the productive features of the palm are considerably influenced by environmental variables; weather factors like sunshine hours, light intensity, ambient temperature, humidity, and rainfall have been reported to play a significant role in fluctuations of coconut yield [5, 6]. The influence of seasonal variations on gas exchange characteristics and biochemical constituents of coconut palms in a particular area have also been reported in earlier studies [7–10]. Weather factors are known to influence crop production especially under rainfall condition and of all the climatic factors, rainfall has maximum influence on the seasonal variation in yield [11]. Variations in environmental factors usually cause genotypes to respond differently from one environment to another resulting in genotype environmental interactions.

The understanding of the distribution of a particular species is also dependent on knowledge of the interrelationship of particular species to their environment. There are wide variations in the crop and cultivation method in different regions.

Ecotypes are cultivars which are grown for a long period to be superior to other local cultivars by the farmers. Wide morphological variability has been observed in these native populations of coconut in various coconut growing countries of the world. Studies have indicated high genetic variability and diversity for whole nut weight, husked nuts weight, kernel weight, copra weight, copra and oil yield of the palms [12]. In India, coconut is generally considered to be a crop of coastal region, even though it is found to be grown well in inlands like Assam State of India [13].

Genetic diversity is thought of as the amount of genetic variability among individuals of a variety, or population of a species [14]. Assessment of genetic diversity of a population provides valuable information that can decide its proper utilization, sustainable management, and design of conservation strategies. There are different techniques available to assess genetic diversity. Among these, molecular markers have the potential to significantly increase the efficiency of coconut genetic improvement, specifically in the areas of germplasm management, genotype identification, and marker-assisted selection of economically important traits [15]. Microsatellites or simple sequence repeats (SSRs) are tandemly repeated motifs of 1–6 bases found in all prokaryotic and eukaryotic genomes analyzed to date. They are present in both coding and noncoding regions and are usually characterized by a high degree of length polymorphism. The origin of such polymorphism is still under debate though it appears most likely to be due to slippage events during DNA replication [16]. The increased number of SSR markers has greatly improved the knowledge about the genetic diversity/relationships between coconut varieties/populations [17–20].

One of the major coconut growing states in India is Kerala, where the most popular variety grown by the farmers is the West Coast Tall (WCT), which occupies over 95 percent of the area under coconut. The WCT palm grows well in all types of soil and is relatively tolerant to moisture stress. It is recommended for large scale cultivation in the coastal regions of Kerala and Karnataka states of India [21]. WCT cultivar has spread inwards from the coastal regions and is now found cultivated even in high ranges. In the process, the cultivar diverged into different ecotypes known by the location where they are cultivated, some of them being Annur, Bedakam, Kuttiyadi, Attingal, and Kanjirappally. These diverse coconut ecotypes have been reported to exhibit morphological or physiological phenotypic differences. The objective of this study was to decipher the genetic diversity among Annur (ANR) and Bedakam (BDK) ecotypes of coconut from Kerala State, India, and compare them with WCT populations, from which they are presumed to have been derived, using molecular markers, which might throw light on the crop's evolutionary diversification.

2. Materials and Methods

2.1. Plant Materials and DNA Extraction. The plant material for this study consisted of 50 leaf samples collected from the two different ecotypes, namely, ANR (17 palms) from Kannur district and BDK (16 palms) from Kasaragod

district, both in Kerala State of India, and WCT cultivar (17 palms) from CPCRI, Kasaragod, Kerala State. Genomic DNA was extracted from spindle leaves of palms following the standardized protocol [22]. The DNA was quantified spectrophotometrically and the bands were checked on 0.8% agarose gel electrophoresis.

2.2. Molecular Analysis. A total of 17 SSR primer pairs specific to coconut were used in the present study to assess the genetic diversity of the respective coconut collections (Table 1). The PCR reactions were carried out in $20\,\mu L$ volume with standardized components: 20 ng genomic DNA, $0.2\,\mu M$ each of forward and reverse primers, $10\,\mu M$ dNTPs (M/s Bangalore Genei Pvt. Ltd.), 10x buffer (10 mM Tris-Hcl (pH 8.3), 50 mM KCl, 1.5 mM $MgCl_2$), and 3 units of Taq DNA polymerase (M/s Bangalore Genei Pvt. Ltd.). After amplification, a volume of $3\,\mu L$ of loading buffer was added to each of the amplified products. The amplified products were run on 3.0% high resolution agarose gel, stained with ethidium bromide following the protocol of Sambrook et al. [23], and were visualized in a gel documentation system.

2.3. Data Analysis. Data analysis was done by scoring of bands. The alleles were scored individually based on comparison in the molecular ladder. The size of the amplicons was compared using a 100 bp ladder. Each band generated by SSR primers was considered as an independent locus. Clearly resolved unambiguous bands were scored visually for their presence or absence with each primer. The scores were obtained in the form of a matrix with "1" and "0" indicating the presence and absence of bands, respectively. Based on the number of polymorphic bands, percentage polymorphism was calculated for each primer.

The genetic associations between varieties were evaluated by calculating Dice's similarity coefficient for pairwise comparisons based on the proportions of shared bands produced by the primers [24]. Similarity matrix was generated using the NTSYS-PC software, version 2.0 [25]. The similarity coefficients were used for cluster analysis and dendrogram was constructed by the Unweighted Pair-Group method (UPGMA) [26]. Shannon's Information Index, expected and observed heterozygosity, unbiased expected heterozygosity, fixation index, principal component analysis (PCA), and analysis of molecular variance (AMOVA) were worked out for the coconut populations using the software GenAlEx 6.5 [27]. For PCA, genetic distance was calculated from the allele data and the genetic distance was plotted as PCA using GenAlEx.

3. Results

In the present study, phylogenetic analysis was carried out using 50 palms that belong to three distinct coconut populations, one a cultivar (WCT) and two ecotypes, namely, ANR and BDK. The 17 SSR primers employed for the study gave clear, unambiguous bands and the data derived were used for further analysis. The details of banding patterns produced in the accessions are given in Table 1. A total of

TABLE 1: SSR primers, their sequences, banding patterns, and percent polymorphism.

Sl. no.	Primer	Primer sequence (5′–3′)	Number of bands	Number of polymorphic bands	Polymorphism (%)
1	CAC2	AGCTTTTTCATTGCTGGAAT CCCCTCCAATACATTTTTCC	4	4	100
2	CAC3	GGCTCTCCAGCAGAGGCTTAC GGGACACCAGAAAAAGCC	3	3	100
3	CAC4	CCCCTATGCATCAAAACAAG CTCAGTGTCCGTCTTTGTCC	4	4	100
4	CAC6	TGTACATGTTTTTTGCCCAA CGATGTAGCTACCTTCCCC	3	3	100
5	CAC8	ATCACCCCAATACAAGGACA AATTCTATGGTCCACCCACA	3	3	100
6	CAC10	GGAACCTCTTTTGGGTCATT GATGGAAGGTGGTAATGCTG	3	3	100
7	CAC13	GGGTTTTTTAGATCTTCGGC CTCAACAATCTGAAGCATCG	4	4	100
8	CNZ1	ATGATGATCTCTGGTTAGGCT AAATGAGGGTTTGGAAGGATT	4	4	100
9	CNZ2	CTCTTCCCATCATATACCAGC ACTGGGGGGATCTTATCTCTG	4	4	100
10	CNZ3	CATCTTTCATCATTTAGCTCT AAACCAAAAGCAAGGAGAAGT	4	4	100
11	CNZ4	TATATGGGATGCTTTAGTGGA CAAATCGACAGACATCCTAAA	4	4	100
12	CNZ5	CTTATCCAAATCGTCACAGAG AGGAGAAGCCAGGAAAGATTT	4	4	100
13	CNZ6	ATACTCATCATCATACGACGC CTCCCACAAAATCATGTTATT	4	4	100
14	CNZ10	CCTATTGCACCTAAGCAATTA AATGATTTTCGAAGAGAGGTC	4	4	100
15	CnCir56	AACCAGAACTTAAATGTCG TTTGAACTCTTCTATTGGG	4	4	100
16	CnCirH9	CACAATCCTTACATCAAA TCTCAAGTTCTTACAGCAGT	3	3	100
17	CnCirG4	AGTATAGTCACGCCAGAAAA AAACCCATAACCAGCAAG	4	4	100
	Total		63	63	
	Average		3.70	3.70	

63 bands were produced by 17 primers, with an average of 3.70 bands/primer. Polymorphism was calculated for all the 17 primers after scoring the bands. All the 17 primers showed 100% polymorphism.

The similarity index, based on Dice's coefficient, was obtained after pair wise comparison of the three coconut populations. The percentage similarity varied between the palms. Maximum similarity was seen between ANR6 and ANR5 (0.97) palms and minimum similarity was seen in BDK16 and ANR1 (0.30) (data not shown). Cluster analysis, based on UPGMA, was performed using NTSYS software in order to obtain a dendrogram. On clustering, it was found that all the three populations formed two major clusters— one containing all WCT and a few BDK palms, while the second cluster exclusively contained BDK palms (Figure 1). In general, palms of ANR, BDK, and WCT grouped together according to the population; however, a few palms did show intergroup affinity. Two palms (ANR1 and WCT8) stood outside their respective clusters. A group of five palms of Annur ecotype clustered separately and this group showed more affinity towards WCT. The palms of BDK ecotype were all clustered together except for one palm that aligned with Annur group. WCT palms were mostly clustered together but one palm aligned with BDK and three palms with ANR ecotype. Among palms studied from BDK, four palms showed affinity towards ANR and WCT palms. Remaining palms of BDK were clustered in a different group. However, one WCT palm (WCT 8) showed variation from all other WCT palms and grouped with BDK ecotype. Other WCT palms were mostly clustered together showing affinity towards ANR palms. The intergroup affinity might be due to sharing of alleles between the three coconut populations studied. The clustering shows that ANR and BDK ecotypes were two separate populations with clear distinction, with just one palm of BDK (BDK 1) showing affinity towards ANR. It is also evident from the clustering that ANR ecotype is

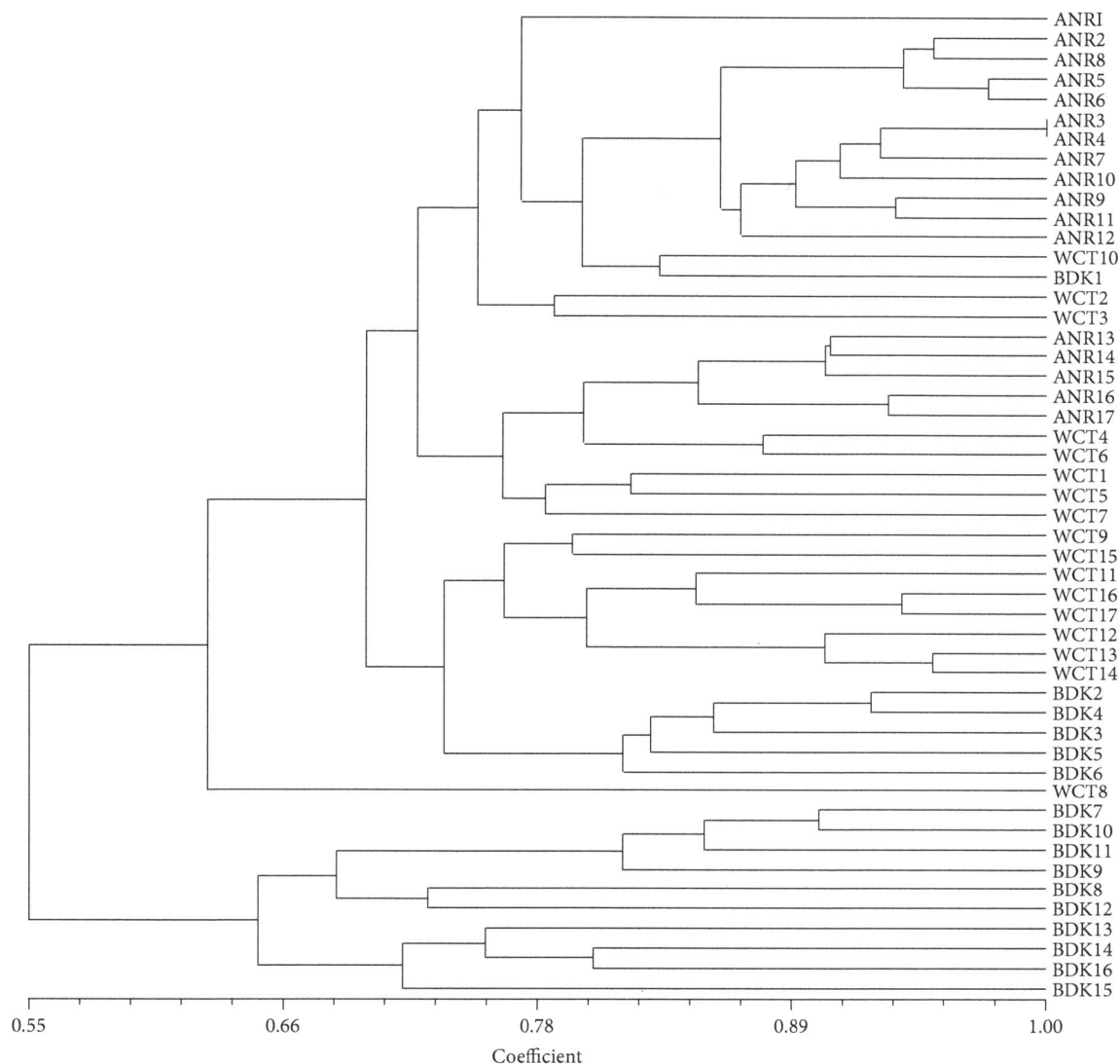

FIGURE 1: UPGMA phylogenetic tree constructed based on Dice's coefficient showing the genetic relationships among the three coconut populations.

closer to WCT than BDK ecotype. When the populations were considered individually, WCT palms clustered at 61% similarity, ANR palms at 70% similarity, and BDK palms at 58% similarity.

Population-wise mean Shannon's Information Index ranged from 0.513 (ANR) to 0.712 (BDK) and the mean observed heterozygosity from 0.257 (ANR) to 0.339 (BDK) (Table 2). The mean unbiased expected heterozygosity (uHe) ranged from 0.325 (ANR) to 0.442 (BDK). Mean fixation index (F_{ST}) ranged from 0.205 to 0.372 (Table 2). Pair-wise population matrix of Nei's genetic identity calculated using GenAlEx program showed a higher average identity between ANR and WCT (0.92) than between WCT and BDK (0.869) and ANR and BDK (0.783) (Table 3). The AMOVA estimation based on 99 permutations using GenAlEx showed a significant ($P = 0.01$) within population variation (79%) compared to among population variation (21%) (Table 4).

For a deeper understanding of the clustering pattern of the coconut populations, we also carried out genetic distance-based principal component analysis (PCA) using GenAlEx. The results showed clear segregation of all the three populations into different quadrates of the PCA (Figure 2). The first and second axes accounted for 85.13 and 14.87% of the total inertia, respectively. As shown in Figure 2, the first axis separated BDK from the other two populations, while the second axis separated WCT from the other two populations.

4. Discussion

Ecotypes are groups of similar populations within the same plant species that are adapted to certain climatic and edaphic conditions. Coconut has been cultivated for centuries in India and the long history of its cultivation along the length

TABLE 2: Shannon's Information Index, expected and observed heterozygosity, unbiased expected heterozygosity, and fixation index for the three populations.

Population	I	H_o	H_e	uH_e	F_{ST}
ANR					
Mean	0.513	0.257	0.315	0.325	0.205
SE	0.066	0.060	0.041	0.043	0.110
WCT					
Mean	0.708	0.339	0.427	0.440	0.244
SE	0.061	0.052	0.037	0.038	0.092
BDK					
Mean	0.712	0.284	0.428	0.442	0.372
SE	0.062	0.057	0.035	0.036	0.105

I = Shannon's Information Index = $-1 * \text{Sum} (pi * \text{Ln} (pi))$.
H_o = observed heterozygosity = Number of Hets/N.
H_e = expected heterozygosity = $1 - \text{Sum } pi^2$.
uH_e = unbiased expected heterozygosity = $(2N/(2N - 1)) * H_e$.
F = fixation index = $(H_e - H_o)/H_e = 1 - (H_o/H_e)$,
where pi is the frequency of the ith allele for the population and Sum pi^2 is the sum of the squared population allele frequencies.

TABLE 3: Pairwise population matrix of Nei's genetic identity.

ANR	WCT	BDK	
1.000			ANR
0.920	1.000		WCT
0.783	0.869	1.000	BDK

TABLE 4: Analysis of molecular variance in the coconut populations.

Source	df	SS	MS	Est. Var.	%
Among pops	2	176.652	88.326	4.330	21%
Within pops	47	760.768	16.187	16.187	79%
Total	49	937.420		20.517	100%

and breadth of the country has resulted in development of many ecotypes, which are generally named or known by the location where they are grown. These ecotypes are likely to have developed certain adaptation to the particular environment conditions in which they have been growing for a long time and possess a rich source of valuable genes for coconut breeding. It is important to identify and conserve these ecotypes, which could withstand the vagaries of nature and still perform better in terms of growth and yield.

West Coast Tall (WCT) coconut population developed on the West coast of India and came to be known by the region where it was cultivated. Though the origin of WCT is not traceable to any area from where it spread, it is obvious that sea journey by nuts was involved in its spread. The cultivar over the years moved from coastal region to the interior areas, which also resulted in further adaptation and diversity in the cultivar. Such adapted WCT populations in certain localities are designated with local names by farmers to differentiate it from other WCT populations. Annur, Bedakam, Kuttiyadi, Attingal, and Kanjirappally are some of these. The variability shown by WCT and ecotypes derived from it might be genetic or due to physiological factors such as climatic change,

Principal coordinates

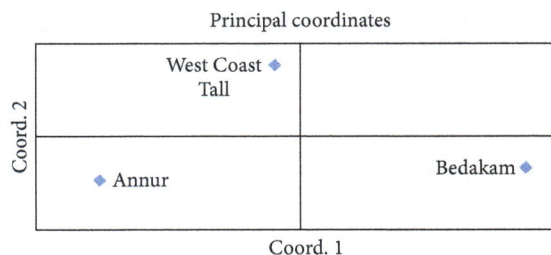

FIGURE 2: Diagram of PCA based on Nei's genetic distance.

pH of the soil, annual rainfall, or any other environmental factors or even human involvement. These coconut ecotypes may possess higher variability which might be an important source for plant breeding, in comparison to the present day cultivars, which have been specifically chosen for their growth performance under certain specific environment.

Molecular markers play an important role in conservation and use of plant genetic resources. Molecular markers are relatively independent of environment. DNA-based markers are a way of exploring the genetic relations between populations and these markers have acted as versatile tools in various fields. Among the various molecular markers currently available, microsatellite markers are the most popular because they are reproducible, enabling their parallel analysis in different laboratories and exchange of the resulting data [28]. Also, microsatellites form an ideal marker system creating complex banding patterns by simultaneously detecting multiple DNA loci. They have been used successfully by many researchers to characterize the genetic diversity of the coconut population [18, 21].

With the objective of deciphering the diversity among and within WCT populations from Kerala region, in the present study, Annur and ecotypes were compared with WCT populations using SSRs. On clustering, it was found that the two ecotypes and WCT formed two major clusters. Annur, Bedakam, and WCT grouped separately, in general, with a few palms showing intergroup affinity. However, most of the Bedakam palms were grouped in separate cluster proving that Annur and Bedakam ecotypes were two separate populations. It is also evident from the clustering that Annur ecotype was closer to WCT than to Bedakam ecotype. Pair-wise population matrix of Nei's genetic identity also revealed a higher average identity between ANR and WCT than between WCT and BDK and ANR and BDK.

The comparison of average observed and expected heterozygosity values did not show great differences between the three coconut populations studied. All the three populations displayed smaller observed than expected heterozygosities-inbreeding may be a factor contributing to this. Among the three populations, palms of ANR ecotype displayed more genetic similarity amongst themselves, with lower observed heterozygosity and fixation index (F_{ST}) compared to the other two populations.

The values of heterozygosities obtained in this study confirm that the coconut ecotypes represent an important reservoir of genetic diversity. The three populations showed significant genetic differentiation, as indicated by F_{ST} values, indicating that a high level of differentiation for the alleles has occurred in members of the subpopulation compared to the total population, and therefore members of the subpopulation tend to carry unique alleles compared to the total population. The among population variation was more than within population variation, based on AMOVA calculations, suggesting very rare genetic exchange between the populations, at least in recent history.

Diversity in the original WCT population allowed it to spread in to different ecoregions resulting in the evolution of ecotypes suitable for the local environment and its subsequent adaptation. The result of molecular characterization revealed significant differences in midland and coastal growing palms. The WCT and ecotypes might have differed due to genetic factors or adaptation resulting from environmental variations like climatic change, pH of the soil, annual rainfall, or any other environmental factors or even human interventions.

One of the important prerequisites for evolutionary change is genetic variation as in its absence populations lack the capacity to evolve. Within a single species, natural selection, coupled with heterogeneity in its habitat, might result in multiple ecotypes, which are genetically distinct [29]. The total genetic variation of a species is likely to be distributed among populations as the impact and direction of natural selection varies from one to another, due to environmental variation and genetic drift [30]. Therefore, with germplasm conservation programmes, it is imperative to accurately measure the amount of genetic diversity and its distribution within and between populations. To these ends, molecular markers provide an efficient and unbiased estimate of these statistics, free of environment effects. Molecular characterization of genetic diversity provides base information which could be utilized in selection of a promising range of accessions for different breeding programs. The microsatellites used in this study appeared to possess a significant potential in this respect. This is one of the first studies to probe the diversity of coconut ecotypes using molecular markers. The results of this study may be used in developing a strategy for conservation of these ecotypes and their utilization in future coconut breeding programmes.

Conflict of Interests

The authors declare that there is no conflict of interests regarding the publication of this paper.

Acknowledgments

The authors thank Director, CPCRI, Kasaragod, for his guidance and facilities provided. The authors also thank the farmers and extension personnel from Annur and Bedakam for their cooperation.

References

[1] L. H. Jones, "Perennial vegetable oil crops," in *Agricultural Biotechnology: Opportunities for International Development*, pp. 213–224, CAB International, Wallingford, UK, 1990.

[2] J. G. Ohler, "Coconut, a tree of life," Plant Production and Protection Paper, 1984.

[3] D. V. Liyanage, "Preliminary studies on the floral biology of the coconut palm," *Tropical Agriculture*, vol. 105, pp. 171–175, 1949.

[4] K. Satyabalan, "Yield variation in west coast tall coconut palms: yield attributes which cause variation in annual yield of nuts in the palms of different yield groups," *Indian Coconut Journal*, vol. 1, pp. 5–8, 1993.

[5] P. Coomans, "Influence of climate factors on seasonal fluctuations of coconut production," *Oleagineux*, vol. 30, no. 4, pp. 153–157, 1975.

[6] D. B. Murray, "Coconut palm," in *Ecophysiology of Tropical Crops*, P. T. Alvin and T. T. Kozlowski, Eds., pp. 24–27, Academic Press, New York, NY, USA, 1977.

[7] K. V. Kasturi Bai and A. Ramadasan, "Changes in the levels of carbohydrates as a function of environmental variables in hybrids and tall coconut palm," in *Coconut Research and Development*, N. M. Nair, Ed., pp. 203–209, Wiley Eastern, New Delhi, India, 1983.

[8] V. Rajagopal, A. Ramadasan, K. V. Kasturi Bai, and D. Balasimha, "Influence of irrigation on leaf water relations and dry matter production in coconut palms," *Irrigation Science*, vol. 10, no. 1, pp. 73–81, 1989.

[9] K. V. Kasturi Bai, V. Rajagopal, C. D. Prabha, M. J. Ratnambal, and M. V. George, "Evaluation of coconut cutivars and hybrids for dry matter production," *Journal of Plantation Crops*, vol. 24, pp. 23–28, 1996.

[10] C. Jayasekara, N. P. A. D. Nainanayake, and K. S. Jayasekara, "Photosythetic characteristic and productivity of coconut palm," *Journal of Plantation Crops*, vol. 24, pp. 538–547, 1996.

[11] K. N. Krishna Kumar, *Coconut phenology and yield response to climate variability and change [Ph.D. thesis]*, Cochin University of Science and Technology, Kerala, India, 2011.

[12] K. Ganesamurthy, C. Natarajan, and M. Jayaramachandran, "Genetic diversity and its exploitation in coconut improvement," in *Proceedings of the National Conference on Agro-Biodiversity*, Conducted by National Biodiversity Authority Chennai, February 2006.

[13] U. Parthasarathy, *A comparative study of coconut cultivation in coastal and inland River Plain Ecosystem of Kasaragod District of Kerala and Kamrup District of Assam [Ph.D. thesis]*, Gauhati University, Assam, India, 2004.

[14] A. H. D. Brown, "Isozymes, plant population genetic structure and genetic conservation," *Theoretical and Applied Genetics*, vol. 52, no. 4, pp. 145–157, 1978.

[15] G. R. Ashburner, W. K. Thompson, G. M. Halloran, and M. A. Foale, "Fruit component analysis of south pacific coconut palm populations," *Genetic Resources and Crop Evolution*, vol. 44, no. 4, pp. 327–335, 1997.

[16] C. Schlotterer and D. Tautz, "Slippage synthesis of simple sequence DNA," *Nucleic Acids Research*, vol. 20, no. 2, pp. 211–215, 1992.

[17] R. Rivera, K. J. Edwards, J. H. A. Barker et al., "Isolation and characterization of polymorphic microsatellites in *Cocos nucifera* L.," *Genome*, vol. 42, no. 4, pp. 668–675, 1999.

[18] B. Teulat, C. Aldam, R. Trehin et al., "An analysis of genetic diversity in coconut (*Cocos nucifera*) populations from across the geographic range using sequence-tagged microsatellites (SSRs) and AFLPs," *Theoretical and Applied Genetics*, vol. 100, no. 5, pp. 764–771, 2000.

[19] L. Perera, J. R. Russell, J. Provan, and W. Powell, "Use of microsatellite DNA markers to investigate the level of genetic diversity and population genetic structure of coconut (*Cocos nucifera* L.)," *Genome*, vol. 43, no. 1, pp. 15–21, 2000.

[20] M. K. Rajesh, P. Nagarajan, B. A. Jerard, V. Arunachalam, and R. Dhanapal, "Microsatellite variability of coconut accessions (*Cocos nucifera* L.) from Andaman and Nicobar Islands," *Current Science*, vol. 94, no. 12, pp. 1627–1631, 2008.

[21] C. Remany, *Cataloging and categorization of unexploited ecotypes of coconut grown in Kerala [Ph.D. thesis]*, Mahatma Gandhi University, Kerala, India, 2003.

[22] M. K. Rajesh, B. A. Jerard, P. Preethi et al., "Development of a RAPD-derived SCAR marker associated with tall-type palm trait in coconut," *Scientia Horticulturae*, vol. 150, no. 4, pp. 312–316, 2013.

[23] J. Sambrook, E. F. Fritsch, and T. Maniatis, *Cloning: A Laboratory Manual*, Cold Spring Harbor laboratory, 1989.

[24] L. Dice, "Measures of the amount of ecologic association between species," *Ecology*, vol. 26, pp. 297–302, 1945.

[25] F. J. Rohlf, "NTSYS-pc numerical taxonomy and multivariate analysis system version 2.0. owner manual," 1997.

[26] P. H. A. Sneath and R. R. Sokal, *Numerical Taxonomy: The Principles and Practice of Numerical Classification*, W. H. Freeman, San Francisco, Calif, USA, 1973.

[27] R. Peakall and P. E. Smouse, "GenAlEx 6.5: genetic analysis in excel. Population genetic software for teaching and research-an update," *Bioinformatics*, vol. 28, pp. 2537–2539, 2012.

[28] P. K. Gupta, H. S. Balyan, P. C. Sharma, and B. Ramesh, "Microsatellites in plants: a new class of molecular markers," *Current Science*, vol. 70, no. 1, pp. 45–54, 1996.

[29] Y. B. Linhart and M. C. Grant, "Evolutionary significance of local genetic differentiation in plants," *Annual Review of Ecology and Systematics*, vol. 27, pp. 237–277, 1996.

[30] M. J. Lawrence and N. Rajanaidu, "The genetic structure of natural populations and sampling strategy," in *Proceedings of the International Workshop on Oil Palm Germplasm and Utilization*, pp. 15–26, Selangor, Malaysia, 1985.

19

Current Population Status and Activity Pattern of Lesser Flamingos (*Phoeniconaias minor*) and Greater Flamingo (*Phoenicopterus roseus*) in Abijata-Shalla Lakes National Park (ASLNP), Ethiopia

Tewodros Kumssa and Afework Bekele

Department of Zoological Sciences, Addis Ababa University, P.O. Box 1176, Addis Ababa, Ethiopia

Correspondence should be addressed to Tewodros Kumssa; tewodroskk@gmail.com

Academic Editor: Alexandre Sebbenn

A study of the population status, habitat preference, and activity pattern of nonbreeding flamingos was carried out in Lakes Abijata, Shalla, and Chitu, part of the Great Rift Valley, Ethiopia, from 2011 to 2013. The current population status and habitat preference of flamingos in the area are still poorly known. Likewise, data on diurnal and seasonal activity pattern of the species are scarce and this leads to the misunderstanding of how Flamingos use local wetlands throughout the different seasons. Data regarding population size and activity pattern were gathered during the wet and dry seasons. Point-count method was used to estimate the population size. Behaviors were recorded using scan sampling techniques. A total of 53671 individuals representing two species of flamingo were counted during both wet and dry seasons from the three lakes. There were more flamingos during the dry season than the wet season in Lake Abijata contrary to Lakes Shalla and Chitu during the wet season. Lesser flamingos (*Phoeniconaias minor*) were the most abundant species comprising 95.39%, while Greater Flamingos (*Phoenicopterus roseus*) accounted for 4.61% of the total population. Lake Abijata is the major stronghold of Lesser Flamingos in the area. There was significant variation in the mean number of both species during the wet and dry season in the different study sites of the lake, respectively. The species were known to use varied habitats within the lakes. The Lesser Flamingo mainly preferred the shoreline and mudflat areas of the lakes. However, Greater Flamingo on several occasions showed preference to offshore area of the lakes. Seasonal average flock sizes were not similar between the species. There was a strong relationship between time allocated to each activity and time of day. Feeding activity varied among daylight hours and was higher in the evening (76.5%) and late morning (74.56%) and least during midday (54%). Some variations in activity breakdown were observed between time blocks and season. Conservation efforts in the park should include the wild flora and fauna not only of the land but also of the aquatic systems. The information in this study will be very useful for the future management of the species in the area.

1. Introduction

The hundreds of thousands of flamingos congregating on the African Rift Valley Lakes is one of the truly spectacular sights of the natural word [1]. Sub-Saharan Africa and India are known for their flamingos but the largest flocks occur not only in the East Africa Rift Valley, particularly in the central section at Lakes Bogoria, Elmenteita, Nakuru, and Magadi in Kenya and Natron and Manyara in Tanzania but also at Lakes Rudolf and Abijata in Ethiopia [2, 3]

and Lakes Chitu and Shalla of Ethiopia. Additionally, a few thousand individuals occur on Ugandan crater lakes [4], while smaller populations are present in southern Africa (55,000–60,000), West Africa (Mauritania/Senegal) (15,000–25,000) and 400,0000 individuals occur in the Rann of Kachchh in northwest India [4, 5].

Flamingos are one of the most easily recognizable birds, with their long necks and legs, unusual bill shape, and their plumage ranging from pale pink to red or orange with black primary and secondary wing feathers [6]. The amount of pink coloration, noticeable particularly on the head and neck, varies greatly amongst individuals, not in relation to age (some birds have a very pink head when otherwise still in immature plumage) but possibly according to diet and an individual's capacity to assimilate carotenes for pigmentation [7]. Lesser flamingos have a highly specialized and fine filter, diet consisting almost entirely of microscopic cyanobacteria and benthic diatoms [8]. However, the Greater Flamingos that occur in the same habitat as the lesser ones (but with lesser density) are generalists consuming copepods, mollusks, annelid worms, small fish, seeds, brine shrimps (*Artemisia* spp.), and other small planktonic and benthic animals in addition to algae [9].

Of the six species of flamingos in the world, two occur in Africa: the Lesser Flamingo (*Phoeniconaias minor*) and the Greater Flamingo (*Phoenicopterus roseus*) [10]. These species overlap in distribution and habitats, occurring mainly in large alkaline or saline lakes, salt pans, and estuaries [11]. Both species are very gregarious and frequently occur in large numbers [12]. The two species mingle freely where they occur, usually at feeding and breeding sites. The Lesser Flamingo is the smallest in size and more numerous in number than the Greater Flamingo one [4]. Both flamingo species are irregular nomadic or partially migratory nomadic and inhabit areas with high seasonal fluctuation in conditions and resources, with a great physical and chemical heterogeneity and geomorphology [13]. Their abundance is associated with variations of water characteristics like conductivity, diversity, and availability of potential food items, alkalinity changes [14], breeding migrations, fresh water requirements and predation pressure [15].

About 60% of the Greater Flamingo population is located in the Mediterranean region [16]. They occur in lower numbers in Africa. In Eastern Africa, they make up about 1% of the average number of flamingo aggregations [17]. This species has a very large range, the population trend appears to be increasing, and the population size is very large and hence does not approach the thresholds for vulnerable. So the species is evaluated as of least concern [5].

Greater Flamingos born in France, Spain, and Italy have been observed to breed at other colonies in the Mediterranean and in West Africa, often moving to a third colony thereafter [18]. It has been suggested that the Mediterranean colonies could function as a metapopulation [18]. Following sustained conservation effort undertaken at the two most important breeding colonies in the western Mediterranean (Camargue in France and Fuente de Piedra in Spain), the Greater Flamingo successfully expanded its range in southern Europe and in North and West Africa [19]. Movements between southern Europe and North Africa have long been known [9], and North Africa has generally been considered to be an important wintering ground for Greater Flamingos and a "nursery" for immature flamingos from Europe [20].

However, the Lesser Flamingos are classified by IUCN as being near threatened, due to its dependence on a limited number of unprotected breeding sites, declining of population, and quality of habitat. It is known to breed in only five sites, two in southern Africa (Makgadikgadi Pans in Botswana and Etosha Pan in Namibia), one in

east Africa (Lake Natron), and two in India (Zinzuwadia and Purabcheria Salt Pans), and on an artificial island at Kamfers Dam in South Africa. Of these breeding sites, only Etosha Pan and the two sites in India are officially protected [21]. More than 75% of breeding individuals are concentrated at only one site (Lake Natron, Tanzania) [22]. A narrow range of breeding conditions is required that occur irregularly resulting in a declining population. An irregular episode of mortality involves tens and sometimes hundreds of thousands of birds in East African Rift Valley Lakes [23]. Africa's flamingo populations are not isolated, and flamingos migrate between the Soda lakes of East Africa and the Etosha and Makgadikgadi Pans in southern Africa [10]. Therefore, flamingo conservation should stretch across many political boundaries, and threats to all key habitats need to be considered in an attempt to conserve the African flamingos.

Abijata-Shalla Lake National Park (ASLNP) in Ethiopia is established primarily to conserve the diverse bird life of the area [24]. A great number of bird species seen in these lakes are seasonal migrants; thus the area remains major place of attraction to bird watchers [25]. The Park supports one of the largest African colonies of flamingos and Great White Pelicans (*Pelecanus onocrotalus*). In addition, a high species richness of mammals, the hot springs, and scenery of the lakes all occur in the protected area [26]. The Park is a candidate wetland of international importance under the Ramsar Convention [24].

Little research has been conducted on the diversity and ecology of avian species in Ethiopia [27]. For instance, the status of the flamingo in Ethiopia has not been well known and the current populationstatus and habitat preference of flamingos in the area are poorly known. Likewise, data on diurnal and seasonal activity pattern of the species are scarce and this leads to the misunderstanding of how flamingos use local wetlands during the different seasons. The present study gives data on the population status of the two species, habitat preference, and their activity pattern in the Park.

2. Materials and Methods

2.1. The Study Area. Abijata-Shalla Lakes region was established as a National Park by the Ethiopian Wildlife Conservation Organization in 1970 with the aim of conserving the biodiversity of the spectacular number of aquatic birds [28]. The park is known as Abijata-Shalla Lakes National Park (ASLNP), deriving its name from the two Lakes Abijata and Shalla [24]. ASLNP is one of Ethiopia's National Parks, located in the Great Rift Valley comprising three lakes: Abijata, Shalla, and Chitu (Figure 1). The site lies between the 7°15′–7°45′N and 38°30′–38°45′E, at about 207 km south of Addis Ababa. ASLNP comprises two types of ecosystems, namely, the water part and land together covering a total area of $887 \, km^2$ of which $405 \, km^2$ is land area while $482 \, km^2$ is water body [29].

Among the three saline lakes in the National Park, Lake Shalla is the deepest (266 m deep) and covering an area of $370 \, km^2$ and Abijata is the shallowest (<7 m deep) and covering an area of $180 \, km^2$ [30]. Lake Chitu is the smallest covering an area of $0.8 \, km^2$ and a maximum depth of 21 m

Current Population Status and Activity Pattern of Lesser Flamingos (Phoeniconaias minor) and Greater Flamingo
(Phoenicopterus roseus) in Abijata-Shalla Lakes National Park (ASLNP), Ethiopia

165

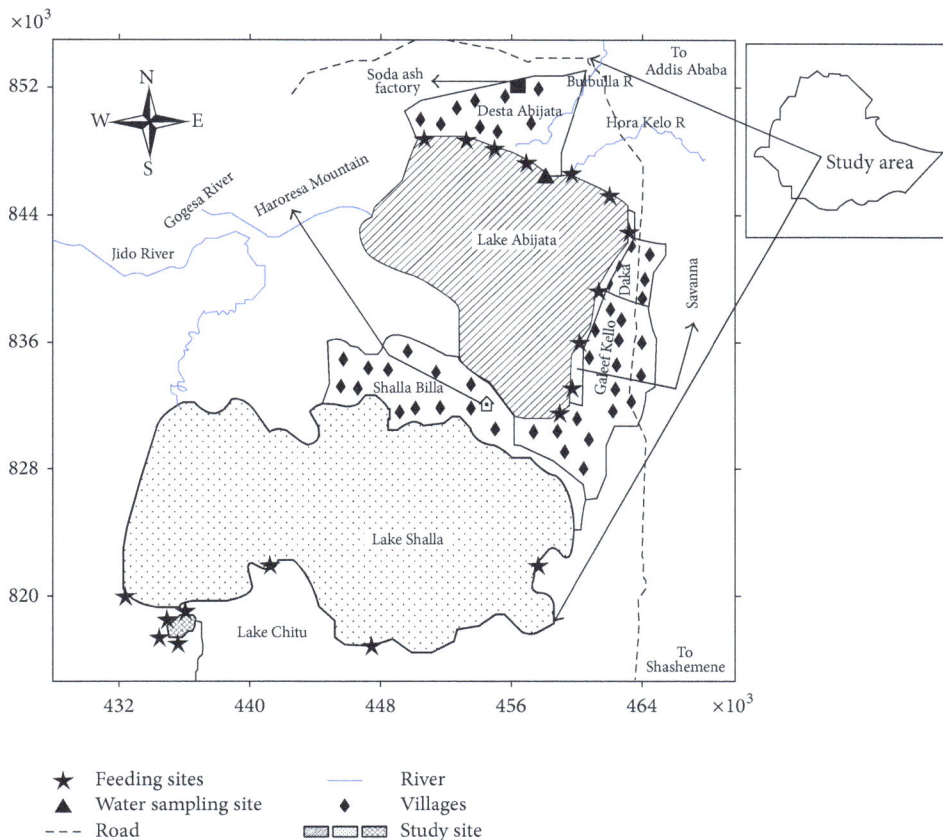

FIGURE 1: Map of the study area with study sites.

and is highly saline. Due to its location in a rainfall deficit area
of the Rift Valley, ASLNP receives an annual rainfall ranging
between 500 and 700 mm and the mean monthly temperature
varies from 18.5°C to 24.6°C with mean annual temperature
of 21°C [29]. The habitat surrounding the lakes in the park
is generally dominated with tree species of *Acacia* and open
scrub rocky slopes [29]. A total of 453 bird species have been
recorded in the park [31]. The park has 6 endemic species to
Ethiopia [28] and holds at least 144 and 292 water-associated
and terrestrial bird species, respectively [29].

The study was carried out from 2011 to 2013. A total of
28 surveys were conducted throughout the study period on
foot. Fourteen were performed during the rainy season (June
to September) and the remaining fourteen during the dry
seasons (November to May). Surveys of the three lake areas
were conducted seasonally to identify and count the presence
of individuals of both flamingo species. During the terrestrial
surveys, depending on the size and shape of the wetland, 5 or
more survey points per lake (in total 36 points) were taken
and systematically all individual flamingos were recorded by
direct count using telescopes (W30x or W22x), binoculars
(10 × 42), and manual counters. Point-count method was
conducted at four locations for 20 minutes in a point using
direct observation through binoculars and telescopes [32].
The mean distance between observers and focal birds was

45 m. For a given wetland, the same census points were used
during each subsequent survey as adopted by [33]. The size
of the census area varied from point to point. The limits of
subareas were determined beforehand and, in order to avoid
double counts, moved to the next point as rapidly as possible
without disturbing the birds. Habitat preference of the bird
species is recorded by the observer each time birds were
detected. For this purpose, onshore, mudflat and offshore
categories were used.

At the time of census, reference points were established to
enable flock identification when counting to avoid repetitions
[34]. Flocks were defined as groups of flamingos in which the
nearest-bird distance averaged <25 m, and where no bird was
>50 m from the nearest flock member as adopted by Bildstein
et al. [35]. For larger number of birds (>4,000 birds), poor
visibility, or both, we counted birds in estimated groups (that
is, 10, 100 [32]. Each count was done by at least two people,
and the average was used as the estimated abundance. The
census was conducted by 10 to 16 people recording the data
together according to the size of the site and bird settlement.
Multi-lake censuses were conducted over short periods of
time at the same time.

To minimize disturbance during counting, silent movement followed by 3 to 5 minutes of waiting period was allowed
to settle down from any disturbance [32]. Census data were

TABLE 1: Number of Lesser and Greater Flamingos on lakes censused from 2011 to 2013.

Site	Species with season								Mean ± SE	
	2011-2012				2012-2013					
	Wet		Dry		Wet		Dry			
	L	G	L	G	L	G	L	G	L	G
Abijata Lake										
Bulbulla and Hora Kelo rivers inlet	9200	278	29700	847	15535	534	22173	464	19152 ± 19.6	530.75 ± 11.7
Gogesa River	345	11	2956	245	482	18	1115	92	1224.5 ± 9.5	91.5 ± 4.2
Savanna	652	—	3516	—	1130	—	2448	—	1936.5 ± 12.6	—
Haroresa mountain	71	—	441	—	168	—	186	—	216.5 ± 8.2	—
Fresh artificial pond of soda ash factory	634	—	2325	—	413	—	1678	14	1262.5 ± 14.2	3.5 ± 0.3
Total	**10902**	**289**	**38938**	**1092**	**17728**	**552**	**27600**	**570**	**23792 ± 15**	**625.75 ± 11**
Lake Shalla	16300	1008	7200	906	19058	3819	6267	548	12206.25 ± 4.3	1570.25 ± 9.3
Lake Chitu	17812	189	13153	313	18513	493	11312	123	15197.5 ± 9.6	279.5 ± 5.1
Total	45014	1486	59291	2311	55299	4864	45179	1241	51195.75 ± 13.4	2475.5 ± 7

L: Lesser Flamingo; G: Greater Flamingo.

collected twice a day, morning (6:30–10:00 a.m.) and late afternoon (10:00 to 6:00 p.m.). These were the periods when most avian species were most active [36].

Various activity patterns of nonbreeding flamingos were recorded during dry and wet seasons of each year at Lake Abijata. Behavioral observations were conducted randomly in two distinct points where flamingos were most numerous. Observations were made using binoculars, at least 35 m from the birds to avoid interference. Since no breeding was observed in the lakes, behaviors associated with the breeding period, such as courting, copulation, and incubation, were not observed. Behaviors were recorded using scan sampling techniques [37] because flock activities of flamingos often are synchronized [38]. Sampling was conducted on three randomly selected days per week with each day divided into four equal time blocks of (1) early morning (06:00–09:00 hr), (2) late morning (09:00–12:00 hr), (3) midday (12:00–14:00), and (4) late afternoon/evening (15:00–18:00) as adopted by Lehner [39]. Data were collected from September 10–14, 2011 (on 4 separate days), October 20–23, 2011 (on 3 days), February 21–27, 2012 (on 6 days), August 12–18, 2012 (on 6 days), March 17–21, 2013 (on 5 days), and May 27–29, 2013 (on 2 days). A total of 303 h of observation was carried out during the hours from 06:00 to 18:00. Each individual of flock was observed for 2 min (focal animal analysis) as used by Altman [37]. The time spent in different activities was calculated and from these values the percentage time spent for each activity during different times on the day was estimated. The activities are divided into five categories as follows:

(i) feeding: stand feeding and walk feeding (feeding was defined as a flamingo holding its head down with the beak either partially or totally submerged),

(ii) moving (walking, swimming, and flight),

(iii) resting (standing, sleep, and grooming),

(iv) preening (feather shaking, wing flapping, tail shaking, and bath),

(v) alert (alarm) [38].

To analyze the data Stata version 12 software was used. ANOVA and chi-square test were performed to find out statistically significant difference among various variables. Behavior data were analyzed with a Kruskal-Wallis test, a nonparametric test.

3. Result

3.1. Population and Habitat Preference of Flamingo. A total of 53671 individuals representing two species of birds were counted during both wet and dry seasons from the three lakes. Lesser Flamingos were the most abundant comprising 95.39%, while Greater Flamingos comprised 4.61%. These constituted 50.32% of the total species during the dry season and 49.68% during the wet season count. There were more flamingos during the dry season than the wet season in Lake Abijata ($x^2 = 10.31$, df = 1, $P < 0.05$). The wet season count significantly outnumbered the dry season count in Lake Shalla ($x^2 = 34.15$, df = 1, $P < 0.05$) and Lake Chitu ($x^2 = 3.76$, df = 1, $P < 0.05$). The difference was also significant between lakes ($x^2 = 24.83$, df = 2, $P < 0.05$) (Table 1).

From the total population, Lake Abijata harbors 46.47% and 25.28% Lesser and Greater Flamingos, respectively, and Lake Shalla supports 23.84% lesser and 63.43% Greater Flamingo, whereas Lake Chitu supports 29.68% and 11.29% of Lesser and Greater flamingos, respectively. Lake Abijata is the major stronghold of Lesser Flamingos in the area. There was significant variation in the mean number of Lesser Flamingo ($F_{4\,101} = 110.42$, $P < 0.005$) and Greater Flamingo ($F_{4\,101} = 130.16$, $P < 0.005$) during the wet and dry season in the different study sites of the lake, respectively. For both species, the highest record was obtained during the dry season in Bulbulla and Hora Kelo rivers inlet site and the lowest was in Haroresa mountain site for Lesser Flamingo and Savanna and Haroresa mountain sites for Greater Flamingo as represented in Table 1.

Current Population Status and Activity Pattern of Lesser Flamingos (Phoeniconaias minor) and Greater Flamingo
(Phoenicopterus roseus) in Abijata-Shalla Lakes National Park (ASLNP), Ethiopia

167

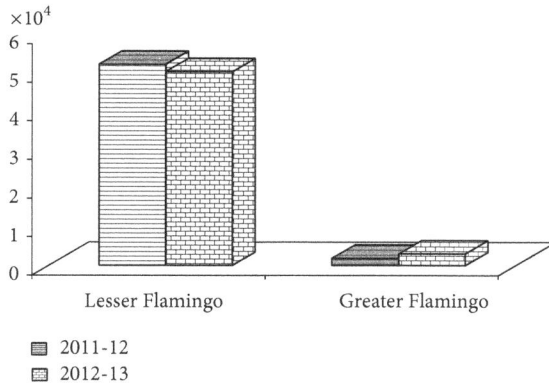

FIGURE 2: Mean number of Lesser and Greater Flamingos from 2011 to 2013 on the whole study sites.

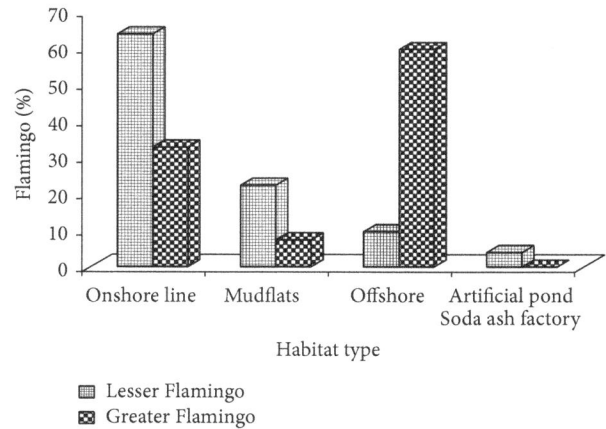

FIGURE 3: Percentage of flamingos observed at different habitat types.

There was no significant difference in the mean number of Lesser Flamingos of the two-year period on the three lakes ($x^2 = 1.13$, df = 1, $P > 0.05$). However, the count of Greater Flamingo showed a significant variation among years ($x^2 = 0.36$, df = 1, $P < 0.05$) (Figure 2). The highest record of Lesser Flamingo was obtained during 2011-2012 year count, while, for Greater Flamingo, the highest record was during 2012-2013 count.

The largest average flock size was for the Lesser Flamingos (10,000 birds) and for Greater Flamingos (14 birds). Lesser Flamingo flocks consisted of greater than 500 birds that are very common and flocks of less than 20 birds that are rare. Seasonal average flock sizes were not similar between the species (Table 2). Flock sizes were the largest during the wet season for both species and were significantly different across seasons, for Lesser Flamingos ($x^2 = 4.14$, df = 1, $P < 0.05$) and for Greater Flamingos ($x^2 = 1.11$, df = 1, $P < 0.05$).

Flamingos were observed in various parts of the lakes during the day. At Lake Abijata, birds were mainly observed along the northwest of the lake during early morning. During the late afternoon, they flew in flocks to the northeastern side and continue feeding during the course of the day. At night, most birds roost back in the northwest side of the lake. Lesser Flamingos were more frequent in shallow area of the lakes and mudflats, occasionally preferring man-made solar salt concentration ponds of Soda ash factories. Lesser Flamingos in Lake Chitu were most of the time concentrated on the southern and western side of the lake shore in relation to human disturbance.

The relative use of the different habitats by flamingo was statistically significant ($F_{463} = 18.49$, $P < 0.005$) as indicated in Figure 3. Lesser Flamingo showed a very high (65.91%) utilization on the shore line. This utilization of habitat by the Lesser Flamingo was very different from the habitat preference of Greater Flamingo (59.76%) which utilized offshore area of the lakes.

3.2. Activity Patterns. The data from Table 3 are used to show different activities of nonbreeding Lesser Flamingo. There was a strong relationship between time allocated to

TABLE 2: Group size of lesser and greater flamingos in the lakes.

Species	Season	Groups	Group size range	Mean group size
Lesser flamingo	Dry	52	2–4400	2100
	Wet	112	2–10,000	7300
Greater flamingo	Dry	26	2–18	15
	Wet	15	2–34	27

each activity and time of day. Feeding activity varied among daylight hours and was higher in the evening (76.5%) and late morning (74.56%) than least during midday (54%). The peaks in moving were similar to the peaks in feeding during all the time blocks of a day and preening and resting were higher during the midday.

Percent time spent in all activity differed significantly except alerting. On average, 68.35% of the bird day time was spent in feeding. Feeding occurred all hours but there was a reduction in feeding activity between 1200 and 1500 hours. Percent time spent in feeding, resting, and preening differed significantly based on seasons (Table 4). Lesser Flamingos fed least during the dry season (62.33%) and most during the wet season (77.91%). Preening activities occurred more during the wet season than the dry season. Alerting and moving showed insignificant variation in season. Resting activity varied between time of day and season. It always peaked during midday and the dry season.

4. Discussion

Flamingos typically are the most prominent and important consumers in the lakes. The largest concentration of flamingo was 24,417 in Lake Abijata. However, drastic and sudden fluctuations in number can occur and there is a very marked exchange between the lakes. There were more flamingos during the dry season than the wet season in Lake Abijata and contrary to Lakes Shalla and Chitu during the wet season. Despite the observed fluctuations, the total flamingo population of the area remained relatively stable, suggesting that

TABLE 3: Percentage of time spent (mean ± SD) on different activities by nonbreeding lesser flamingos based on time of day.

Activity	Time of the day (hours)				H	P
	0600–0900	0900–1200	1200–1500	1500–1800		
Feeding	68.33 ± 7.21	74.56 ± 4.2	54 ± 3.2	76.5 ± 4.5	51.12	0.000*
Moving	6.5 ± 5.11	7.17 ± 1.43	3.95 ± 0.21	10.67 ± 0.6	78.13	0.000*
Resting	8.83 ± 2.3	2.1 ± 0.3	20.55 ± 1.4	3.83 ± 0.2	42.11	0.000*
Preening	10.01 ± 3.4	8.67 ± 2.03	17.5 ± 2.1	2.33 ± 1.3	28.34	0.000*
Alerting	6.33 ± 1.62	7.5 ± 072	4 ± 0.78	6.67 ± 0.4	16.47	0.34

*Differ significantly (Kruskal-Wallis test, $P < 0.05$) between time blocks.

TABLE 4: Percentage of time spent (mean ± SD) on different activities by nonbreeding lesser flamingos based on seasons.

Activity	Season		H	P
	Dry	Wet		
Feeding	62.33 ± 3.1	77.91 ± 2.4	37.23	0.000*
Moving	5.27 ± 2.3	7.11 ± 1.4	7.21	0.121
Resting	18.74 ± 1.7	1.27 ± 0.4	22.3	0.000*
Preening	5.36 ± 0.6	10.90 ± 2.2	27.12	0.000*
Alerting	8.3 ± 1.3	2.81 ± 0.3	3.45	0.141

*Differ significantly (Kruskal-Wallis test, $P < 0.05$) between seasons.

the flamingos responded to the effect of seasonal variation by moving between lakes. This aspect of behavior makes flamingos nonresident in any one saline lake, moving and exploiting various nearby lakes as their home range [40]. A seasonal pattern of abundance was observed which was positively correlated with weather and water level quality has been shown for many species of flamingos [41]. The algal food resources are not stable and decline from time to time [42]. Hence, any lake that provides food in suitable quality and quantity makes a valuable contribution to the survival of these birds. Therefore, seasonal abundance of diet might be the primary cause for the difference in the number of individuals within and among lakes. Diet abundance within lakes is associated with variations of water characteristics like conductivity and salinity [43].

The relative difference in the number of flamingos in Lake Abijata sites might be due to the difference in quantity of their diet among sites. In particular, differences in concentration of flamingos at the rivers inlet sites might be due to the concentration of algae as a result of fresh water nutrients provided by the rivers for algal growth and access to use the fresh water to drink. Vareschi [14] and Tuite [41] stated a striking characteristic of flamingos to strong fluctuations on Rift Valley Soda Lakes. Variations could also be caused by availability of fresh water [44].

The Greater Flamingos which occur at low density in the same habitat with the Lesser Flamingo are mostly found (63%) in Lake Shalla. This might be related to the feeding habit of the birds. Greater Flamingos are generalists consuming copepods, mollusks, and other small planktonic and benthic animals in addition to algae [4]. Greater Flamingos feed mainly on invertebrates which they filter from water or mud over a large range of habitats [45]. Shalla has poor phytoplankton and yet supports dense number of benthic organisms [46].

Flamingos are gregarious birds and form very large feeding groups. They form large foraging flocks, which can be interpreted as a consequence of the high patchiness of food distribution. On several occasions, Greater Flamingos showed preference to offshore area of the lakes. In contrast, Lesser Flamingo mainly preferred the shoreline and mudflat of the lakes. This preference for shallow water bodies may be related to the species' foraging behavior, characterized by feeding on small diatoms and *Spirulina* near the lake banks. Lesser Flamingos require shallow eutrophic wetlands and waterbodies which are more saline than those used by the Greater Flamingos, because it feeds mainly blue-green algae that bloom under these conditions [47]. During the wet season, Lesser Flamingos at Lake Abijata were observed in mudflat habitat where diatoms may be plentiful [48].

Feeding usually was the major activity of flamingo, probably because of the small size of their prey items that flamingos are forced to spend more time on feeding than on other forms of behavior. More feeding activity of flamingo in the late afternoon may reflect their need to obtain energy for overnight energetic requirements. The general pattern of flamingo feeding in different regions worldwide is that they fed mainly during morning and late afternoon to early evening and roosted during the midday [35]. The rate of resting was the highest in the middle of the day, between 012:00 and 15:00. Probably, the main purpose of this rest is to avoid the heat of the day to conserve water loss. Flamingos spend more time in feeding activities during the wet than dry seasons. There are several possible explanations. It is their breeding season which necessitates an increase in forage intake regardless of the ambient conditions. However, the main effect was the decline of availability of their diet in the area (Lake Abijata) during wet season [48]. Preening showed a significant variation with season and time of a day (high during midday and low during late afternoon). It plays an important role in deparasitisation and feather adjustment [49]. Similar to preening, moving showed marked seasonal and time of day changes. Individuals spent more time moving in the afternoon partly because a fraction of the population left the site to go to another one after having taken the required food intake and also disturbance by raptors (African fish eagle) occurs mainly in the afternoon.

5. Conclusion

Seasonal and annual fluctuations are observed in the total number of birds counted. Such changes in flamingo counts at individual lakes may indicate changes in the importance of each of these Soda lakes both within years and between years. The distribution and abundance of flamingos are related to food supply. Thus, changes in the numbers of flamingos at a particular lake during the year and between years may reflect fluctuations in the availability of food supply and may at least in part be a result of anthropogenic activities. Conservation efforts in the park should include not only the wild flora and fauna not only of the land but also of the aquatic systems because both places represent one integrated system. Presently, there is no effective protection of flamingo feeding areas or enforcement of laws protecting the bird. Ethiopian Wildlife Conservation Authority should take a more positive position action to protect flamingos.

Conflict of Interests

The authors declare that there is no conflict of interests regarding the publication of this paper.

Acknowledgments

The authors wish to thank Addis Ababa University and Animal Diversity Research Project for providing the required facilities for carrying out the above research work.

References

[1] J. del Hoyo, A. Elliot, and J. Sargatal, *Handbook of the Birds of the World*, vol. 1 of *Ostrich to Ducks*, Lynx Edicions, Barcelona, Spain, 1992.

[2] M. A. Ogilvie and C. Ogilvie, *Flamingos*, Alan Sutton Publishing, Gloucester, UK, 1986.

[3] C. Mlingwa and N. Baker, "Lesser Flamingo *Phoenconaias minor* counts in Tanzanian soda lakes," in *Implications for Conservation*, G. C. Boere, C. A. Galbraith, and D. A. Stroud, Eds., pp. 260–267, The Stationery Office, Edinburgh, Scotland, 2006.

[4] B. Childress, S. Nagy, and B. Hughes, *International Single Species Action Plan for the Conservation of the Lesser Flamingo (Phoeniconaias Minor)*, AEWA Technical Series, no. 34, Agreement on the Conservation of African-Eurasian Migratory Waterbirds, Bonn, Germany, 2008.

[5] S. Delany and D. Scott, *Waterbird Population Estimates*, Wetlands International, Wageningen, The Netherlands, 2006.

[6] K. Unger and J. J. Elston, "Successful ex-situ breeding of lesser flamingo (*Phoeniconaias minor*)," in *Flamingo Bulletin of the IUCE-SSC-Wetlands International Flamingo Specialist Group*, B. Childress, F. Arengo, and A. Bechet, Eds., no. 17, pp. 64–67, Wildfowl & Wetlands Trust, Slimbridge, UK, 2009.

[7] A. R. Johnson, F. Cezilly, and V. Boy, "Plumage development and maturation in the greater flamingo *Phoenicopterus ruber*," *Ardea*, vol. 81, pp. 25–34, 1993.

[8] L. Krienitz, P. K. Dadheech, and K. Kotut, "Mass developments of a small sized ecotype of *Arthrospira fusiformis* in Lake Oloidien, Kenya, a new feeding ground for lesser flamingos in east Africa," *Fottea*, vol. 13, pp. 215–225, 2013.

[9] A. R. Johnson and F. Cezilly, *The Greater Flamingo*, T. & A. D. Poyser, London, UK, 2007.

[10] R. E. Simmons, "Population declines, viable breeding areas, and management options for flamingos in southern Africa," *Conservation Biology*, vol. 10, no. 2, pp. 504–514, 1996.

[11] O. Nasirwa, "Conservation status of flamingos in Kenya," *Waterbirds*, vol. 23, pp. 47–51, 2000.

[12] R. E. Simmons, "Declines and movements of Lesser Flamingos in Africa," *Waterbirds*, vol. 23, pp. 40–46, 2000.

[13] T. P. Boyle, S. M. Caziani, and R. G. Waltermire, "Landsat TM inventory and assessment of waterbird habitat in the southern altiplano of South America," *Wetlands Ecology and Management*, vol. 12, no. 6, pp. 563–573, 2005.

[14] E. Vareschi, "The ecology of Lake Nakuru (Kenya). I. Abundance and feeding of the lesser flamingo," *Oecologia*, vol. 32, no. 1, pp. 11–15, 1978.

[15] C. H. Tuite, "Population size, distribution and biomass density of the Lesser Flamingo in the Eastern Rift Valley, 1974–1976," *Journal of Applied Ecology*, vol. 16, no. 3, pp. 765–775, 1979.

[16] B. Childress, "New flamingo population estimates for Africa and Southern Asia," in *Flamingo, Bulletin of the IUCN-SSC/Wetlands International Flamingo Specialist Group*, B. Childress, A. Béchet, F. Arengo, and N. Jarrett, Eds., no. 13, pp. 18–21, Wildfowl & Wetlands Trust, Slimbridge, UK, 2005.

[17] L. A. Bennun, "Threats to lesser flamingos in East Africa," in *Conservation of the Lesser Flamingo in Eastern Africa and Beyond*, G. W. Howard, Ed., pp. 31–33, IUCN East Africa Programme, Lake Bogoria Conference, Nairobi, Kenya, 1994.

[18] O. Balkiz, A. Béchet, L. Rouan et al., "Experience-dependent natal philopatry of breeding greater flamingos," *The Journal of Animal Ecology*, vol. 79, no. 5, pp. 1045–1056, 2010.

[19] N. Baccetti, L. Panzarin, F. Cianchi, L. Puglisi, M. Basso, and E. Arcamone, "Two new greater flamingo (*Phoenicopterus roseus*) breeding sites in Italy," in *Flamingo, Bulletin of the IUCN-SSC/Wetlands International Flamingo Specialist Group*, B. Childress, F. Arengo, and A. Bechet, Eds., no. 16, pp. 24–27, Wildfowl & Wetlands Trust, Slimbridge, UK, 2008.

[20] M. Smart, H. Azafzaf, and H. Dlensi, "Analysis of the mass of raw data on Greater Flamingos Phoenicopterus roseus on their wintering grounds, particularly in North Africa," in *Flamingo, Bulletin of the IUCNSSC/ Wetlands International Flamingo Specialist Group, Special Publication 1: Proceedings of the 4th International Workshop on the Greater Flamingo in the Mediterranean Region and Northwest Africa*, A. Béchet, M. Rendón-Martos, J. Amat, N. Baccetti, and B. Childress, Eds., pp. 58–61, Wildfowl & Wetlands Trust, Slimbridge, UK, 2009.

[21] B. Childress, B. Hughes, D. Harper, and W. van den Bossche, "East African flyway and key site network of the Lesser Flamingo (*Phoenicopterus minor*) documented through satellite tracking," *Ostrich*, vol. 78, no. 2, pp. 463–468, 2007.

[22] R. Koenig, "The pink death: die-offs of the lesser flamingo raise concern," *Science*, vol. 313, no. 5794, pp. 1724–1725, 2006.

[23] BirdLife International, "*Phoeniconaias minor*," IUCN red list of threatened species, 2008, http://www.iucnredlist.org.

[24] EWNSH, *Important Bird Areas of Ethiopia: A First Inventory*, Ethiopian Wildlife and Natural History Society, Addis Ababa, Ethiopia.

[25] J. G. Stephenson, *An Appraisal of the Conservation of Nature in the Lakes Abijiata and Shalla Locality with Recommendations*, Ethiopian Wildlife Conservation Organisation, Addis Ababa, Ethiopia, 1978.

[26] EWNHS, *Important Bird Areas Program Site Account: Abijata-Shalla Lakes National Park*, Ethiopia Wildlife and Natural History Society, Addis Ababa, Ethiopia, 2000.

[27] J. S. Ash and T. M. Gullick, "The present situation regarding endemic breeding birds of Ethiopia," *Scopus*, vol. 13, pp. 90–96, 1989.

[28] J. C. Hillman, *Ethiopia: Compendium of Wildlife Conservation Information*, vol. 1 & 2, New York Zoological Society, The Wildlife Conservation Society International, New York, NY, USA; Ethiopian Wildlife Conservation Organization, Addis Ababa, Ethiopia, 1993.

[29] F. Senbeta and F. Tefera, "Environment crisis in the Abijiata-Shalla Lakes national park," *Walia*, vol. 22, pp. 1–13, 2001.

[30] T. Ayenew, "Environmental implications of changes in the levels of lakes in the Ethiopian Rift since 1970," *Regional Environmental Change*, vol. 4, no. 4, pp. 192–204, 2004.

[31] R. Almaw, *A Checklist of the Birds of the Abijata-Shalla Lakes National Park (Central Rift Valley)*, Ethiopian Wildlife Conservation Authority, Addis Ababa, Ethiopia, 2012.

[32] C. Bibby, N. D. Burgess, and D. A. Hill, *Bird Census Techniques*, Academic Press, London, UK, 1992.

[33] W. J. Sutherland, *Ecological Census Techniques: A Handbook*, Cambridge University Press, London, UK, 1996.

[34] N. D. Niemuth, M. E. Estey, R. E. Reynolds, C. R. Loesch, and W. A. Meeks, "Use of wetlands by spring-migrant shorebirds in agricultural landscapes of North Dakota's drift prairie," *Wetlands*, vol. 26, no. 1, pp. 30–39, 2006.

[35] K. L. Bildstein, P. C. Frederick, and M. G. Spalding, "Feeding patterns and aggressive behavior in juvenile and adult American flamingos," *The Condor*, vol. 93, no. 4, pp. 916–925, 1991.

[36] W. J. Sutherland, *The Conservation Handbook Research Management and Policy*, Cambridge University Press, Cambridge, UK, 2000.

[37] J. Altmann, "Observational study of behavior: sampling methods," *Behaviour*, vol. 49, no. 3, pp. 227–267, 1974.

[38] M. P. Kahl, "Distribution and numbers a summary," in *Flamingos*, J. Kear and H. Duplaix-Hall, Eds., pp. 93–149, T. & A. D. Poyser, Berkhamsted, UK, 1975.

[39] P. N. Lehner, *Handbook of Ethological Methods*, Cambridge University Press, Cambridge, UK, 2nd edition, 1996.

[40] G. McCulloch, A. Aebischer, and K. Irvine, "Satellite tracking of flamingos in southern Africa: the importance of small wetlands for management and conservation," *Oryx*, vol. 37, no. 4, pp. 480–483, 2003.

[41] C. H. Tuite, "The distribution and density of Lesser Flamingos in east Africa in relation to food availability and productivity," *Waterbirds*, vol. 23, pp. 52–63, 2000.

[42] E. Vareschi and J. Jacobs, "The ecology of Lake Nakuru. VI. Synopsis of production and energy flow," *Oecologia*, vol. 61, pp. 70–82, 1985.

[43] S. M. Caziani, O. R. Olivio, E. R. Ramírez et al., "Seasonal distribution, abundance, and nesting of Puna, Andean, and Chilean flamingos," *The Condor*, vol. 109, no. 2, pp. 276–287, 2007.

[44] E. Vareschi, "The ecology of Lake Nakuru (Kenya). III. Abiotic factors and primary production," *Oecologia*, vol. 55, no. 1, pp. 81–101, 1982.

[45] G. Zweers, F. de Jong, and H. Berkhoudt, "Filter feeding in flamingos (*Phoenicopterus ruber*)," *Journal of Avian Biology*, vol. 97, pp. 1–28, 1995.

[46] C. Tudorancea and A. D. Harrison, "The benthic communities of the saline lakes: Abijata and Shala (Ethiopia)," *Hydrobiologia*, vol. 158, no. 1, pp. 117–123, 1988.

[47] L. H. Brown, E. K. Urban, and K. Newman, *The Birds of Africa*, vol. I, Academic Press, London, UK, 1982.

[48] T. Kumssa and A. Bekele, "Phytoplankton composition andphysico-chemicalparameters study in water bodies of Abijata-Shalla Lakes National Park (ASLNP), Ethiopia," *Greener Journal of Biological Sciences*, vol. 4, no. 2, pp. 69–76, 2014.

[49] P. Cotgreave and D. H. Clayton, "Comparative analysis of time spent grooming by birds in relation to parasite load," *Behaviour*, vol. 131, no. 3-4, pp. 171–187, 1994.

20

Genetic Diversity Assessment and Identification of New Sour Cherry Genotypes Using Intersimple Sequence Repeat Markers

Roghayeh Najafzadeh,[1] Kazem Arzani,[1] Naser Bouzari,[2] and Ali Saei[3]

[1] Department of Horticultural Sciences, Tarbiat Modares University (TMU), P.O. Box 14115-336, Tehran, Iran
[2] Horticultural Section, Stone Fruit Research Group, Seed and Plant Improvement Research Institute of Karaj (SPII), P.O. Box 31585-4119, Karaj, Iran
[3] Genomics Section, Agricultural Biotechnology Research Institute of Iran (ABRII), P.O. Box 85135-487, Isfahan, Iran

Correspondence should be addressed to Roghayeh Najafzadeh; roghayehnajafzadeh@yahoo.com and Kazem Arzani; arzani_k@modares.ac.ir

Academic Editor: Alexandre Sebbenn

Iran is one of the chief origins of subgenus *Cerasus* germplasm. In this study, the genetic variation of new Iranian sour cherries (which had such superior growth characteristics and fruit quality as to be considered for the introduction of new cultivars) was investigated and identified using 23 intersimple sequence repeat (ISSR) markers. Results indicated a high level of polymorphism of the genotypes based on these markers. According to these results, primers tested in this study specially ISSR-4, ISSR-6, ISSR-13, ISSR-14, ISSR-16, and ISSR-19 produced good and various levels of amplifications which can be effectively used in genetic studies of the sour cherry. The genetic similarity among genotypes showed a high diversity among the genotypes. Cluster analysis separated improved cultivars from promising Iranian genotypes, and the PCoA supported the cluster analysis results. Since the Iranian genotypes were superior to the improved cultivars and were separated from them in most groups, these genotypes can be considered as distinct genotypes for further evaluations in the framework of breeding programs and new cultivar identification in cherries. Results also confirmed that ISSR is a reliable DNA marker that can be used for exact genetic studies and in sour cherry breeding programs.

1. Introduction

The sour cherry belongs to the family of Rosaceae, subfamily Prunoideae, genus *Prunus,* and subgenus *Cerasus* [1]. It is an allotetraploid species ($2n = 4x = 32$) resulting from a natural hybridization between *Prunus avium* L. (Sweet Cherry) and *Prunus fruticosa* Pall. (Ground Cherry) [2]. This species is reported to have originated from the area that comprises Asia Minor, Iran, Iraq, and Syria [3] and has been used as rootstock and also in breeding programs for developing new commercial cultivars, dwarf, and resistant rootstocks [4–6]. According to the FAO database, Iran ranked third worldwide in 2011 for cherry production after Turkey and the USA with a total of 241 thousand tons produced [7].

Genetic variability is a prerequisite for any plant breeding program [8]. As an origin of the subgenus *Cerasus*, Iran has rich cherry germplasm resources. Using diverse *Cerasus* subgenus resources to broaden the genetic base of cherry cultivars and rootstocks and improving them for development of the cherry industry are important goals for cherry breeders in Iran [9, 10]. Therefore, it is necessary to characterize and preserve these genotypes and cultivars [11]. DNA markers are very useful in distinguishing between accessions and in investigations of genetic diversity or relatedness [12]. Different DNA markers have been broadly used to analyze genetic variations in *Prunus*, such as RAPD [13], AFLP [14], and SSR [15]. The major limitations of these methods are low reproducibility of RAPD, high cost of AFLP, and the need to know the flanking sequences to develop species specific primers for SSR polymorphism [16]. In comparison, intersimple sequence repeats (ISSRs) have been developed that overcome most limitations [17]. These markers involve

TABLE 1: Accessions of genotypes used in this study and their origins.

Number	Accession	Origin
1	KaThLa1SSGe21	Lavasan
2	Hamedan	Hamedan
3	KaTaJo2Ge9	Taleghan
4	KaThMe3Ge19	Chalus
5	KaThLa8Ge31	Lavasan
6	KrRIV4C20	Kerman
7	EsASC1V1SS1	Esfahan
8	KaThLa3Ge23	Lavasan
9	Bulgar	Bulgaria
10	Montmorency	France
11	Erdi Jubileum	Hungary
12	Erdi Botermo	Hungary

PCR amplification of DNA by a single, 16–18 bp long primer composed of a repeated sequence [18]. ISSR gives multilocus patterns which are very reproducible, abundant, and polymorphic in plant genomes [19]. It is useful in areas of cultivar identification, germplasm characterization, phylogenetic relationship analysis, and genetic linkage mapping in a wide range of plant species [20, 21], including cherries [20–24].

In the present study, ISSR analysis was used to evaluate the genetic variation new Iranian sour cherries, with the aim of using the ISSR technique to these genotypes for the use them in cherries breeding programs as well as for conservation management of subgenus Cerasus germplasm in Iran. It is hoped that with supplementary tests and analyses we can identify and introduce new sour cherry cultivars to the fruit industry.

2. Materials and Methods

2.1. Plant Materials. During the breeding programs, collection, and evaluation of local sour cherry germplasms from different regions in Iran in order to achieve proper cultivars and rootstocks, after five years visual observations, it was found that some of the genotypes had quite superior growth characteristics and fruit quality that they can be considered for the introduction of new cultivars [25]. These superior genotypes were chance seedlings, so they were selected according to the 5-year visual observations and grafted onto "Mahlab" rootstocks which were available in the fruit research collection of the Seed and Plant Improvement Institute in Kamal Abad, Karaj, Iran. Then 2-year determination of the genotypes also approved it and showed that these selected genotypes had such superior growth characteristics and fruit quality [26]. In this study, these selected genotypes (see Table 1) and 4 improved cultivars (Bulgar, Montmorency, Erdi Jubileum, and Erdi Botermo) were analyzed. These genotypes were 5 years old and had been planted at 4×5 m. The young leaves of these genotypes were collected during May 2012. The characterization of the accessions is shown in Table 1.

2.2. DNA Extraction and PCR Amplification. Total genomic DNA was extracted from the young leaves using a CTAB (hexadecyl trimethyl ammonium bromide) method, according to the protocol described in Saunders et al. [27]. Then, the DNA extract was suspended in 50 μL 1X TE buffer (1 M pH 8.0 Tris-HCl; 0.5 M pH 8.0 EDTA) and kept at −20°C. The quality and concentration of each DNA sample was determined using a NanoDrop spectrophotometer at 260, 280 nm (ND-1000, Co, USA) and running 3 μL DNA in 0.7% (w/v) agarose gels in 0.5X TAE buffer. 30 ISSR primers which were selected by Agricultural Biotechnology Research Institute Laboratory, Esfahan, Iran, according to [21, 22, 24, 28, 29] studies and synthesized by Metabion Co. (Germany) were used. These synthesized ISSR primers were initially screened and finally 23 primers were selected to be used in this study. The list of primers and their information are presented in Table 2. For PCR analysis, approximately 25 ng of genomic DNA was used in a 25 mL reaction containing 1X PCR reaction buffer, 2 mM MgCl$_2$, 0.8 mM dNTPs, 5 pmol of each primer, 1U Taq DNA polymerase (Fermentas, Lithuania), and DNA-free water. Amplifications were performed in a thermocycler (Applied Biosystems, Veriti, USA) programmed for a first denaturation step of 3 min at 94°C, followed by 40 cycles of 30 s at 94°C for denaturation, 30 s at 30–57°C (varied for each primer according to Table 2) for annealing, 1 min at 72°C for elongation, and final extension at 72°C for 5 min. They were then held at 4°C until the tubes were removed. Amplified products were separated by electrophoresis in 1.5% (w/v) agarose gels at constant voltage (95 V) in 1X TAE buffer for approximately 90 min, stained with gelred, and photographed with UV light (Figure 1). The size of produced fragments was defined according to size marker (GeneRuler 1 kb DNA ladder, SM0311, Fermentas).

2.3. Data Analysis. Only reproducible and well-defined alleles were considered potential polymorphic markers. The alleles scored as present (1) or absent (0) with Phoretix Pro, ver. 10.4 software. In order to increase the accuracy of the scoring, each gel electrophoresis was scored in three replications. The cophenetic correlation coefficient (CCC) was calculated and the similarity matrix was used for the cluster analysis and construction of dendrogram using the Simple Coefficient and Unweighted Pair-Group Method with Arithmetic average (UPGMA) [30] and for genetic relationships among subgenera and sections, Principal Coordinate Analysis (PCoA) and three-dimensional projection of genotypes (3D) [31] were used using the NTSYS software ver. 2.02 [32]. The following parameters were calculated for each primer: number of total alleles per locus, number of polymorphic alleles, polymorphism percentage, Polymorphism Information Content (PIC), average gene diversity (Hi), and fragment size. Polymorphism percentage was calculated using the ratio of number of polymorphic alleles to total alleles [33]. PIC was calculated as described by Warburton and Crossa [34] (PIC $= 1 - \sum (P_i/P_k)^2$), where P_i is the proportion of the population carrying the ith allele calculated for each microsatellite locus and P_k is total alleles. Average gene diversity was calculated by direct counts for the putative locus, identified by each primer described by the IPGRI and Cornell University [33] method ($H_i = 1 - p2 - q2$), where p and q are the frequency of the ith allele.

Number	Primers	Sequence (5'-3')	TA[1] (°C)	Total number of alleles (a)	Number of polymorphic alleles (b)	% Polymorphism (b/a) × 100	PIC[2]	Average gene diversity (Hi)	Band size range (bp)
1	ISSR-1	GTGGTGGTGGC	30	16	16	100	0.90	0.45	600–1500
2	ISSR-2	GAGAGAGAGAGAGAGAT	48	21	21	100	0.94	0.40	400–1500
3	ISSR-3	CTCTCTCTCTCTCTCTG	47	25	24	96.00	0.94	0.43	700–2500
4	ISSR-4	GAGAGAGAGAGAGAGATG	52	26	26	100	0.94	0.37	400–1400
5	ISSR-5	AGAGAGAGAGAGAGAGTT	52	21	21	100	0.94	0.36	400–2100
6	ISSR-6	CTCTCTCTCTCTG	39	29	29	100	0.94	0.40	400–2200
7	ISSR-7	CTCTCTCTCTCTCTCTTG	47	18	17	94.44	0.93	0.37	700–1700
8	ISSR-8	CACACACACACAAC	39	25	24	96.00	0.95	0.40	700–2300
9	ISSR-9	CACACACACACAGT	39	18	18	100	0.89	0.39	700–2800
10	ISSR-10	CACACACACACAGG	36	21	21	100	0.94	0.37	400–1700
11	ISSR-11	CACACACACACAAG	30	23	23	100	0.94	0.43	700–3000
12	ISSR-12	CACACACACACACACAGG	47	15	14	93.33	0.88	0.43	500–2100
13	ISSR-13	CTCTCTCTCTCTCTCTRG	53	29	29	100	0.96	0.34	400–2800
14	ISSR-14	DBDACACACACACAC	55	29	29	100	0.95	0.42	400–2800
15	ISSR-15	HVHTCCTCCTCCTCCTCCTCC	57	9	7	77.77	0.85	0.45	600–1800
16	ISSR-16	GAGAGAGAGAGAGAGAC	53	25	25	100	0.95	0.45	400–3200
17	ISSR-17	ACACACACACACACACC	53	17	17	100	0.90	0.44	600–2200
18	ISSR-18	GAGAGAGAGAGAGAGAYC	47	15	14	93.33	0.90	0.45	600–1800
19	ISSR-19	AGAGAGAGAGAGAGAGYT	52	29	29	100	0.95	0.41	280–1600
20	ISSR-20	ACACACACACACACACYG	53	16	16	100	0.90	0.43	400–2800
21	ISSR-21	CACACACACACACACART	48	24	24	100	0.94	0.39	600–1700
22	ISSR-22	GAGAGAGAGAGAGAGAYG	47	20	20	100	0.93	0.40	400–2200
23	ISSR-23	GAGAGAGAGAGAGAGACG	55	18	18	100	0.94	0.42	400–3200
	Mean	—	—	21.26	20.95	98.45	0.93	0.41	—
	Total	—	—	489	482	—	—	—	—

Note: D = (G, A, T), B = (G, T, C), H = (A, T, C), V = (G, A, C), R = (A, G), Y = (C, T), and N = (A, T, G, C).
[1]Annealing temperature. [2] Polymorphism Information Content.

3. Results

The results of ISSR fingerprinting of 12 sour cherry genotypes using 23 primers are given in Table 2. Results showed that 489 alleles were generated at 23 ISSR loci, 482 of which were polymorphic. The number of total alleles per locus varied from 9 (ISSR-15 loci) to 29 (ISSR-6, ISSR-13, ISSR-14, and ISSR-19 loci) alleles with an average of 21.26 across the genotypes. Other primers with a high number of alleles per locus were ISSR-4 (26 alleles) followed by ISSR-3, ISSR-8, and ISSR-16 (25 alleles). The average polymorphic alleles per primer were 20.95, and of the 23 loci, ISSR-6, ISSR-13, ISSR-14, and ISSR-19 loci had the highest polymorphic alleles (29 alleles), followed by ISSR-4 (26 alleles), ISSR-16 (25 alleles), and ISSR-3 and ISSR-8 (24 alleles). The polymorphism percentage for primers ranged from 77.77% to 100.0% with an average of 98.45% (Table 2).

PIC values ranged from 0.85 to 0.96 with an average of 0.93. The highest PIC, that is, an indicator of effectiveness of primers used in genetic diversity studies, was ISSR-13 (0.96) with 29 alleles, followed by ISSR-8, ISSR-14, ISSR-16, and ISSR-19 (0.95). The lowest PIC value belonged to ISSR-15 (0.85) with (7 alleles) (Table 2). Average gene diversity ranged from 0.34 to 0.45 with a mean of 0.41. Among the loci, the highest H_i value belonged to the ISSR-1, ISSR-15, ISSR-16, and ISSR-18 locus (0.45), while the lowest values belonged to the ISSR-13 loci (0.34) (Table 2). Fragment size of the 23 primers ranged from 280 to 3200 bp. The lowest ranges were those of ISSR-19 (280–1600) and ISSR-4 (400–1400), and the highest range was related to ISSR-16 and ISSR-23 (400–3200) bp (Table 2).

The highest cophenetic correlation coefficient based on ISSR data with a simple similarity coefficient was ($r = 0.91$) (Table 3). The high value of this coefficient indicates the

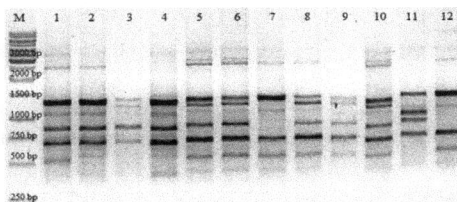

FIGURE 1: ISSR band profiles generated by the primer ISSR-16. Genotype number: see Table 1. M: weight marker: 1 kb DNA.

TABLE 3: Comparison of different methods for constructing similarity matrices and dendrograms.

Similarity coefficient	Cluster algorithm		
	UPGMA	UPGMC	Complete linkage
D	r = 0.72	r = 0.42	r = 0.71
J	r = 0.74	r = 0.47	r = 0.73
SM	r = 0.91	r = 0.24	r = 0.79

D: Dic [36]; J: Jaccard [37]; SM: Simple Matching [30].
UPGMA: Unweighted Pair-Group Method with Arithmetic average;
UPGMC: Unweighted Pair-Group Method using Centroids.

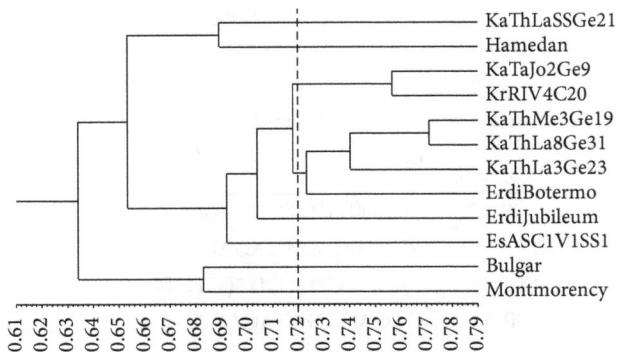

FIGURE 2: Dendrogram of ISSR analysis on sour cherry genotypes used in this study.

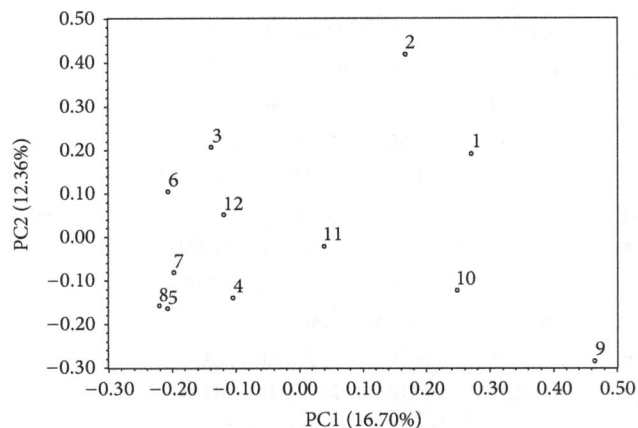

FIGURE 3: Principle coordinate analysis (PCoA) for 23 ISSR primers applied on sour cherry genotypes. Numbers represent the genotypes according to Table 1.

suitability of the grouping method. The CCC is considered to be a good representation of the data matrix in the dendrogram if it is $0.80 \leq$ CCC [35]. Thus, the similarity matrix was used for the cluster analysis and construction of dendrogram using the simple coefficient and UPGMA.

Genetic similarity between genotypes was estimated using the simple similarity coefficient. The genetic similarity ranged from 0.56 to 0.77 with an average of 0.72. It showed a high diversity among genotypes. The EsASC1V1SS1 and Bulgar showed the lowest similarity (0.56), and the KaThMe3Ge19 and KaThLa8Ge31 showed a high similarity (0.77). Other genotypes such as Bulgar with KaTaJo2Ge9 (0.58), KrRIV4C20 (0.59), and Erdi Botermo (0.59) had low similarity. Moreover, KaThLa8Ge31 with KrRIV4C20 (0.75), Erdi Jubileum (0.75) and KaThLa3Ge23 (0.76), KaTaJo2Ge9 with KrRIV4C20 (0.76), and Ka ThMe3Ge19 with Erdi Botermo (0.76) had high similarity (Table 4).

A dendrogram based on the simple similarity coefficient and UPGMA analysis is presented in Figure 2. According to the dendrogram, genotypes were separated into two main clusters. Promising Iranian genotypes were separated from the improved cultivars and were further divided into eight subclusters with a genetic similarity of 0.72. One (I) included the KaThLa1SSGe21 genotype (from Lavasan), two (II) included Hamedan (Hamedan), and subcluster three (III) contained KaTaJo2Ge9 (Taleghan) and KrRIV4C20 (Kerman). These two genotypes were similar to the matrix 0.76, so it seemed that they had the same genetic origin. Subcluster 4 (IV) which was the biggest subcluster included KaThMe3Ge19 (Chalus), KaThLa8Ge31 (Lavasan), KaThLa3Ge23 (Lavasan), and Erdi Botermo (Hungary). Five (V) included the Erdi Jubileum cultivar (Hungary); six (VI) included EsASC1V1SS1 (Esfahan); seven (VII) included the Bulgar (Bulgaria); eight (III) included Montmorency cultivar

(France). According to these results, it seems that the genetic diversity of the sour cherry genotypes was not entirely a function of geographical variation; thus, the genotypes of the 4th subgroup with different distribution centers were placed in adjacent genetic groups.

The genetic relationship between these genotypes was also visualized by performing PCoA that showed two significant axes, which explained 16.70% and 12.36% of the total variance, respectively. The first two eigenvalues accounted for 29.06% of the variation observed in the genotypes (Table 5).

The two-dimensional plot generated from PCoA also supported the clustering pattern of the UPGMA dendrogram (Figure 3). This reflected a higher genetic diversity in the studied collection, which was confirmed by a principle component analysis of the genotype data. Results of this analysis showed a wider genetic distribution of genotypes in the studied collection.

In the three-dimensional PCoA plot, generally, similar groupings with the UPGMA dendrogram and additional information were also revealed. The first three principal axes accounted for 16.70%, 12.36%, and 10.52% of the total variation, respectively, indicating the complex multidimensional nature of ISSR variation (Figure 4).

TABLE 4: The simple similarity matrix for sour cherry genotypes based on ISSR data.

	1	2	3	4	5	6	7	8	9	10	11	12
1	1.00											
2	0.69	1.00										
3	0.67	0.70	1.00									
4	0.67	0.64	0.71	1.00								
5	0.64	0.64	0.72	0.77	1.00							
6	0.63	0.67	0.76	0.70	0.75	1.00						
7	0.64	0.63	0.66	0.69	0.72	0.69	1.00					
8	0.63	0.63	0.73	0.72	0.76	0.71	0.73	1.00				
9	0.64	0.60	0.58	0.64	0.61	0.59	0.56	0.60	1.00			
10	0.66	0.66	0.66	0.66	0.69	0.64	0.64	0.66	0.68	1.00		
11	0.70	0.68	0.70	0.70	0.75	0.70	0.68	0.70	0.64	0.70	1.00	
12	0.65	0.67	0.73	0.76	0.72	0.71	0.67	0.69	0.59	0.67	0.68	1.00

Genotype number: see Table 1.

TABLE 5: Eigen values, percentage, and cumulative proportions for 11 principal coordinate axes, derived from ISSR markers application on sour cherry genotypes.

Axes	Eigen value	Percent	Cumulative
1	1.15	16.70	16.70
2	0.94	12.36	29.06
3	0.90	10.52	39.59
4	0.88	9.84	49.43
5	0.82	8.98	58.42
6	0.77	8.60	67.02
7	0.75	8.00	75.03
8	0.72	7.35	82.39
9	0.66	6.56	88.95
10	0.63	5.99	94.94
11	0.61	5.05	100.00

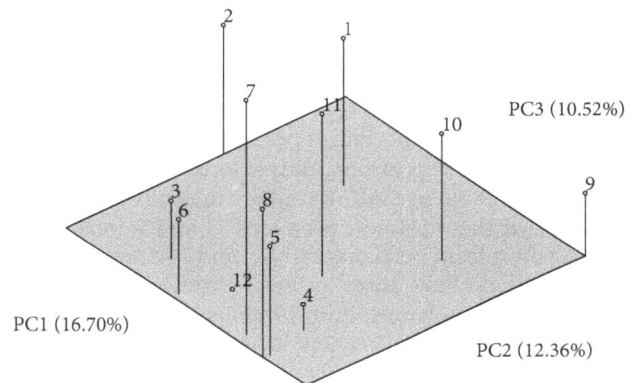

FIGURE 4: 3D plot for 23 ISSR primers applied on sour cherry genotypes. Numbers represent the genotypes according to Table 1.

4. Discussion

Iran is accepted as an origin and diversity center for cherries. In this study, we report for the first time the use of ISSR markers to assess the genetic characterization and to determine genetic relationships between promising Iranian sour cherry genotypes selected from different regions of Iran and improved cultivars. In the present study, 23 ISSR loci in sour cherry genotypes were assayed. The results obtained showed that ISSR primers can be effectively used for genetic diversity studies as well as genetic identification of sour cherries, which was also found in other investigations of cherries [20–24]. In fact, primers tested in this study produced good and various levels of amplifications.

A total of 489 amplified products were obtained using 23 ISSR primers. The average number of total alleles per locus identified in this study (21.26) was higher than the number identified in other studies of cherries. Average polymorphism percentage across all genotypes was 98.45% indicating a high level of polymorphism. Among the 23 loci, ISSR-6, ISSR-13, ISSR-14, and ISSR-19 had the highest polymorphic alleles. Shahi-Gharahlar et al. [24] in their study of 12 ISSR primers

tested on 39 accessions of Iranian wild *Prunus* sub-*cerasus* reported that these ISSR primers generated 156 alleles and that the number of alleles per locus ranged from 9 to 19 with an average of 13 alleles, and the polymorphism percentage was 81.80–100% with an average of 96.46%. Ganopoulos et al. [20] studied 10 ISSR primers on 19 Greek traditional sweet cherries and two international cultivars. They reported that these ISSR primers generated 91 alleles and that the number of alleles per locus ranged from 2 to 10 with an average of 9.1 alleles, and the polymorphism percentage was 25–75% with an average of 57.7%. Moreover, Li et al. [22] studied 18 ISSR primers on 10 Chinese sour cherries and reported that these ISSR primers generated 150 alleles and that the number of alleles per locus ranged from 4 to 13 with an average of 8.33 alleles, and the polymorphism percentage was 1.33–32% with an average of 18.67%. The high number of generated alleles in our study may be due to the use of several different genotypes that had high genetic diversity.

The PIC values ranged from 0.85 to 0.96 with a mean value of 0.93. The high value of PIC represents the larger number of alleles and polymorphism [18]. Yilmaz et al. [21] in their study of 20 ISSR primers tested on 16 genotypes from genus *Prunus* reported that PIC ranged from 0.35 to 0.93 with an average of 0.74. The average gene diversity ranged from 0.34 to 0.45 with a mean of 0.41. Ganopoulos et al. [20] reported that gene diversity ranged from 0.29 to 0.48 with an average of 0.36. This particular average gene diversity value (0.41) was higher than that (0.36) identified in another survey [20]. The high h_i showed a high diversity among sour cherry genotypes. Fragment size of the 23 primers ranged from 280 to 3200 bp. Band size ranges using ISSR markers were also found in other investigations of cherries with the range of 530–3100 bp [20], 231–1986 bp [22], 400–1950 bp [23], and 200–2100 bp [24].

According to these results, primers tested in our study produced good and various levels of amplifications as compared to other studies. For example, Shahi-Gharahlar et al. [24] in their study on subgenus *Cerasus* used ISSR-2 primer and reported that this primer produced 12 total alleles with polymorphism percentage of 91.70% and fragment size of 250 to 1200 bp, while in our study this primer produced 21 total alleles with polymorphism percentage of 100% and size range of 400–1500 bp. Also Li et al. [22] used ISSR-20 primer in sour cherries and reported that this primer produced 7 total alleles with size range of 281–1458 bp, while this primer produced 16 total alleles with polymorphism percentage of 100% and size range of 400–2800 bp in our study. Moreover, Yilmaz et al. [21] tested ISSR-3 and ISSR-18 primers in genus *Prunus* and reported that these primers produced 13 and 9 total alleles with polymorphism percentage of 100 and 89%, respectively, while we found that these primers produced 25 and 15 total alleles with polymorphism percentages of 96 and 93%, respectively. Also these primers have been tested by other researchers in various species. For example, Sofalian et al. [28] studied ISSR-2, ISSR-3, and ISSR-5 primers on wheat accessions. They reported that these primers produced 3, 15, and 10 total alleles, respectively, while the total alleles produced by these primers in our study were higher than their study. Moreover, Acharya and Sharma [29] in their study on genus *Papaver* tested these primers and reported that ISSR-1, ISSR-6, ISSR-8, ISSR-10, ISSR-11, and ISSR-19 did not produce any band, while in our study these primers produced good and various levels of amplifications. So ISSR-6 and ISSR-19 had the highest number of alleles (29 alleles). The highest number of alleles in their study related to ISSR-2 (4 alleles) and ISSR-3 (6 alleles). Our results indicated a high level of polymorphism of the genotypes based on these markers specially primers with high numbers of polymorphic alleles, polymorphism percentage, PIC values, and gene diversity were to ISSR-4, ISSR-6, ISSR-13, ISSR-14, ISSR-16, and ISSR-19, which can be used effectively in genetic diversity studies of the sour cherry.

Autochthonous varieties, cultivars, and wild genotypes are rich resources for genes for breeding objectives [10, 11]. The mean number of alleles produced by Iranian genotypes was higher than the number of alleles produced by foreign genotypes. These results indicated that since the Iranian genotypes were not selected for breeding programs, they were more likely to have a more diverse genetic background, and they can be used to select different genotypes in order to produce new cultivars. The high average gene diversity with a mean of 0.41 and the high number of alleles observed with a mean of 21.26 for all loci showed that these ISSR markers are highly polymorphic and can be useful in the study of genetic diversity.

Genetic similarity between genotypes ranged from 0.56 to 0.77 with an average of 0.72, which showed a high diversity among the genotypes. Cluster analysis with a genetic similarity of 0.72 divided the genotypes into eight distinct groups that separated Iranian genotypes from improved cultivars, and PCoA supported the cluster analysis results. These genotypes grouped within the same cluster in the dendrogram also occupied the same positions in two-dimensional scaling. Shahi-Gharahlar et al. [24] reported that the genetic similarity measured within 39 accessions of *subcerasus* ranged from 0.04 to 0.85 with an average of 0.28. A dendrogram constructed according to ISSR data of these genotypes divided them into 11 subclusters in which improved cherry cultivars were separated from wild genotypes, and this is consistent with the results of our study.

5. Conclusions

In summary, the good discrimination efficiency and high reproducibility of ISSR markers make them particularly suitable to identify the closely related and unknown sour cherry genotypes. In addition, the high genetic diversity observed within superior Iranian sour cherry genotypes and improved cultivars reflects the necessity for the conservation of this germplasm. Since the Iranian genotypes were superior to the improved cultivars and were separated from them in most of the groups, these genotypes can be considered as distinct genotypes for further evaluations in the framework of breeding programs and new cultivar identification in cherries. Hence, it is expected that the results of this study will assist current sour cherrybreeding efforts in Iran and will maintain the genetic integrity of the genetic resources. It is hoped that with supplementary tests and analyses we can identify and introduce new sour cherry cultivars to the fruit industry.

Conflict of Interests

The authors declare that there is no conflict of interests regarding the publication of this paper.

Acknowledgments

Materials for this study were provided and partially supported by Grant no. (0-100-120000-04-0000-84104) with the support of the Seed and Plant Improvement Institute, Horticultural Section, Karaj, Islamic Republic of Iran. The authors would like to thank Tarbiat Modares University (TMU) for providing the facilities.

References

[1] N. E. Looney and A. D. Webster, *Cherries: Crop Physiology, Production and Uses*, CAB International Press, Oxfordshire, UK, 1996.

[2] A. F. Lezzoni, H. Schmidt, and A. Albertini, "Cherries," in *Genetic Resources of Temperate Fruit and Nut Crops*, J. N. Moore

and J. R. Ballington, Eds., pp. 109–175, Society for Horticultural Science, Wageningen, The Netherland, 1991.

[3] N. I. Vavilov, *The Origin, Variation, Immunity and Breeding of Cultivated Plants*, Ronald Press, New York, NY, USA, 1951.

[4] G. Charlot, M. Edin, F. Flochlay, P. Soing, and C. Boland, "Tabel Edabriz: a dwarf rootstock for intensive cherry orchards," *Acta Horticulturae*, vol. 667, no. 1, pp. 217–222, 2005.

[5] K. Hrotko, L. Magyar, and M. Gyeviki, "Evaluation of native hybrids of *Prunus fruticosa* Pall. As cherry interstocks," *Acta Agriculturae Serbica*, vol. 13, pp. 41–45, 2008.

[6] L. Magyar and K. Hrotkó, "*Prunus cerasus* and prunus fruticosa as interstocks for sweet cherry trees," *Acta Horticulturae*, vol. 795, pp. 287–292, 2008.

[7] FAOSTAT, FAOSTAT, FAO Statistical Databases (United Nations), FAO, 2013, http://faostat.fao.org/.

[8] G. S. Khush, "Molecular genetics-plant breeder's perspective," in *Molecular Techniques in Crop Improvement*, S. M. Jain, D. S. Brar, and B. S. Ahloowalia, Eds., pp. 1–8, Kluwer Academic, Dordrecht, The Netherlands, 2002.

[9] E. Ganji-Moghadam and A. Khalighi, "Relationship between vigor of Iranian *Prunus mahaleb* L. selected dwarf rootstocks and some morphological characters," *Scientia Horticulturae*, vol. 111, no. 3, pp. 209–212, 2007.

[10] A. Shahi-Gharahlar, Z. Zamani, M. R. Fatahi, and N. Bouzari, "Assessment of morphological variation between some Iranian wild *Cerasus* sub-genus genotypes," *Horticulture, Environment and Biotechnology*, vol. 51, no. 4, pp. 308–318, 2010.

[11] H. Demirsoy and L. Demirsoy, "Characteristics of some local sweet cherry cultivars from Homeland," *Journal of Agronomy*, vol. 3, no. 2, pp. 88–89, 2004.

[12] G. Lacis, I. Rashal, S. Ruisa, V. Trajkovski, and A. F. Iezzoni, "Assessment of genetic diversity of Latvian and Swedish sweet cherry (*Prunus avium* L.) genetic resources collections by using SSR (microsatellite) markers," *Scientia Horticulturae*, vol. 121, no. 4, pp. 451–457, 2009.

[13] Y. L. Cai, D. W. Cao, and G. F. Zhao, "Studies on genetic variation in cherry germplasm using RAPD analysis," *Scientia Horticulturae*, vol. 111, no. 3, pp. 248–254, 2007.

[14] D. Struss, R. Ahmad, S. M. Southwick, and M. Boritzki, "Analysis of sweet cherry (*Prunus avium* L.) cultivars using SSR and AFLP markers," *Journal of the American Society for Horticultural Science*, vol. 128, no. 6, pp. 904–909, 2003.

[15] M. Bouhadida, A. M. Casas, M. J. Gonzalo, P. Arús, M. Á. Moreno, and Y. Gogorcena, "Molecular characterization and genetic diversity of Prunus rootstocks," *Scientia Horticulturae*, vol. 120, no. 2, pp. 237–245, 2009.

[16] A. Belaj, Z. Satovic, G. Cipriani et al., "Comparative study of the discriminating capacity of RAPD, AFLP and SSR markers and of their effectiveness in establishing genetic relationships in olive," *Theoretical and Applied Genetics*, vol. 107, no. 4, pp. 736–744, 2003.

[17] K. Wu, R. Jones, L. Danneberger, and P. A. Scolnik, "Detection of microsatellite polymorphisms without cloning," *Nucleic Acids Research*, vol. 22, no. 15, pp. 3257–3258, 1994.

[18] W. Powell, G. C. Machray, and J. Proven, "Polymorphism revealed by simple sequence repeats," *Trends in Plant Science*, vol. 1, no. 7, pp. 215–222, 1996.

[19] B. Bornet and M. Branchard, "Nonanchored Inter Simple Sequence Repeat (ISSR) markers: reproducible and specific tools for genome fingerprinting," *Plant Molecular Biology Reporter*, vol. 19, no. 3, pp. 209–215, 2001.

[20] I. V. Ganopoulos, K. Kazantzis, I. Chatzicharisis, I. Karayiannis, and A. S. Tsaftaris, "Genetic diversity, structure and fruit trait associations in Greek sweet cherry cultivars using microsatellite based (SSR/ISSR) and morpho-physiological markers," *Euphytica*, vol. 181, no. 2, pp. 237–251, 2011.

[21] K. U. Yilmaz, S. Ercişli, B. M. Asma, Y. Doğan, and S. Kafkas, "Genetic relatedness in prunus genus revealed by inter-simple sequence repeat markers," *HortScience*, vol. 44, no. 2, pp. 293–297, 2009.

[22] M. M. Li, Y. L. Cai, Z. Q. Qian, and G. F. Zhao, "Genetic diversity and differentiation in Chinese sour cherry *Prunus pseudocerasus* Lindl., and its implications for conservation," *Genetic Resources and Crop Evolution*, vol. 56, no. 4, pp. 455–464, 2009.

[23] A. Lisek and E. Rozpara, "Identification and genetic diversity assessment of cherry cultivars and rootstocks using ISSR-PCR technique," *Journal of Fruit and Ornamental Plant Research*, vol. 17, no. 2, pp. 95–106, 2009.

[24] A. Shahi-Gharahlar, Z. Zamani, R. Fatahi, and N. Bouzari, "Estimation of genetic diversity in some Iranian wild *Prunus* subgenus *Cerasus* accessions using inter-simple sequence repeat (ISSR) markers," *Biochemical Systematics and Ecology*, vol. 39, no. 4–6, pp. 826–833, 2011.

[25] N. Bouzari, E. Ganji-Moghadam, F. Karami et al., *National Project Collection and Evaluation of Local Sour Cherry Germplasms in Order To Achieve Proper rootsTock and Cultivars, Project No. 0-100-120000-04-0000-84104*, Seed and Plant Improvement Institute, Horticultural Section, Karaj, Iran; Agricultural Research and Educatin Organization Press, Tehran, Iran, 2010.

[26] N. Najafzadeh, K. Arzani, and N. Bouzari, "Assesment of morphological and pomological variation of some selected Iranian sour cherry (*Prunus cerasus* L.) genotypes," *Seed and Plant*, vol. 1, no. 2, pp. 123–137, 2014.

[27] J. A. Saunders, M. J. Pedroni, L. D. J. Penrose, and A. J. Fist, "AFLP analysis of opium poppy," *Crop Science*, vol. 41, no. 5, pp. 1596–1601, 2001.

[28] O. Sofalian, N. Chaparzadeh, A. Javanmard, and M. S. Hejazi, "Study the genetic diversity of wheat landraces from northwest of Iran based on ISSR molecular markers," *International Journal of Agriculture and Biology*, vol. 10, no. 4, pp. 466–468, 2008.

[29] H. S. Acharya and V. Sharma, "Molecular characterization of opium poppy (*Papaver somniferum*) germplasm," *American Journal of Infectious Diseases*, vol. 5, no. 2, pp. 148–153, 2009.

[30] P. H. A. Sneath and R. R. Sokal, *Numerical Taxonomy, the Principles and Practice of Numerical Classification*, Freeman WH, San Francisco, Calif, USA, 1973.

[31] S. A. Mohammadi and B. M. Prasanna, "Analysis of genetic diversity in crop plants-salient statistical tools and considerations," *Crop Science*, vol. 43, no. 4, pp. 1235–1248, 2003.

[32] F. G. Rohlf, *NTsys-Pc Numerical Taxonomy and Multivariate System Version 2.0 Applied*, vol. 12, Biostatistics, New York, NY, USA, 2000.

[33] IPGRI and Cornell University, *Genetic Diversity Analysis With Molecular Marker Data: Learning Module, Measures of Genetic Diversity*, IPGRI, Maccarese, Rome; Cornell University, Ithaca, NY, USA, 2003.

[34] M. Warburton and J. Crossa, *Data Analysis in the CIMMYT. Applied Biotechnology Center for Fingerprinting and Genetic Diversity Studies*, CIMMYT, Mexico City, Mexico, 2002.

[35] H. C. Romesburg, *Cluster Analysis for Researchers*, Krieger, Malabar, Fla, USA, 1990, Reprint of 1984 edition.

[36] N. M. Nei and W. Li, "Mathematical model for studying genetic variation in terms of restriction endonucleases," *Proceedings of the National Academy of Sciences of the United States of America*, vol. 76, no. 10, pp. 5269–5273, 1979.

[37] P. Jaccard, "Nouvelles recherches sur la distribution florale," *Bulletin de la Societe Vaudoise des Sciences Naturelles*, vol. 44, pp. 223–270, 1908.

21

Endophytic Fungal Diversity in Medicinal Plants of Western Ghats, India

Monnanda Somaiah Nalini,[1] Ningaraju Sunayana,[2] and Harischandra Sripathy Prakash[2]

[1] *Department of Studies in Botany, University of Mysore, Manasagangotri, Mysore, Karnataka 570 006, India*
[2] *Department of Studies in Biotechnology, University of Mysore, Manasagangotri, Mysore, Karnataka 570 006, India*

Correspondence should be addressed to Monnanda Somaiah Nalini; nmsomaiah@gmail.com

Academic Editor: Raeid Abed

Endophytes constitute an important component of microbial diversity, and in the present investigation, seven plant species with rich ethnobotanical uses representing six families were analyzed for the presence of endophytic fungi from their natural habitats during monsoon (May/June) and winter (November/December) seasons of 2007. Fungal endophytes were isolated from healthy plant parts such as stem, root, rhizome, and inflorescence employing standard isolation methods. One thousand five hundred and twenty-nine fungal isolates were obtained from 5200 fragments. Stem fragments harbored more endophytes (80.37%) than roots (19.22%). 31 fungal taxa comprised of coelomycetes (65%), hyphomycetes (32%), and ascomycetes (3%). *Fusarium, Acremonium, Colletotrichum, Chaetomium, Myrothecium, Phomopsis,* and *Pestalotiopsis* spp. were commonly isolated. Diversity indices differed significantly between the seasons ($P < 0.001$). Species richness was greater for monsoon isolations than winter. Host specificity was observed for few fungal endophytes. UPGMA cluster analysis grouped the endophytes into distinct clusters on the basis of genetic distance. This study is the first report on the diversity and host-specificity of endophytic fungal taxa were from the semi evergreen forest type in Talacauvery subcluster of Western Ghats.

1. Introduction

The microbes residing in the internal parts of plant tissues called "endophytes" constitute a group of plant symbionts and are a component of microbial diversity. Endophytes offer plethora of unknown advantages to the host with immense applications in agriculture and medicine [1, 2]. Recently, challenging hypotheses related to endophyte diversity [3], their role in oxidative stress protection [4], heavy metal tolerance [5], and as components of tropical community ecology [6, 7] have emerged. A perusal of the literature over the past decades indicated many ethnomedicinal plant species with rich botanical history, sampled from unique ecological niches species are known to harbor potential endophytic microbes [8].

There has been an increasing surge of interest among the research groups for the isolation of endophytes from the tropical plant species [9, 10], owing to high plant diversity. One such region represents the Western Ghats, stretching a length of 1,600 Km from the river Tapti in the state of Gujarat to the Southern tip of Kerala, recognized as one of the 34 hot spots of biodiversity. The Western Ghats represent rich flora with enormous species diversity as well as endemic taxa and are therefore recognized as one among the hot spots of the world [11]. Western Ghats are divided into seven subclusters. A proposal to include and declare 39 sites in this region as the World Natural Heritage Cluster Site by UNESCO is underway (http://www.atree.org/wg_unesco_whs).

India has many regions of unique ecological niche harboring variety of medicinal plants. One such region in the peninsular India is Kodagu District, the land of coffee cultivation. Kodagu is situated in the Western Ghats of peninsular India and is known for its majestic mountain ranges, coffee plantations, and teak wood forests. The Talacauvery subcluster (12°17′ to 12°27′N and 75°26′ to 75°33′E) of the Western Ghats is situated in Kodagu. The altitude ranges from 1525 above mean sea level. Annual precipitation of 3525 mm is largely restricted during May to October,

TABLE 1: Details of medicinal plants collected from the natural habitats of Talacauvery subcluster of Western Ghats.

Plant species*	Common name	Ayurvedic name	Family	Habit	Parts collected
Tylophora asthmatica (W. and A.)	Indian ipecacuanha	Anthrapachaka	Asclepiadaceae	Twiner	Stem
Rubia cordifolia L.	Indian madder	Majith	Rubiaceae	Climber	Stem
Plumbago zeylanica L.	White leadwort	Chitramool	Plumbaginaceae	Shrub	Stem
Phyllanthus amarus (Schum. and Thonn.)	Niruri	Bhoomyamalaki	Euphorbiaceae	Herb	Stem and root
Eryngium foetidum L.	Fit weed/spiny/serrated coriander	Bhandhanya	Apiaceae	Herb	Roots and stem
Centella asiatica L.	Asiatic pennywort	Gotu Kola	Apiaceae	Runner	Stolon, roots, and inflorescence
Zingiber sp.	Wild ginger	Shunti	Zingiberaceae	Herb	Rhizome, aerial stem, and root

Ten plants* were sampled, pooled, and used for isolation of endophytes [13].

although premonsoon showers are not uncommon during February to April. The average temperature is 23°C. Kodagu has a reservoir of forest belts and diverse vegetation ranging from tropical wet evergreen forests to scrub jungles. Several tribes residing in the forests still use medicinal plants of ethnopharmacological importance as the source of natural medication for their ailments [12]. Ethnomedicinal plants are often used in ayurvedic medicinal system in India for the treatment of various diseases.

Despite the reports of ethnomedicinal plants of this region, the biodiversity and the endophytic microbes of this region remain unexplored. Therefore, in the present investigation, seven medicinal plants representing six families were subjected to diversity studies on fungal endophytes during two seasons.

2. Methodology

2.1. Plant Materials and Study Site.
Plant parts such as stem, root, rhizome, and inflorescence were collected from seven healthy medicinal plant species: Tylophora asthmatica, Rubia cordifolia, Plumbago zeylanica, Phyllanthus amarus, Eryngium foetidum, Centella asiatica, and Zingiber sp. inhabiting the natural vegetation of the Talacauvery Region of Western Ghats, located at 012°17′ to 012°27′N and 075°26′ to 075°33′E of Kodagu, Karnataka, during the monsoon (May to June) and winter seasons (November-December) of 2007 (Table 1). The natural vegetation is an evergreen/semievergreen type of forests. The mean temperature was 23°C and mean annual precipitation is 3525 mm. Herbarium specimens of the plants were prepared and submitted to the herbarium collections in the DOS in Botany, University of Mysore. Ten individual plants from each were pooled for isolations. The samples were placed in polyethylene bags, labeled, transported in ice box to the laboratory, and placed in a refrigerator at 4°C until isolation. All samples were processed within 24 h of collection.

2.2. Isolation and Identification of Endophytic Fungi.
Samples were washed thoroughly in distilled water, blot dried, and first immersed in 70% ethanol (v/v) for one min followed by second immersion in sodium hypochlorite (3.5%, v/v) for three minutes. They were rinsed three times in changes of

sterile distilled water and dried on sterile blotters under the airflow to ensure complete drying. Bits of 1.0 × 0.1 cm size were excised with the help of a sterile blade. A total of 5200 segments from stem, roots, inflorescence, and rhizomes of plant species were placed on water agar (2.5%) supplemented with the antibiotic streptomycin sulphate (100 mg/L). Forty segments were plated per plate. The plates were wrapped in clean wrap cling film and incubated at 22°C with 12 h light and dark cycles for up to 6 to 8 weeks. The effectiveness of surface sterilization of tissues was checked by placing the aliquots of sterilants on agar plates and observing fungal colonies if any for two weeks [14].

Periodically the bits were examined for the appearance of fungal colony and each colony that emerged from segments was transferred to antibiotic-free potato dextrose agar medium (PDA, 2%) to aid identification. The morphological identification of the isolates was done based on the fungal colony morphology and characteristics of the reproductive structures and spores [15–17]. Sporulation was induced by inoculating cultures onto sterilized banana leaf bits (one cm^2) impregnated on potato dextrose agar in petri dishes. All fungal mounts were made on microscopic glass slides in lactophenol-cotton blue and sealed with nail polish. Cultures which failed to sporulate were grouped as mycelia sterilia. All the fungal isolates have been catalogued as DST# series with plant code and maintained as culture collections of the department by cryopreservation on PDA overlaid with 15% glycerol (v/v) at −20°C in a deep freezer.

2.3. Data Analysis.
Isolation rate (IR), the measure of fungal richness of a sample, was calculated as the number of isolates obtained from tissue segments, divided by the total number of segments, and expressed as fractions but not as percentages [18]. The colonization frequency (CF), expressed as percentage, was calculated according to Kumaresan and Suryanarayanan [19] as follows:

$$\%\text{CF} = \frac{\text{Number of tissue segments colonized by a fungus}}{\text{Total number of tissue segments plated}} \times 100.$$

(1)

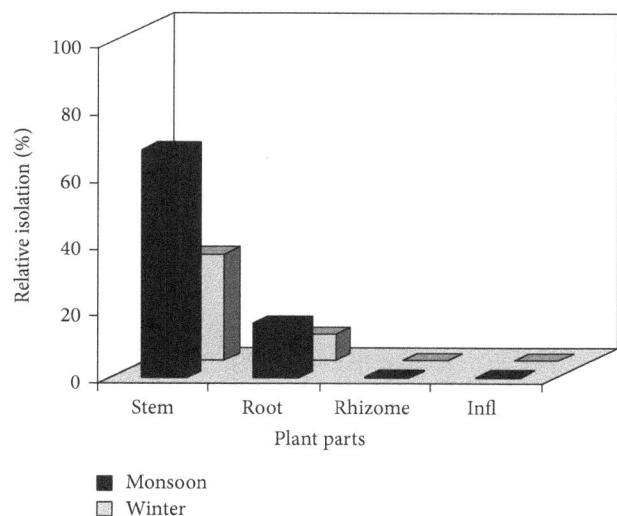

FIGURE 1: Relative seasonal isolations of fungal endophytes from plant parts of medicinal species.

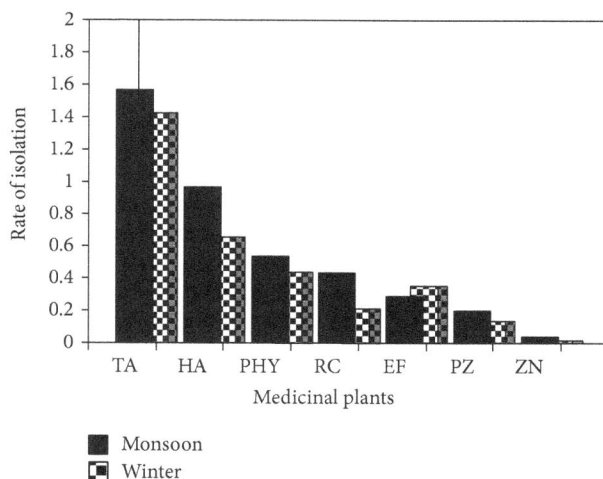

FIGURE 2: Seasonal isolation rates of fungal endophytes from medicinal species. Endophytic fungi were isolated from seven medicinal plant species during monsoon and winter seasons, respectively.

The percentage of dominant endophytes (D) was calculated based on the %CF divided by the total number of endophytes × 100 [20].

Differences in the extent of colonization of the samples were analyzed by univariant analysis of variance (one-way ANOVA) and Tukey's honestly significant difference (HSD) as post hoc test using the statistical software SPSS16.0. The fungal isolations were considered for analysis of ANOVA and Tukey's HSD. Simpson and Shannon diversity indices were calculated for endophytic fungi from different seasons with Estimate S, software (version 6, http://viceroy.eeb.uconn.edu/estimates/). Species richness was calculated using the online web page rarefactor calculator (http://www2.biology.ualberta.ca/jbrzusto/rarefact.php).

Rarefaction indices were employed to compare the species richness among the plant species during two seasons. The expected number of species in N isolations was calculated [21]. Unweighted pair group method with arithmetic mean (UPGMA) cluster analysis was applied for all the isolates from plant species based on the number of isolates recovered from each plant species using a dendrogram constructed based on Nei's genetic distances [22] using tools for population genetics analysis (TFPGA) software [23].

3. Results

A total of 1529 isolates were obtained from 5200 tissue fragments from seven medicinal plant species. The extent of endophytes colonization varied in plant parts with stem fragments harboring 80% of endophytic isolates followed by root (19.22%). In other plant parts, colonization was lower. Isolations of endophytes from various plant parts showed greater numbers of endophytes during monsoon than winter (Figure 1). The high isolation rates (IR) of fungal endophytes were recorded as 1.41 to 1.58 for *T. asthmatica* in both seasons, while in *Zingiber* sp., low rates of isolations

were obtained (Figure 2). Thirty-one fungal taxa were identified which consisted of coelomycetes (65%), hyphomycetes (32%), and ascomycete isolations of 3%. The frequency of fungal colonization (%CF) differed among the seven plant species (Table 2). *Fusarium* sp., *Acremonium*, *Chaetomium*, and *Phoma* are some of the endophytes with high colonization frequency. The dominant fungal genera include *Fusarium* spp. (%D = 10.64) and *Acremonium* (%D = 9.48). Few endophytic fungi such as *A. strictum* had wide distributions in host plants and were isolated from most plants with the exception of *Zingiber* sp. and *Plumbago zeylanica*, whereas species of *Fusarium*, *Trichoderma*, *Curvularia*, and *Penicillium* were isolated from more than three plant species.

Host-specificity was observed for few of the fungal endophytes isolated from two of the seven medicinal plants (Table 2). *Colletotrichum dematium*, *Nigrospora oryzae*, *Heinesia rubi*, *Pestalotiopsis guepinii*, and unidentified red pycnidial form were isolated from the stem segments of *T. asthmatica* only, while in *Rubia cordifolia* one endophytic *Periconia* exhibited specificity. *P. islandicum* and *T. viride* were isolated from root segments of *Phyllanthus amarus*.

Diversity indices of fungal endophytes varied within plant species as well as between seasons (Table 3). High Shannon-Weiner diversity index was recorded for *T. asthmatica* (H^1 = 2.60) and *P. amarus* (H^1 = 2.27) during monsoon and winter seasons, whereas low indices were recorded for *E. foetidum* and *Zingiber* during monsoon and winter seasons, respectively. 42% of the total 31 taxa were found in monsoon season, while 55% of them colonized in both seasons. Simpson index (1/l) was high for *T. asthmatica* with a richness of 19 fungal species during monsoon season, while *P. amarus* recorded highest richness of species during winter season. Rarefaction curves calculated for the endophytic fungal isolations indicated maximum species richness for *T. asthmatica* and *P. amarus* during monsoon and winter isolations, respectively

TABLE 2: Colonization frequency* of endophytic fungi isolated from plant parts of seven medicinal plant species.

Seasons	Monsoon														Winter													D (%)
Plant species	TA	RC	P	PHY		EF			CA		ZN				TA	RC	PZ	PHY		EF		CA		ZN				
Plant Parts	Stem	Stem		S	R	S	R	I	St	R	S	Rh	R		Stem	Stem		S	R	S	R	St	R	S	Rh	R		
Endophytes																												
Acremonium strictum	25.0	16.0	—	10.0	9.0	8.0	9.0	—	2.0	9.0	—	—	—		12.5	7.5	15.0	—	5.0	2.5	6.0	5.0	3.5	—	—	—		9.48
Alternaria alternata	—	—	—	—	—	—	—	—	—	—	—	—	—		—	8.5	—	—	—	1.5	—	1.5	—	—	—	—		0.65
Aspergillus terreus	11.5	—	5.0	—	—	—	—	—	—	—	—	—	—		—	—	—	—	7.0	—	—	—	—	—	—	—		1.54
Red pycnidia	5.0	—	—	—	—	—	—	—	—	—	—	—	—		—	—	—	—	—	—	—	—	—	—	—	—		0.33
Botryodiplodia theobromae	2.5	7.5	—	2.5	—	—	—	—	—	—	—	—	—		—	5.0	—	—	—	2.5	1.5	—	—	—	—	—		1.41
Cladosporium herbarum	—	—	—	2.5	—	—	—	—	—	—	—	—	—		—	—	—	—	—	—	—	—	—	—	—	—		0.20
Colletotrichum dematium	—	—	—	—	—	—	—	—	—	—	—	0.5	—		20.0	—	—	—	—	—	—	—	—	—	—	—		1.31
C. lindemuthianum	—	—	—	—	—	—	—	—	—	16.0	—	—	—		10.0	3.5	—	—	—	—	6.0	5.0	—	—	—	—		2.32
Chaetomium globosum	27.5	—	—	2.5	—	—	16.0	—	6.0	4.0	—	1.0	—		16.0	10.0	3.5	7.5	—	5.0	—	2.5	—	—	—	—		4.51
Curvularia lunata	—	—	1.5	—	—	—	—	—	—	—	—	—	—		—	—	—	1.5	—	—	—	—	—	—	—	—		2.12
Fusarium sp.	—	—	—	—	—	2.5	—	9.5	—	—	—	—	—		—	—	—	—	—	—	—	1.0	—	—	—	—		0.20
F. graminearum	7.5	20.0	6.0	—	—	—	—	1.0	3.5	—	—	0.5	—		—	1.5	2.0	—	—	—	1.0	4.0	—	—	1.0	—		0.69
F. oxysporum	5.0	10.0	—	—	4.5	—	5.0	—	—	10.0	—	4.0	—		—	—	—	4.5	—	—	—	—	—	—	—	—		3.01
F. solani	—	—	—	—	2.5	—	—	—	—	—	—	0.5	—		—	—	11.0	—	—	—	—	—	—	—	—	—		2.81
F. pseudonygamai	—	—	5.5	10.0	—	7.5	—	—	—	—	2.0	—	—		—	—	—	—	—	—	—	—	—	—	—	—		0.20
F. verticillioides	21.0	—	—	2.0	—	—	—	—	—	—	—	—	—		—	—	—	—	—	—	—	—	—	—	—	—		3.73
Gelasinospora spp.	2.5	—	—	—	—	—	—	—	—	—	—	—	—		—	—	—	—	—	—	—	—	—	—	—	—		0.29
Heinesia rubi	1.5	—	—	—	—	—	—	—	—	—	—	—	—		—	—	—	—	—	—	—	—	—	—	—	—		0.10
Memnoniella spp.	8.5	2.5	—	—	—	—	—	—	—	—	—	—	—		—	—	—	—	—	—	—	—	—	—	—	—		0.69
Myrothecium verrucaria	3.0	7.5	—	—	—	—	—	—	—	—	—	—	—		—	—	—	6.0	—	—	—	—	—	—	—	—		1.08
Nigrospora oryzae	5.0	—	—	—	—	—	—	—	—	—	1.0	—	—		—	—	—	—	—	—	—	—	—	—	—	—		0.33
Periconia sp.	—	17.5	—	—	—	—	—	—	—	—	—	—	—		2.0	—	—	—	—	—	—	—	—	—	—	—		1.14
Pestalotiopsis guepinii	2.0	—	—	5.0	—	—	—	—	—	—	—	—	—		2.0	—	—	—	—	—	—	—	—	—	—	—		0.13
Pestalotiopsis sp.	—	—	—	—	—	—	—	1.0	—	3.0	—	—	—		9.0	—	—	—	—	—	—	—	—	—	—	—		0.59
Phoma spp.	6.0	10.0	—	—	—	—	—	—	—	—	—	—	—		—	—	—	—	—	—	—	—	—	—	—	—		1.63
Phomopsis sp.	5.0	—	—	—	2.0	—	1.0	—	—	—	—	—	—		—	—	—	—	—	—	—	1.0	—	—	—	—		0.39
Penicillium sp.	10.0	4.5	—	—	1.0	—	—	—	—	—	—	—	—		—	—	2.0	5.0	2.5	5.0	—	1.5	—	—	—	—		2.06
P. islandicum	—	—	7.5	7.5	—	—	—	—	—	—	—	—	—		—	—	—	—	—	—	—	—	—	—	—	—		0.55
Trichoderma harzianum	—	—	7.5	—	4.5	3.0	—	—	—	—	—	1.5	—		—	—	—	5.0	1.5	5.0	—	5.0	—	—	2.0	—		1.96
T. viride	—	—	5.0	5.0	—	—	—	—	—	—	—	—	—		—	—	—	—	1.0	—	—	—	—	—	—	—		0.39
Sphaeronema sp.	2.5	—	3.5	12.0	—	—	—	—	—	—	1.0	—	—		—	—	—	—	—	—	—	—	—	—	—	—		0.82
Verticillium albo-atrum	—	—	3.5	—	5.0	—	—	—	—	—	—	—	—		—	—	—	—	—	—	—	—	—	—	—	—		0.29
Pycnidial forms	—	7.5	15.0	—	—	—	—	—	—	—	—	—	—		—	—	—	—	—	—	—	—	—	—	—	—		1.47
Sterile mycelia	6.0	—	3.5	5.0	—	1.0	—	—	—	—	—	—	—		2.0	—	—	4.5	—	—	—	—	—	—	—	—		1.44
Total	314	191	107	126	46	47	68	2	44	79	0	6	6	10	143	72	63	73	34	18	29	20	25	0	0	6	0	1529
	314	191	107	172		117			123		22				143	72	63	107		47		45		6				1529

*200 segments were plated for frequency analysis during monsoon and winter seasons; TA: *Tylophora asthmatica*; RC: *Rubia cordifolia*; PZ: *Plumbago zeylanica*; PHY: *Phyllanthus amarus*; EF: *Eryngium foetidum*; CA: *Centella asiatica*; ZN: *Zingiber* sp;; S: stem; St: stolon; R: root; Rh: rhizome; I: inflorescence.

TABLE 3: Diversity indices (H^1) and species richness of the medicinal plants.

Plant species	Monsoon				Winter			
	Total isolates	Total species richness	Diversity indices		Total isolates	Total species richness	Diversity indices	
			Simpson	Shannon			Simpson	Shannon
Tylophora asthmatica	314	19	0.90	2.60	143	7.0	0.78	1.66
Rubia cordifolia	191	9	0.86	2.05	72	6.0	0.79	1.62
Plumbago zeylanica	107	8	0.84	1.96	63	4	0.63	1.14
Phyllanthus amarus	172	14	0.87	2.29	107	11.0	0.89	2.27
Eryngium foetidum	117	9	0.77	1.68	47	6	0.75	1.53
Centella asiatica	123	8	0.83	1.88	45	6	0.76	1.57
Zingiber sp.	22	8	0.79	1.81	6	2	0.44	0.64

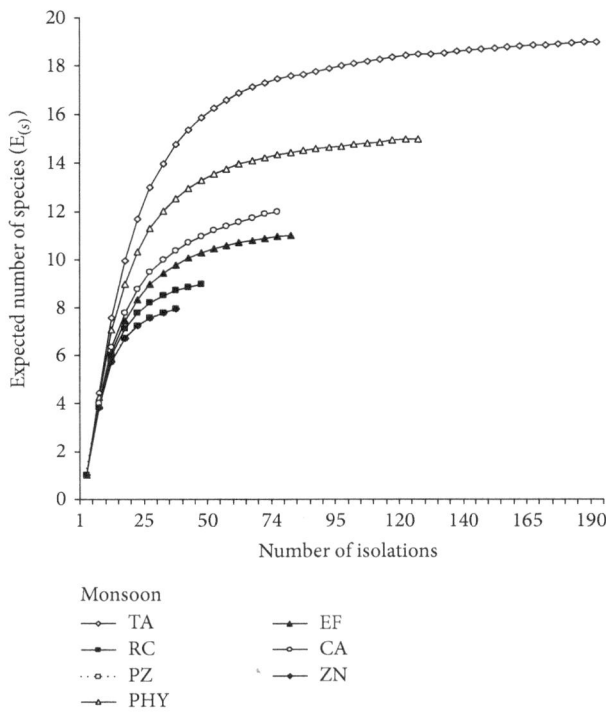

FIGURE 3: Rarefaction curvesof fungal endophytes from monsoon isolations (number of isolations versus expected number of species $E_{(s)}$). TA: *Tylophora asthmatica*; RC: *Rubia cordifolia*; PZ: *Plumbago zeylanica*; PHY: *Phyllanthus amarus*; EF: *Eryngium foetidum*; CA: *Centella asiatica*; and ZN: *Zingiber* sp.

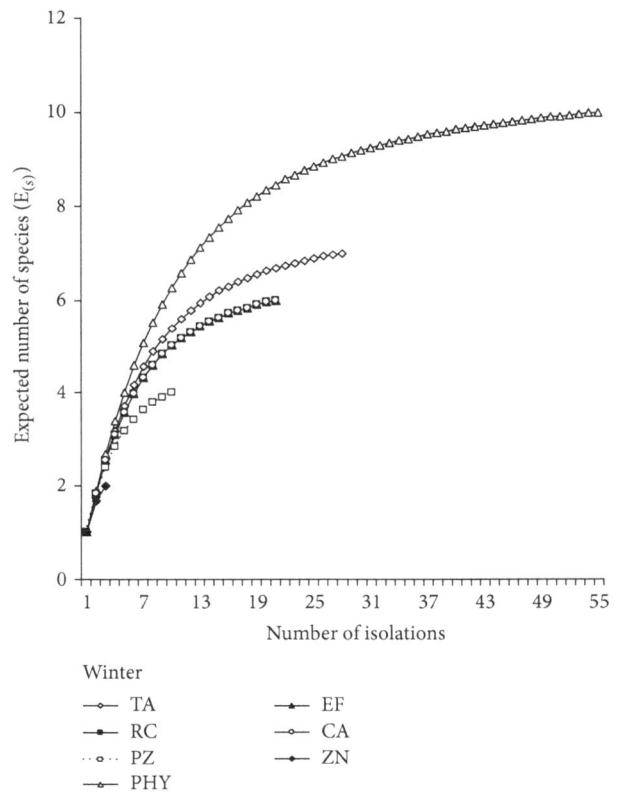

FIGURE 4: Rarefaction curves of fungal endophytes from winter isolations (number of isolations versus expected number of species $E_{(s)}$). TA: *Tylophora asthmatica*; RC: *Rubia cordifolia*; PZ: *Plumbago zeylanica*; PHY: *Phyllanthus amarus*; EF: *Eryngium foetidum*; CA: *Centella asiatica*; and ZN: *Zingiber* sp.

(Figures 3 and 4). Differences in the number of isolates and colonization frequency differed significantly between seasons ($P < 0.001$) as indicated in Table 4.

Nei's genetic distance between endophytes isolated from plant species ranged from 0.3185 (between populations of *Zingiber* and *C. asiatica*) to 0.9116 (between populations of *Zingiber* and *T. asthmatica*) (Table 5) which was widely ranged. This indicates a closer relationship between endophytic fungal patterns of both plants. In general, fungal species from *T. asthmatica* is most distanced from the other

plants studied. In order to represent the relationships among plant species, cluster analysis (UPGMA) was used to generate a dendrogram based on Nei's genetic distances between populations (Figure 5). In this dendrogram, all plants form a distinct cluster. When the transect line was placed at approximately 0.4 on the distance scale, two distinct groups were formed. The first cluster was formed by CA-ZN, while

TABLE 4: ANOVA table of seasonal variation of endophytic fungi analyzed from seven medicinal plant species.

	Sum of squares	Degrees of freedom	Mean^2	F	Significance
Between seasons	11931.9	26	458.919	34.382	0.001
Within seasons	1001.062	75	13.346	—	—
Total	**12932.96**	**101**	—	—	—

TABLE 5: Nei's genetic distance of plant species analyzed for endophytic fungi.

	TA	RC	PZ	PHY	EF	CA	ZN
TA	* * *	0.7885	0.5008	0.6061	0.4055	0.7885	0.9116
RC		* * *	0.5008	0.7239	0.4055	0.3610	0.6061
PZ			* * *	0.6633	0.3610	0.5008	0.4520
PHY				* * *	0.4520	0.5008	0.6633
EF					* * *	0.4055	0.3610
CA						* * *	0.3185
ZN							* * *

*Indicates that genetic distance between the same genus does not exist, it is shown for different genera.

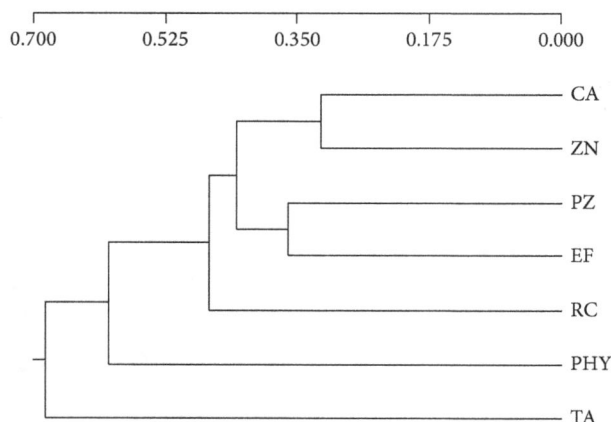

FIGURE 5: UPGMA cluster analysis of fungal endophytes representing seven medicinal plant species. Plant codes are TA: *Tylophora asthmatica*; RC: *Rubia cordifolia*; PZ: *Plumbago zeylanica*; PHY: *Phyllanthus amarus*; EF: *Eryngium foetidum*; CA: *Centella asiatica*; and ZN: *Zingiber* sp.

PZ-EF formed the second cluster. PHY, RC, and TA are found outside the cluster.

4. Discussion

Medicinal plants are considered as a repository of "endophytic microbes" living in the internal tissues of plants. The quest for identifying novel bioactives from the endophytic fungi has resulted in the sampling of host plants such as herbs, shrubs, tree species, and vines in unique places of ecological adaptations around the rainforests of the world. Such niches harbor great species diversity, unintervened by

human activities. Efforts in this direction to sample plants located in the rainforests around the world with potential ethnomedicinal values have resulted in the isolation of fungal endophytes, unique to a particular plant species with distinct bioactivity.

4.1. Endophyte Colonization in Medicinal Species. The medicinal plant species were sampled from the Talacauvery subcluster situated in the Kodagu District of Western Ghats of Southern India. This region is among one of the 34 hot spots of biodiversity. Recently, a proposal to include this biodiversity spot in the list of UNESCO Heritage cluster site is underway (http://www.atree.org/wg_unesco_whs). The natives as well as the ethnic tribes inhabiting this region still depend on the plants as a source of medicine to cure some of the ailments [12]. Seven medicinal plant species assigned to six plant families were selected for the study in natural populations in two seasons from a single location from the study area stretching over an area of 25 kilometers. Sampling was conducted during monsoon and winter seasons, as two of the herbaceous species, *E. foetidum* and *Zingiber* sp., grow only till the second half of the year (June to December) and their nonavailability during summer (March to May) makes it difficult to consider the summer season for endophytic analysis.

From 5200 segments of plant materials a total of 1529 isolates were obtained; these were grouped into 31 taxa. Mycelia sterilia, the fungal taxa that failed to sporulate, were also reported from this study. This fungal group is prevalent in endophytic studies [24]. The fungal endophytes were analyzed from four plant parts, namely, stem, root, rhizome, and inflorescence; however, their occurrence in root and inflorescence was investigated for few plant species only, as the phenology and sampling of plants never correlated with seasons. The leaves were not considered for isolations since some of the plants were climbers and stragglers with delicate hairy surfaces and stringent surface sterilization techniques would render them unsuitable for plating on agar medium. Relative percentages of endophytic isolations from stem segments were greater (80.37%) than isolations from roots (19.22%). Our results are supported by the earlier work of Huang et al. [25] on 29 traditional Chinese medicinal plants that fungal endophytes are more frequent in stem tissues than roots. Among the fungal taxa, coelomycete isolations were more dominant than hyphomycetes and have been found in earlier studies in endophytes of tree species [26].

Endophytes such as *Colletotrichum*, *Phoma*, *Acremonium*, *Chaetomium*, *Botryodiplodia*, and *Trichoderma* were isolated with %D > 2.0. Few fungal taxa that are less frequently isolated are *Pestalotiopsis*, *Penicillium islandicum*, *Cladosporium herbarum*, *Alternaria alternata*, *F. graminearum*, *Phomopsis*, and *Sphaeronema*. *Colletotrichum* spp. are the most frequently encountered endophytic fungi [18], whereas *Pestalotiopsis* spp. are well documented as endophytes of many rainforest plants [27, 28], tropical tree species, namely, *Terminalia arjuna* [29], *Azadirachta indica* [30], and many herbs and shrubs [25, 31, 32]. It is necessary to screen newer plant species for the isolation of fungal endophytes, as Hawksworth and Rossman [33] estimate that there are still millions of species of fungi yet to be identified. Differences in the colonization frequencies of endophytes during two seasons were observed and more isolation during monsoon season is due to the fact that the slimy conidia of fungal spores are dispersed better by rain splashes and germination of conidia is influenced by climatic factors [34].

4.2. Host-Specificity of Fungal Endophytes. We observed that some fungal taxa exhibited host-specificity, a phenomenon often associated with endophytes. Three plant species, *T. asthmatica*, *R. cordifolia*, and *P. amarus*, were host-specific to endophytes. The red pycnidial endophyte (TA-005) was isolated from the stem fragments of *T. asthmatica* only, suggesting the host-specificity of this endophyte. *Pestalotiopsis guepinii* was isolated from the stem segments of *T. asthmatica*. It has been reported as an endophyte of *Wollemia nobilis*, growing in Sydney, Australia [35]. *Heinesia rubi*, *P. islandicum*, and TA-005 are new reports of fungi as endophytes. Host-specificity of endophytic fungi has been observed earlier for grasses [36], orchids [37], and forest tree species [38, 39]. Recently, Sun et al. [40] reiterated the term "host-specificity" as taxa that occur exclusively on a stated host but not on other hosts in the same habitat [41]. Our studies also indicate the host-specificity of endophytes as the plant species were sampled from a single habitat.

4.3. Seasonal Diversity of Fungal Endophytes. Diversity indices for fungal endophytes as analyzed by Shannon-Weiner (H^1) and Simpson ($1/l$) indices indicated differences in seasonal variation and species richness. High indices were noted for *T. asthmatica* ($H^1 = 2.6$) and *P. amarus* ($H^1 = 2.27$) during monsoon and winter seasons, respectively. The fungal species did not differ significantly between plant species, whereas they differed between seasons ($P < 0.001$). Seasonal variation in fungal isolates and colonization frequency has been reported for many host plants [42, 43]. High colonization frequency as well as the species richness of endophytic fungi is limited to leaf segments rather than stem or bark segments of host plants sampled from five medicinal species of Kudremukh Region of Western Ghats [44]. Species richness in our study is limited to stem fragments among the plant parts considered for analysis.

Most studies on fungal endophytes in tropics have revealed remarkable patterns of endophyte colonization and estimates of diversity in foliages of forest tree species representing various sites such as Panamanian Forest [45] and Iwokrama Forest Reserve, Guyana [39]. In the Nilgiri Biosphere Reserve, Western Ghats, India, 75 dicotyledonous species in three different tropical forest types were sampled to study foliar endophytes and diversity [10]. The endophyte

diversity in forest types was limited due to loose host affiliations among endophytes. Studies on foliar endophytes from the sampling of herbaceous and shrubby medicinal plant species have revealed differences in the colonization rates as well as seasonal diversity in Malnad Region of Bhadra Wildlife Sanctuary in Southern India [32, 46].

The present study provides firsthand information on the diversity and seasonal influence on the colonization frequencies of endophytic fungi from selected medicinal plants from one of the subclusters of biodiversity hot spots in the Western Ghats of Southern India. Although the isolation and analysis of endophyte communities in herbs, shrubs, and trees are not uncommon, each of the studies is unique with reference to number of hosts, species of fungal endophytes, and their specificity. The fungal endophytes have been subjected to fermentation studies, and extracts are being tested for biological activities.

5. Conclusion

The study provides firsthand information on the diversity and seasonal influence on the colonization frequencies of endophytic fungi from seven medicinal plants from one of the subclusters of biodiversity hot spots in the Western Ghats of Southern India. The present investigation is the first isolation of endophytes from the medicinal species and their plant parts. Though isolation of endophytes has been accomplished from various forest types and locations around the globe, each study is unique in documenting newer endophytic taxa. We are currently working on the fermentation of fungal endophytes to obtain newer antioxidants with therapeutic applications.

Conflict of Interests

The authors declare no conflict of interests regarding the publication of this paper.

Acknowledgments

This work was carried out with the financial assistance from the Department of Science & Technology (DST)-SERC Division, Government of India, under the Women Scientist Scheme [DST-WOS (A)] awarded to Monnanda Somaiah Nalini (DST sanction no. SR/WOS (A)/LS-76/2006 dt. 02.08.2007). Monnanda Somaiah Nalini is thankful to the Chairman, DOS in Biotechnology, for providing the necessary facilities for the completion of the project.

References

[1] K. Clay, J. Holah, and J. A. Rudgers, "Herbivores cause a rapid increase in hereditary symbiosis and alter plant community composition," *Proceedings of the National Academy of Sciences of the United States of America*, vol. 102, no. 35, pp. 12465–12470, 2005.

[2] P. Álvarez-Loayza, J. F. White Jr., M. S. Torres et al., "Light converts endosymbiotic fungus to pathogen, influencing seedling

survival and niche-space filling of a common tropical tree, *Iriartea deltoidea*," *PloS ONE*, vol. 6, no. 1, 2011.

[3] R. Linnakoski, H. Puhakka-tarvainen, and A. Pappinen, "Endophytic fungi isolated from Khaya anthotheca in Ghana," *Fungal Ecology*, vol. 5, no. 3, pp. 298–308, 2012.

[4] J. F. White Jr. and M. S. Torres, "Is plant endophyte-mediated defensive mutualism the result of oxidative stress protection?" *Physiologia Plantarum*, vol. 138, no. 4, pp. 440–446, 2010.

[5] H.-Y. Li, D.-W. Li, C.-M. He, Z.-P. Zhou, T. Mei, and H.-M. Xu, "Diversity and heavy metal tolerance of endophytic fungi from six dominant plant species in a Pb-Zn mine wasteland in China," *Fungal Ecology*, vol. 5, no. 3, pp. 309–315, 2012.

[6] A. E. Arnold, "Understanding the diversity of foliar endophytic fungi: progress, challenges, and frontiers," *Fungal Biology Reviews*, vol. 21, no. 2-3, pp. 51–66, 2007.

[7] K. D. Hyde and K. Soytong, "The fungal endophyte dilemma," *Fungal Diversity*, vol. 33, pp. 163–173, 2008.

[8] G. Strobel and B. Daisy, "Bioprospecting for microbial endophytes and their natural products," *Microbiology and Molecular Biology Reviews*, vol. 67, no. 4, pp. 491–502, 2003.

[9] A. E. Arnold and F. Lutzoni, "Diversity and host range of foliar fungal endophytes: are tropical leaves biodiversity hotspots?" *Ecology*, vol. 88, no. 3, pp. 541–549, 2007.

[10] T. S. Suryanarayanan, T. S. Murali, N. Thirunavukkarasu, M. B. Govinda Rajulu, G. Venkatesan, and R. Sukumar, "Endophytic fungal communities in woody perennials of three tropical forest types of the Western Ghats, southern India," *Biodiversity and Conservation*, vol. 20, no. 5, pp. 913–928, 2011.

[11] R. A. Mittermeier, N. Myers, P. R. Gil, and C. G. Mittermeier, *Hotspots: Earth's Biologically Richest and Most Endangered Terrestrial Ecoregions*, Cemex Conservation International, Washington, DC, USA, 2000.

[12] R. D. Kshirsagar and N. P. Singh, "Some less known ethnomedicinal uses from Mysore and Coorg districts, Karnataka state, India," *Journal of Ethnopharmacology*, vol. 75, no. 2-3, pp. 231–238, 2001.

[13] N. Yoganarasimhan, *Medicinal Plants of India*, Karnataka Interline Publishing, Bangalore, India, 1996.

[14] B. Schulz, U. Wanke, S. Draeger, and H. J. Aust, "Endophytes from herbaceous plants and shrubs: effectiveness of surface sterilization methods," *Mycological Research*, vol. 97, no. 12, pp. 1447–1450, 1993.

[15] H. Barnett and B. Hunter, *Illustrated Genera of Imperfect Fungi*, Burgess Publishing, Minneapolis, Minn, USA, 1998.

[16] K. H. Domsch, W. Gams, and T. Anderson, *Compendium of Soil Fungi*, Academic Press, New York, NY, USA, 2003.

[17] J. F. Leslie and B. A. Summerell, *The Fusarium Laboratory Manual*, Blackwell Publishing, London, UK, 2006.

[18] W. Photita, P. W. J. Taylor, R. Ford, K. D. Hyde, and S. Lumyong, "Morphological and molecular characterization of *Colletotrichum* species from herbaceous plants in Thailand," *Fungal Diversity*, vol. 18, pp. 117–133, 2005.

[19] V. Kumaresan and T. S. Suryanarayanan, "Occurrence and distribution of endophytic fungi in a mangrove community," *Mycological Research*, vol. 105, no. 11, pp. 1388–1391, 2001.

[20] V. Kumaresan and T. S. Suryanarayanan, "Endophyte assemblages in young, mature and senescent leaves of *Rhizophora apiculata*: evidence for the role of endophytes in mangrove litter degradation," *Fungal Diversity*, vol. 9, pp. 81–91, 2002.

[21] J. A. Ludwig and J. F. Reynolds, *Statistical Ecology: A Primer on Methods and Computing*, John Wiley & Sons, New York, NY, USA, 1988.

[22] M. Nei, "Genetic distance between populations," *American Naturalist*, vol. 106, no. 949, pp. 283–293, 1972.

[23] M. P. Miller, *Tools For Population Genetic Analyses (TFPGA) V L.3: A Windows program For the Analysis of Allozyme And moleculAr Genetic data*, Department of Biological Sciences, Northern Arizona University, Flagstaff, Ariz, USA, 1997.

[24] D. C. Lacap, K. D. Hyde, and E. C. Y. Liew, "An evaluation of the fungal 'morphotype' concept based on ribosomal DNA sequences," *Fungal Diversity*, vol. 12, pp. 53–66, 2003.

[25] W. Y. Huang, Y. Z. Cai, K. D. Hyde, H. Corke, and M. Sun, "Biodiversity of endophytic fungi associated with 29 traditional Chinese medicinal plants," *Fungal Diversity*, vol. 33, pp. 61–75, 2008.

[26] M. V. Tejesvi, B. Mahesh, M. S. Nalini et al., "Fungal endophyte assemblages from ethnopharmaceutically important medicinal trees," *Canadian Journal of Microbiology*, vol. 52, no. 5, pp. 427–435, 2006.

[27] N. Raj, *Coelomycetous Anamorphs With Appendage Bearing Conidia*, Edwards Brothers, Ann Harbor, Michigan, 1993.

[28] G. A. Strobel, "Microbial gifts from rain forests," *Canadian Journal of Plant Pathology*, vol. 24, no. 1, pp. 14–20, 2002.

[29] M. V. Tejesvi, B. Mahesh, M. S. Nalini et al., "Endophytic fungal assemblages from inner bark and twig of *Terminalia arjuna* W. & A. (Combretaceae)," *World Journal of Microbiology and Biotechnology*, vol. 21, no. 8-9, pp. 1535–1540, 2005.

[30] B. Mahesh, M. V. Tejesvi, M. S. Nalini et al., "Endophytic mycoflora of inner bark of *Azadirachta indica* A. Juss," *Current Science*, vol. 88, no. 2, pp. 218–219, 2005.

[31] K. Rajagopal, S. Kalavathy, S. Kokila et al., "Diversity of fungal endophytes in few medicinal herbs of southern India," *Asian Journal of Experimental Biological Sciences*, vol. 1, pp. 415–418, 2010.

[32] Y. L. Krishnamurthy, S. B. Naik, and S. Jayaram, "Fungal communities in herbaceous medicinal plants from the malnad region, Southern India," *Microbes and Environments*, vol. 23, no. 1, pp. 24–28, 2008.

[33] D. L. Hawksworth and A. Y. Rossman, "Where are all the undescribed fungi?" *Phytopathology*, vol. 87, no. 9, pp. 888–891, 1997.

[34] D. Wilson and G. C. CarrolL, "Infection studies of *Discula quercina*, an endophyte of *Quercus garryana*," *Mycologia*, vol. 86, no. 5, pp. 635–647, 1994.

[35] G. A. Strobel, W. M. Hess, J.-Y. Li et al., "*Pestalotiopsis guepinii*, a taxol-producing endophyte of the wollemi pine, *Wollemia nobilis*," *Australian Journal of Botany*, vol. 45, no. 6, pp. 1073–1082, 1997.

[36] K. Clay, "Fungal endophytes of grasses," *Annual Review of Ecology and Systematics*, vol. 21, no. 1, pp. 275–297, 1990.

[37] P. Bayman, L. L. Lebrón, R. L. Tremblay, and D. J. Lodge, "Variation in endophytic fungi from roots and leaves of *Lepanthes* (Orchidaceae)," *New Phytologist*, vol. 135, no. 1, pp. 143–149, 1997.

[38] A. E. Arnold, Z. Maynard, and G. S. Gilbert, "Fungal endophytes in dicotyledonous neotropical trees: patterns of abundance and

diversity," *Mycological Research*, vol. 105, no. 12, pp. 1502–1507, 2001.

[39] P. F. Cannon and C. M. Simmons, "Diversity and host preference of leaf endophytic fungi in the Iwokrama Forest Reserve, Guyana," *Mycologia*, vol. 94, no. 2, pp. 210–220, 2002.

[40] X. Sun, Q. Ding, K. D. Hyde, and L. D. Guo, "Community structure and preference of endophytic fungi of three woody plants in a mixed forest," *Fungal Ecology*, vol. 5, no. 5, pp. 624–632, 2012.

[41] P. Holliday, *A Dictionary of Plant Pathology*, Cambridge University Press, Cambridge, UK, 1998.

[42] J. Collado, G. Platas, and F. Peláez, "Identification of an endophytic *Nodulisporium* sp. from Quercus ilex in central Spain as the anamorph of *Biscogniauxia mediterranea* by rDNA sequence analysis and effect of different ecological factors on distribution of the fungus," *Mycologia*, vol. 93, no. 5, pp. 875–886, 2001.

[43] X.-X. Gao, H. Zhou, D.-Y. Xu, C.-H. Yu, Y.-Q. Chen, and L.-H. Qu, "High diversity of endophytic fungi from the pharmaceutical plant, *Heterosmilax japonica* Kunth revealed by cultivation-independent approach," *FEMS Microbiology Letters*, vol. 249, no. 2, pp. 255–266, 2005.

[44] N. S. Raviraja, "Fungal endophytes in five medicinal plant species from Kudremukh Range, Western Ghats of India," *Journal of Basic Microbiology*, vol. 45, no. 3, pp. 230–235, 2005.

[45] A. E. Arnold, Z. Maynard, G. S. Gilbert, P. D. Coley, and T. A. Kursar, "Are tropical fungal endophytes hyperdiverse?" *Ecology Letters*, vol. 3, no. 4, pp. 267–274, 2000.

[46] B. S. Naik, J. Shashikala, and Y. L. Krishnamurthy, "Diversity of fungal endophytes in shrubby medicinal plants of Malnad region, Western Ghats, Southern India," *Fungal Ecology*, vol. 1, no. 2-3, pp. 89–93, 2008.

22

Genetic Divergence, Implication of Diversity, and Conservation of Silkworm, *Bombyx mori*

Bharat Bhusan Bindroo and Shunmugam Manthira Moorthy

Central Sericultural Research and Training Institute, Srirampura, Mysore, Karnataka 570 008, India

Correspondence should be addressed to Shunmugam Manthira Moorthy; moorthysm68@gmail.com

Academic Editor: Alexandre Sebbenn

Genetic diversity is critical to success in any crop breeding and it provides information about the quantum of genetic divergence and serves a platform for specific breeding objectives. It is one of the three forms of biodiversity recognized by the World Conservation Union (IUCN) as deserving conservation. Silkworm *Bombyx mori*, an economically important insect, reported to be domesticated over 5000 years ago by human to meet his requirements. Genetic diversity is a particular concern because greater genetic uniformity in silkworm can increase vulnerability to pests and diseases. Hence, maintenance of genetic diversity is a fundamental component in long-term management strategies for genetic improvement of silkworm which is cultivated by millions of people around the worlds for its lusture silk. In this paper genetic diversity studies carried out in silkworm using divergent methods (quantitative traits and biochemical and molecular markers) and present level of diversity and factors responsible for loss of diversity are discussed.

1. Introduction

Sericulture is a unique field of agriculture, because silkworms are reared on an extensive scale in rearing houses and their silk cocoons are utilized as fine material for clothing. Like agriculture, sericulture also requires a continuous flow of productive silkworm breeds and host plant varieties to meet the ever-changing demand of people involved in the industry besides the consumer sector. To meet all these requirements, the breeder needs very wide and inexhaustible genetic resources to meet the ever-changing demands from various sectors. Considering the great economic importance of *Bombyx mori*, silk producing countries, such as China, Japan, India, Russia, Korea, Bulgaria, and Iran, have collected number of silkworm breeds suitable for a wide range of agroclimatic conditions. More than 4000 strains are maintained in the germplasm of *B. mori* and 46 institutes are involving silkworm genetic resources maintenance, which includes univoltine, bivoltine, and polyvoltine strains. These different genotypes display large differences in their qualitative and quantitative traits that ultimately control silk yield. It was estimated that silkworm genome consists of about $4.8' 108$ bp; its genetic information volume is about one-sixth of human being. There are over 450 morphological, physiological, and biochemical characters recorded at present, among them 300 (including multiallele) had been located on 27 groups of the total 28 chromosomes [1]. Apart from a rich biodiversity of geographical races, there are also a large number of mutants for a variety of characters present in *B. mori* [2].

2. Genetic Divergence in Silkworm

Study on genetic diversity is critical to success in any crop breeding and it provides information about the quantum of genetic divergence and serves as a platform for specific breeding objectives [3]. Genetic diversity is usually thought of as the amount of genetic variability among individuals of a variety or population of a species [4]. It results from the many genetic differences between individuals and may be manifest in differences in DNA sequence, in biochemical characteristics (e.g., in protein structure or isoenzyme properties), in physiological properties (e.g., abiotic stress resistance or growth rate), or in morphological characters [5]. Genetic diversity has been conventionally estimated on the basis of different biometrical techniques (Metroglyph, D^2 divergence analysis, and principal component analysis) such as phenotypic diversity index (H), or coefficient of parentage utilizing morphological, economical, and biochemical data [6–9].

The genetic diversity of B. mori is derived from hybridization of different geographical origins, mainly the Japanese, Chinese, European, and Indian strains, which have distinct traits. Among these four geographical strains, silkworm of temperate origin produces a higher quantity of good, finer, stronger silk fiber, whereas the tropical strains are hardy, tolerant to pathogen load, and resistant to diseases. However, the tropical strains produce low amounts of silk, which is coarser and weaker [10]. To help the breeders in the process to identify the parents that nick better, several methods of divergence analysis based on quantitative traits have been proposed to suit various objectives. As most of the desirable characters in silkworm are quantitative nature, multivariate statistical methods have been employed to measure the genetic diversity among the stocks. Among them, D^2 analysis of Mahalanobis [10] using Tocher's optimization method [11] occupies a unique place and an efficient method to gauge the extent of diversity among genotypes, which quantify the difference among several quantitative traits. It is being used by most of the workers and has been found as an extremely useful tool for estimating genetic divergence in the silkworm [12–16]. Table 1 summarizes genetic divergence study carried out by several workers in silkworm. Using this method, silkworm genotypes were formed into different clusters indicating presence of distinct divergence among the genotypes. Though divergence reported to exist among genotypes and mixed trend of clustering observed. Jolly et al. [13] subjected forty-nine silkworm breeds for this analysis and reported that these breeds were found to form three distinct clusters indicating the presence of distinct diversity among the breeds. Subba Rao et al. [14] and Govindan et al. [15] reported that breeds derived from the same parents were included in different clusters showing variation among the breeds derived from the same source. On the other hand, the breeds derived from the same source were included in the same cluster showing close affinity between advanced sister lines [12, 14, 15] and those of differing genetic background occupied in single cluster indicated uniformity in selection procedures [12]. However, genotypes of temperate and tropical origin formed into separate clusters indicating environmental influence on the expression of characters [17]. Though theoretically geographical diversity is important factor, it is not the whole determining factor for genetic divergence [18, 19]. All these studies were aimed to identify suitable parents for breeding programme and recommended to cross the genotypes from different clusters [20–22] for yield improvement. Though many characters in silkworm are subjected for divergence study, characters, namely, fecundity larval weight, single cocoon weight, cocoon shell weight, and filament length only, contributed about 97% to the total genetic divergence [12, 13, 16, 23, 24].

3. Diversity in Silkworm

Genetic diversity is most often characterized using data that depict variation in either discrete allelic states or continuously distributed (i.e., quantitative) characters, which lead to different possible metrics of genetic diversity [25]. Genetic diversity can be assessed among different accessions/individuals within the same species (intraspecific), among species (interspecific). and between genus and families [26]. It plays an important role in any breeding either to exploit heterosis or to generate productive recombinants. The choice of parents is of paramount importance in any kind of breeding programme; hence, the knowledge of genetic diversity and relatedness in the germplasm is a prerequisite for crop improvement programmes. Genetic diversity is also an essential aspect in conservation biology because a fundamental concept of natural selection states that the rate of evolutionary change in a population is proportional to the amount of genetic diversity present in it [27]. Decreasing genetic diversity increases the extinction risk of populations due to a decline in fitness. Genetic diversity also has the potential to affect a wide range of population, community, and ecosystem processes both directly and indirectly. However, these effects are contingent upon genetic diversity being related to the magnitude of variation in phenotypic traits [28].

In general, cocoon colour and cocoon shape, larval, marking, and quantitative traits have been used for differentiation of silkworm genotypes and, based on that, parents are being selected. However, recent advent of different molecular techniques led breeders to estimate genetic diversity on the basis of data generated by different molecular markers, which provided a means of rapid analysis of germplasm and estimates of genetic diversity, which were often found to corroborate phenotypic data. These molecular markers are broadly categorized as biochemical and molecular markers.

3.1. Biochemical Markers. Application of isoenzymes and other molecular markers helps to estimate genetic diversity much more accurately than that of morphological traits. Electrophoresis identifies variation (alleles) at loci that codes for enzymes (usually termed isozymes or allozymes). One advantage of allozyme loci is that they are codominant and heterozygotes can be scored directly. Understanding the genetic constitution of an individual in the population of races and allelic variations through isozyme studies is known to reflect the differential catalytic ability of allelic genes and their significant role in the adaptive strategy of the genotypes [29].

The diversity study carried out in silkworm through protein profiles, enzymes, and isozymes are summarized in Table 2. Isozymes like esterase, acid phosphatase, alkaline phosphatase, amylase, phosphoglucomutase, aspartate aminotransferase, malate dehydrogenase, glucose 6 phosphate dehydrogenase, and carbonic anhydrase have been used by various authors to study diversity in silkworm genotypes [30–43]. Among the different isoenzymes analyzed, esterase was most preferred because of its diverse substrate specificity and polymorphic expression followed by acid phosphatase [36, 37, 44]. Eguchi et al. [32] found four

TABLE 1: Genetic divergence study reported in silkworm.

SL number	Reference number	Number of genotypes used and clusters formed	Measures of genetic diversity	Conclusion
1	[89]	49 and 3	Mahalanobis (1936) and Tocher (1956)	(1) Presence of distinct diversity. (2) Breeds derived from the same parents were included in different clusters. (3) Breeds derived from the same source were included in the same cluster.
2	[23]	32 and 7	Mahalanobis (1936) and Tocher	Geographical diversity did not contribute much to genetic diversity.
3	[14]	15 and 5	Mahalanobis (1936) and Tocher (1956)	(1) Enough diversity present. (2) Suggested for making crosses between different clusters.
4	[116]	50 and 5	Mahalanobis (1936) and Tocher (1956)	Cluster III was the largest, consisting of 34 strains. The clusters are compared for various features influencing silk production.
5	[15]	18 and 8	Mahalanobis (1936) and Tocher (1956)	Breeds derived from the same ancestry were included in different clusters and those of different genetic background occupied a single cluster.
6	[20]	25 and 6	Mahalanobis (1936) and Tocher (1956)	The genetically divergent parents were grouped into four classes.
7	[18]	30 and 5	Mahalonobis' D^2 values (Ward's minimum variance)	Geographical diversity though important is not the determining factor for genetic divergence.
8	[24]	24 and 7	Mahalanobis (1936) and Tocher	Genotypes of temperate and tropical origin formed separate clusters.
9	[117]	11 and 3	Mahalanobis (1936) and Tocher (1956)	The intracluster distance ranged from 0.00 to 1689.37 implying the prevalence of substantial amount of intracluster diversity.
10	[21]	22 and 6	Mahalonobis' D^2 values (Ward's minimum variance.)	There is no relation between geographical diversity and genetic diversity.
11	[118]	65 and 9	Mahalonobis' D^2 values (Ward's minimum variance)	Breeds in the optimum distance obtained cluster can be used in the conventional silkworm breeding programme to improve silk quality.
12	[119]	47 and 12	Mahalonobis' D^2 values (Ward's minimum variance)	Geographic diversity had no association with genetic diversity.
13	[120]	51 and 2	UPGMA	Clusters of individuals exhibited high internal (within clusters) homogeneity and high external (between clusters) heterogeneity.
14	[121]	16 and 3	Mahalanobis (1936), UPGMA	The strains of the same origin did not group together, demonstrating they can have different biological and development performance.
15	[122]	8 and 5	Mahalanobis (1936), UPGMA	Genetic distance and not the geographic diversity is to be considered while identifying parents for hybridization programme.
16	[16]	21 and 7	Mahalanobis (1936) and Tocher (1956)	Silkworm genotypes originating from the same geographical regions fell in one cluster.
17	[17]	56 and 8	Mahalanobis (1936) and Tocher (1956)	Silkworm genotypes originating from different geographical regions fell in one cluster while those originating from a single region fell in different clusters.
18	[123]	51 and 4	Hierarchical agglomerative clustering UPGMA	Inclusion of genotypes of the same origin in different clusters clearly indicates the presence of considerable genetic diversity among the populations.
19	[19]	19 and 3	The hierarchical cluster analysis using Euclidian distance	Cluster analysis and conformity with the variability in the performance of the genotypes for different traits. Geographic diversity had no association with genetic diversity.
20	[124]	4 and 2	UPGMA method (Sokal and Michener)	The optimum level of genetic distance is necessary to obtain heterosis.

TABLE 2: Genetic divergence study reported in silkworm using enzymes, protein, and isozymes.

SL number	Reference number	Number of genotypes and clusters	Measures of genetic diversity	Conclusion
1	[32]	—	Esterase used for polymorphism	Polymorphism noticed among genotypes.
2	[125]	—	Protein profiles used for genetic diversity	Divergence arises internally after a relatively long amino terminal sequence which appears to be conserved. A plausible explanation for the observed genetic variability is the occurrence of relatively large unequal crossing-over exchanges in the repetitive domain of the fibroin gene.
3	[126]	—	Esterase used for polymorphism	Polymorphism noticed among genotypes.
4	[127]		Acid phosphatase used for polymorphism	Polymorphism observed among genotypes.
5	[128]	20	Enzymes	Rich genetic diversity among genotypes.
6	[129]	10	Esterase used for polymorphism	Polymorphism noticed among genotypes.
7	[36]	12 and 6	Nei and Li (1978) [130] and Yeh et al. (1999) [131]	Rich genetic diversity among genotypes.
8	[132]	8 and 2	Enzymes and UPGMA	Genetic diversity noticed among genotypes.
9	[133]		Nei and Li (1978)	The protein profile of different breeds has indicated the polymorphism and genetic diversity among silkworm breeds.
10	[37]	15	—	Esterase exhibited polymorphism among the bivoltine breeds.
11	[134]	6 and 3	Protein	Genetic differentiation among populations of different races.
12	[44]	12 and 2	Nei and Li (1978) UPGMA	The mean value of FST (0.2224) calculated on the base of the established polymorphism showed that 22.24% of the genetic variability was observed between the different strains, which corresponds to the level of the interstrain genetic differentiation.
13	[40]	21 and 8	Nei and Li (1978) UPGMA	Genetic variations were observed and they can be identified by relating with their morphology and geographical origins
14	[135]		Nei and Li (1978) UPGMA	Protein profiles studied and presence of rich genetic diversity among germplasm stocks. Different origin accessions established a close relationship indicating close affinity in protein pattern.
15	[41]	15	Esterase used for polymorphism	Variation in esterase pattern was observed among genotypes.
16	[43]	10 and 2	Nei (1978) by UPGMA dendrogram (Sneath and Sokal, 1973)	A perusal of genetic diversity within and among strains indicated that 34.72% of the observed variation occurred among strains and the rest of the variation (65.28%) within strains. Their rich genetic diversity needs to be exploited in conservation and breeding programme.
17	[136]	4	(Swofford and Selander, 1981)	The lower degree of observed heterozygosity and the higher degree of homozygotes proved the inbreeding effect.
18	[42]	15 and 3	Nei (1978) and UPGMA	Japanese and Chinese strains could not be totally separated by the isoenzyme system analysis. The results indicate that, in spite of the genetic distance and differentiation among the lineages, they cannot be separate just with the isozymes alleles. The high FST value (0.6128) allows the conclusion that the lineages are differentiated.

fundamental types of esterase and about 70% of the Japanese, Chinese, and European races investigated belong to A type and 20% to 0 type, while B type was found only in Chinese races. Yoshitake et al. [45] analyzed polymorphism pattern of esterase and acid phosphates in 300 strains of silkworm and concluded that distribution of acid phosphatase and esterase was similar in European and Japanese strains and there was resemblance between Chinese and European strains. A higher degree of interstrain variability was reported on the acid phosphatase [43, 44] and esterase [36, 37, 41, 42]. Acid phosphatase is also found to be a suitable marker for analyzing the inter- and intrastrain diversity and the strain differentiation [44]. Isozyme analysis in different silkworm genotypes by different authors indicated rich genetic diversity between the genotypes and results were mainly used to separate populations and strains in order to use them in selection programs.

3.2. Molecular Markers. Molecular diversity studies assess all levels of genetic structure and species specific complex components [46]. The detection and exploitation of naturally occurring DNA sequence polymorphisms have wide potential applications in animal and plant improvement programmes as a means for varietal and parentage identification facilitate genetic diversity and relatedness estimations in germplasm [47]. The results obtained from different molecular markers may themselves be quite different from those obtained by using biochemical markers such as isozymes or morphological characters. The molecular markers, namely, RAPD, RFLP, ISSR, and SSR, have been effectively utilized in analyzing the genetic diversity and phylogenetic relatedness in the domesticated silkworm *Bombyx mori* [48–56]. Details of diversity study carried out in silkworm through molecular markers are summarized in Table 3. RAPD based dendrogram resulted in a clear separation of two groups, one comprising of diapausing and other comprising of nondiapausing genotypes [49, 57–60]. Among the diapausing genotypes, all the "Chinese type" genotypes which spin oval cocoons grouped separately, while the "Japanese type" genotypes which spin peanut shaped cocoons were found in another group. Further genotypes, which share the same geographical origin, were grouped in the same cluster [57, 61]. SSR and mtDNA markers analysis revealed considerable genetic diversity among the nondiapausing silkworm genotypes that were developed in India, China, and Bangladesh [62]. The dendrogram constructed analysing RFLP markers revealed two distinct groups as Khorasan native (Iran) and Japanese commercial lines. The distinct clustering of these two sets of strains and lines reflects differences of the geographical origin and morphological, qualitative, and quantitative traits associated with them [54]. Kim et al. [63] made phylogenetic analysis using the individual or the nine concatenated intronic sequences which showed no clustering on the basis of known strain characteristic such as voltinism, moultinism, egg colour, blood colour, cocoon colour, or cocoon shape. Furthermore, the tree obtained by them using the nine concatenated intronic sequences comprising 5,897 bp including indels resulted in a similar conclusion. However, Tunca et al. [64] stated moderately low level of diversity among genotypes studied. Supporting this argument recently, Jagadeesh Kumar [65] reported the low level of genetic distance between the breeds on the basis of gene frequency evidenced by the boot strap values in the constructed dendrogram with the help of molecular markers.

On the whole, the diversity study conducted using phenotypic characters and molecular markers had reported adequate genetic variation between genotypes. But these differentiations mostly based on voltinism and geographical origin indicating narrow genetic base between the available genotypes.

4. Status of Genetic Diversity in Silkworm

Zhang et al. [51] reported that genetic distances within Japanese strains are closer than those of Chinese strains and within a strain; the individual polymorphism is significantly higher in wild silkworm than those of domesticated silkworm. According to Liu et al. [66] at the species level, *Antheraea pernyi* and *Bombyx mori* showed high levels of genetic diversity, whereas *Samia cynthia ricini* showed low level of genetic diversity. However, at the strains level, *Antheraea pernyi* had relatively the highest genetic diversity and *B. mori* had the lowest genetic diversity. Analysis of molecular variance (ANOVA) suggested that 60% and 72% of genetic variation resided within strains in *Antheraea pernyi* and *Samia cynthia ricini*, respectively, whereas only 16% of genetic variation occurred within strains in *B. mori*. Similarly, genetic variation was measured using the population size scaled mutation rate which was significantly smaller in domesticated strains (0.011), when compared to the wild strains (0.013) of *B. mori*. The rate of heterozygosity in domesticated strains was reported to be two times lower than that in wild varieties (0.003 and 0.008, resp.). Recently, Yukuhiro et al. [67] analyzed PCR amplified carbamoyl-phosphate synthetase 2, aspartate transcarbamylase, and dihydroorotase (CAD) gene fragments from 146 *Bombyx mori* native strains and found extremely low levels of DNA polymorphism. CAD haplotype analysis of 42 samples of Japanese *B. mandarina* revealed four haplotypes. No common haplotype was shared between the two species and at least five base substitutions were detected. These results suggesting that low levels of gene flow between the two species. Further extremely low level of DNA polymorphism in *B. mori* compared to its wild relatives suggested that the *CAD* gene itself or its tightly linked regions are possible targets for silkworm domestication. This information clearly indicates narrow level of genetic diversity in silkworm.

5. Causes for Loss of Genetic Diversity

The existence of genetic variation within a population is crucial for its ability to evolve in response to novel environmental challenges. Genetically variable populations are expected to evolve morphological, physiological, or behavioural mechanisms to cope with the novel conditions [68]. This sorting process not only results in populations that are better adapted to their local environments, but may also, at least in theory, cause a reduction in the genetic variation. Forces that affect genetic variation within populations are effective population size, mutation, genetic drift, gene flow, inbreeding depression, out breeding depression, and natural selection. In silkworm, reduction in genetic diversity might be mainly due to domestication, breeding systems, selection, genetic drift, and inbreeding. In maize, too, selection and drift due to the domestication are the principal factors that influence

TABLE 3: Molecular diversity reported in silkworm.

Sl number	Reference number	Number of genotypes and Cluster	Measures of genetic diversity	Conclusion
1	[48]	13 and 2	RAPD	Silkworm genotypes were clustered into two groups, one consisting of six diapausing and the other of seven nondiapausing genotypes. RAPD technique could be used as a powerful tool to generate genetic markers that are linked to traits of interest in the silkworm.
2	[57]	13	RAPD and banded krait minor satellite DNA	The RAPD based dendrogram resulted in a clear separation of two groups, one comprising diapausing and the other nondiapausing genotypes. The clustering pattern of RFLP obtained was comparable to the phenogram resulting from RAPD analysis.
3	[49]	5 and 3	RAPD	Some of the DNA fragments were strain specific and some could differentiate the multivoltine from the bivoltine strains or vice versa.
4	[50]	13 and 2	SSR	Detailed analysis of silkworm strains with microsatellite loci revealed a number of alleles ranging from 3 to 17 with heterozygosity values of 0.66–0.90. Along with strain specific microsatellite markers, diapause and nondiapause strain-specific alleles were also identified
5	[137]	13 and 2	ISSR and RAPD	The highest diversity index was observed for ISSRPCR (0.957) and the lowest for RAPDs (0.744). Differentiated diapause and nondiapause strains
6	[138]	31 and 7	SSR	The average heterozygosity value for each SSR locus ranged from 0 to 0.60, and the highest one was 0.96 (Fl0516 in 4013). The mean polymorphism index content (PIC) was 0.66 (range of 0.12–0.89). SSR markers are an efficient tool for fingerprinting cultivars and conducting genetic-diversity studies in the silkworm
7	[139]	20 and 6	RAPD	Multivoltine Silkworm has more genetic diverse than bivoltine
8	[51]	12	SSR	Within a strain, the individual polymorphism of wild silkworm was significantly higher in abundance than those of domesticated silkworm
9	[140]	5	RAPD	The genetic distances between the clusters and within the clusters estimated 6 percent variability between the 4 races and Nistari. RAPDs are very efficient in the estimation of genetic diversity in populations that are closely related and acclimatized to local environmental conditions.
10	[52]	29 and 4	CAP	Considerable genetic diversity observed. Grouped strains roughly according to their geographical origin.
11	[55]	96	SSR	The mean polymorphism index content was 0.71 (range of 0.299–0.919). UPGMA cluster analysis of Nei's genetic distance grouped silkworm strains based on their origin.
12	[141]	6 and 2	AFLP	Higher degree of genetic similarity within Japanese commercial lines than the Iranian native strains. The distinct clustering of these two sets of strains and lines reflects differences of the geographical origin and morphological, qualitative, and quantitative traits associated with them.
13	[61]	7 and 2	AFLP	The genetic similarity estimated within and among silkworms could be explained by the pedigrees, historical and geographical distribution of the strains, effective population size, inbreeding rate, selection intensity, and gene flow.
14	[64]	6	RAPD	The genetic diversity in studying strains was moderately low. Estimates of gene diversity in populations were higher in total (Ht) as compared to those within population diversity (Hs).
15	[142]	20 and 6	ISSR	In selected mutant genetic stocks, the average number of observed alleles was (1.7080 ± 0.4567), effective alleles (1.5194 ± 0.3950), and genetic diversity (Ht) (0.2901 ± 0.0415). ISSR is a valuable method for determining the genetic variability among mutant silkworm strains.

TABLE 3: Continued.

Sl number	Reference number	Number of genotypes and Cluster	Measures of genetic diversity	Conclusion
16	[63]	25 and 3	Intronic sequences	The degree of sequence divergence in some introns is very variable, suggesting the potential of using intronic sequences for strain identification.
17	[58]	12	RAPD, ISSR, and RFLP-STS	RAPD generated 93.6%, ISSR was 84.62, and RFLP was 75.6% polymorphism. Ability to discriminate bivoltine and multivoltine.
18	[143]	3 and 2	RAPD	The diversity within the populations (Hs) was 0.1334 and the magnitude of differentiation among the populations (GST) was 0.2968.
19	[56]	14	ISSR	ISSR markers has generated 92 percent were polymorphic,diapausing and non-diapausing silkworm stocks could be distinguished by specific marker.
20	[144]	8 and 2	ISSR, RAPD, and isozymes	Sufficient polymorphism and genetic diversity observed.
21	[145]	30 and 2	ISSR	PCA analysis helped to visualize the two major clusters which included the multivoltines and bivoltines separately. The grouping of bivoltines in the PCA analysis clearly showed higher similarity among bivoltines as compared to the multivoltines.
22	[62]	13 and 2	SSR and mtDNA	The heterozygosity generated by the seven pairs of SSR primers varied from 0.098 to 0.396. Considerable genetic diversity is present among the 13 silkworm genotypes.
23	[146]	30 and 2	ISSR	The grouping of bivoltines in the PCA analysis clearly showed higher similarity among bivoltines as compared to the multivoltines.
24	[66]	A. Pernyi-3 S. cynthia ricini-12 B. mori-12	RAPD	At the species level, A. pernyi and B. mori showed high levels of genetic diversity, whereas S. cynthia ricini showed low level of genetic diversity. However, at the strain level, A. pernyi had relatively the highest genetic diversity and B. mori had the lowest genetic diversity.
25	[147]	14 and 2	RAPD, ISSR	High polymorphisms (70.91 and 74.70%) were revealed by ISSR and RAPD markers.
26	[59]	4 and 2	RAPD	Multivoltine silkworm races are genetically more distant than the two bivoltine silkworm. Genetic distances among the multivoltine and bivoltine silkworm were 0.52 and 0.27, respectively.
27	[60]	9 and 3	RAPD	The average genetic distance between the samples was 0.53. The average genetic distance from analyzed samples proved to be relatively high.
28	[148]	8 and 6	RAPD	Genetic distances varied from 0.28889 (B75.2-C1.4) to 0.92437 (A1.2-A1.3) with an average of 0.58497. Silkworms group a high genetic diversity.
29	[65]	5 and 2	ISSR	Artificial selection during seven continuous generations generally caused lesser genetic distance between the breeds.
	[149]	6 and 3	ISSR	This marker could not discriminate same geographical races correctly.
30	[150]	10 and 3	RAPD	The genotypes were grouped based on voltinism and bivoltines are subgrouped based on silk productivity nature of silkworm breeds.
31	[151]	10	SSR	Sufficient polymorphism and genetic diversity observed. The genotypes were grouped based on voltinism and subdivided based on cocoon shape and cocoon colour.

the amount and distribution of genetic variation in crop genomes as compared to their wild progenitors in maize [69].

5.1. Domestication. Over the past 12 000 years, humans have sampled, selected, cultivated, travelled through, and colonized new environments, thus inducing a plethora of bottlenecks, drifts, and selection. Plant breeders have accelerated the whole process by selecting preferred genotypes [46]. In the broadest sense, alteration and narrowing of crop genetic diversity began with the first domestication of wild plants/animals. Domestication represents a relatively recent evolutionary event, occurring over the past 13,000 years after the Neolithic revolution [70, 71]. This process frequently leads to the improvement of economically important traits and the diversification of morphological traits in domesticated species compared to their wild ancestors. Silkworm domestication, which is a relatively recent evolutionary event, may have generated a large number of alterations and diversification in the structure of an evolutionarily conserved morphogenetic gene [72]. There is an assumption that the process of domestication and selection has resulted in drastic narrowing of the genetic variation and homozygosity in mulberry silkworm which has been domesticated over 5000 years ago. Xia et al. [73] compared the whole genome sequencing of 29 B. mori strains and 11 Chinese B. mandarina individuals by 1.50 billion short reads and concluded that B. mori was clearly genetically differentiated from B. mandarina. At the same time, based on the high level of conservation of genetic variability, the authors estimated that a large number of B. mandarina individuals were used for domestication (i.e., the population bottleneck during silkworm domestication might not have been severe.) Therefore, gene flow limited to B. mori could have occurred for many genes during silkworm domestication. Recently, Yu et al. [74] and Guo et al. [75] reported decreased level of genetic variation in B. mori genes or regions compared to those in Chinese B. mandarina in the domestication targeted gene. About 40.7% or 49.2% of the genetic diversity of wild silkworm was lost in domesticated silkworm [74]. Study conducted with B. Mandarina and B. mori by Guo et al. [75] reveals that diversity of B. mori is significantly lower than that of B. mandarina. Further gene DefA showed signature of artificial selection by all analysis methods and might experience strong artificial selection in B. mori during domestication resulting less diversity [75]. However, when analysing the carotenoid binding protein (CBP) genes in B. mori identified large copy number variations and retrotransposon associated structural differences in CBP from B. mori, which were absent from B. mandarina, and concluded that domestication can generate significant diversity of gene copy number and structure over a relatively short evolutionary time.

5.2. Breeding Too Causes Loss of Variability. Breeding systems and life history traits govern the transmission of genes between generations and have been long recognized as impacting the genetic diversity and population genetic structure [76–78]. Breeding is a strong force in the reduction of genetic diversity [79] and views the introduction of modern

varieties as evidence of genetic erosion [80]. Silkworm breeding by definition is the selection of superior genotypes and/or phenotypes over a period of time. During the last decades, development and increased focus on more efficient selection programmes have accelerated genetic improvement in a number of breeds. As a result, highly productive silkworm breeds have replaced local ones across the world [81–84]. This development has led to growing concerns about the erosion of genetic resources. As the genetic diversity of low-production breeds is likely to contribute to current or future traits of interest [85, 86], they are considered essential for maintaining future breeding options.

Selection naturally results in a narrowing of the genetic base of the genotype. Even if the breeder has introduced alleles from indigenous races to his target genotype, he/she must then begin the process of "weeding out" the alleles that are undesirable. This weeding out of undesirable alleles is once again narrowing the genetic base of the line. Practically, a breeder typically uses the best genotypes available and selects superior progeny. The continual use of the best genotypes as parents naturally narrows the gene pool to only those alleles that are available from the elite parents and therefore tends to decrease the genetic variation of the population [87]. There is also a threat or loss of genetic diversity as a result of replacement of wild species by exotic high-yielding varieties. Typically, population size is also a major source of loss of genetic diversity.

5.3. Effects of Selection on Diversity. Patterns of diversity in any populations are likely to be affected by selection. Balancing selection due to overdominance (heterozygote advantage), or to frequency dependent selection, may maintain variants in populations, and environmental differences may select for different genotypes in different populations [3]. Purifying selection, however, removes deleterious variants that arise by mutation; such variants are expected to be present at frequencies lower than predicted for the neutral equilibrium. Another form of directional selection occurs when advantageous mutations rapidly reach high frequencies whether they spread throughout a species to fixation or just within a population undergoing adaptation to its local environment [88, 89]. Artificial selection has been widely utilized in the breeding programmes concerning B. mori, which is of commercially important insect. Nevertheless, the genetic diversity of silkworm is greatly reduced during systematically extensive selection for a few target traits. In general, selection of superior individuals results in genetic gain, but also loss in genetic diversity and it is strongly dependent on selection method and selection intensity [90]. Selection will have two important consequences: (1) the genetic average value will be changed, thus conventionally it measured a gain, and (2) there will be change in diversity, and this will be measured by relative effective number of families. It is a well-known fact that diversity is affected by directional selection. Directional or disruptive selection will ultimately fix one allele and thereby deplete genetic variation. It has been suggested that directional selection decreases the level of developmental precision or developmental stability [91] because it may prevent the evolution of canalisation and possibly favour those mechanisms that increase the phenotypic variation [92, 93] showing that systematic selection of parents' results in reduced genetic variation among their offspring. After 4-5 generations with the same selection intensity, the reduction will stabilize. Class example of diversity changes through directional selection and inbreeding in silkworm was reported by Pradeep et al. [94]. They have separated larval populations of Nistari strain based on the shortest larval duration (SLD) and the longest larval duration (LLD) and maintained for 4 more generations. RAPD and ISSR primers generated polymorphic profiles in LLD and SLD lines. Distinct markers specific to LLD individuals were observed from the 3rd generation and indicated selection induced differentiation of allelic variants for longer larval duration. This finding implies that selection combined with inbreeding could result in lines with different genetic properties following separation from the original parental populations. According to Strunnikov [95], continuous selection and inbreeding could have induced a homozygous state of the recessive gene for longer larval duration, where shorter larval duration is the dominant and fitness character. Though it introduced diversity, because of losing its dominant or fitness characters, chances of survival become vulnerable. Further, as reported by Seidavi [96], the genetic performance of selected population of silkworm for cocoon weight trait after the fourth generation shown increased sensitivity towards environment resulting in poor survival due to selection based on productivity traits indicating effect of selection on diversity.

5.4. Genetic Drift.
Genetic drift is the chance changes in allele frequency that result from the random sampling of gametes from generation to generation in a finite population. It has the same expected effect on all loci in the genome [97]. In a large population, on the average, only a small chance change in the allele frequency will occur as the result of genetic drift. On the other hand, if the population size is small, then the allele frequency can undergo large fluctuations in different generations in a seemingly unpredictable pattern and can result in chance fixation (going to a frequency of 1.0) or the loss (going to a frequency of 0.0) of an allele. A classic illustration of how finite population size affects allele frequency was provided by Buri [98]. He looked at the frequency of two alleles at the brown locus that affects eye color in Drosophila melanogaster in randomly selected populations of size 16. However around 107 number of populations had 0 to 32 bw^{75} genes in different (19) generations. The total number of populations fixed for one of the two alleles increased at nearly a linear rate after generation 4 and in generation 19 it is nearly equal for the two alleles, with 30 populations fixed for bw and 28 fixed for bw^{75}. In silkworm germplasm maintenance centers, at every cycle only 40–60 cocoons are selected from each strain/breed for the next generation, from which around 20 layings are prepared and subsequently only 5-6 layings are brushed for next generation. This size of population is small which may be lead to change in allele frequency as explained by Buri [98].

5.5. Effects of Inbreeding.
Loss of genetic diversity among populations occurs due to the synergetic effects of inbreeding and environmental stressors [99]. The negative interaction between inbreeding and environmental stress reflects on population growth rates and inbreeding and environmental effects may interact in their effects on population dynamics. Inbreeding is characterized by an increase in homozygosity resulting in increased expression of recessive deleterious alleles (partial dominance hypothesis) [100] and/or reduced opportunity to express heterozygote superiority (overdominance hypothesis) [101]. Selfing has direct genetic consequences, including its effect on the intensity of inbreeding depression [102] and the partitioning of genetic diversity within and among populations [103]. A consequence of inbreeding is that it makes it much more likely that an individual is homozygous for a rare gene because it is more likely that two related parents simultaneously possess a rare allele and transmit it to their inbred offspring than the two unrelated individuals independently transmit the same rare allele to noninbred offspring. Thus inbreeding seems to reduce fitness because it reveals harmful genes in homozygotes [104].

Sericulture practicing countries maintain hundreds of inbred lines of silkworm in germplasm centres for several decades by selection and inbreeding. Sibling mating of the progenies derived from a single brood is preferred for pure stocks so that the original traits of the races are maintained through generations. Generally, breeders try to maintain the original characters of the races/breeds through selection with care to avoid inbreeding depression. However, the effects of inbreeding can accumulate over many generations, as the frequency of slightly deleterious alleles can gradually increase over time due to genetic drift [105, 106]. This is a particular concern in small populations, where natural selection can be inefficient for alleles that have only slight effects on fitness [107]. The rate at which genetic diversity is lost will depend on the population's size and degree of isolation; small, isolated populations can lose genetic diversity within a few generations, whereas large, continuous populations may not lose significant amounts of diversity over thousands of years [108]. In small populations where genetic drift is most rapid, the fixation of common alleles will result in the reduction of genetic diversity. This phenomenon is applicable in silkworm, as different silkworm strains are maintained with small population leading to genetic drift thereby may be reducing genetic diversity. Further, Li et al. [109] when analysing genetic diversity in B. mandarina and B. mori concluded that the polymorphism level ($\theta\pi$) of mt sequence among Chinese wild population (6.20×10^{-3} nucleotide differences per site) is more than six times that among domesticated varieties (1.14×10^{-3}) and pointing out that the relative larger reduction in polymorphism is most likely caused by inbreeding or population bottlenecking.

6. Broadening the Genetic Diversity

Continuous breeding and selection of silkworm breed for uniformity narrows genetic diversity. One of the approaches to broaden the genetic diversity is the use of recent advances in molecular biology and biotechnology, which allow the transfer of specific genes from diverse sources to target genotypes. In [110], through transgenic approach, by gene addition, subtraction, and pathway redirection, the genetic constituents of crops can be modified and broadened, resulting in new and improved traits. Another approach of broadening the genetic diversity is by the use of exotic germplasm [111]. It will create genotypes with a diverse range of desirable characteristics. The genetic variation that breeders need to introduce these characteristics is often available only through the exchange of genetic resources. This exchange is necessary because some areas of the world have richer resources of genetic diversity, which will be useful in creation of variation.

7. Germplasm Conservation

Conservation of genetic diversity is essential to the long-term survival of any species, particularly in light of changing environmental conditions. Reduced genetic diversity may negatively impact the adaptive potential for a species. Increasing population size and maximizing genetic diversity are among the primary goals of conservation management [112]. The silkworm germplasm maintenance centres generally follow brushing of "composite population" type of all strains to avoid inbreeding depression as well as genetic erosion and maintain the gene pool as far as possible. Composite laying is defined as collection of a known number of eggs from a known number of individual laying sources that represents the whole population. Though composite layings method can retain gene pool, there is a concern regarding populations, as even slight selection has a drastic effect on genetic variability when the effective population size N is large [113]. In this method, only 250–500 larvae are retained in a strain; improper selection can lead to inbreeding depression and natural selection can be inefficient for alleles that have only slight effects on fitness [107].

8. Strategies Required for Conservation of Silkworm Genetic Resources

(1) The curator of the germplasm bank should carefully verify the available genetic resources and avoid duplicates before collection and introduction of new material.

(2) Development of cost-effective, viable, and cost-economic conservation practices through modification or development of long-term preservation of silkworm genetic resources to reduce the number of crop cycles is required.

(3) Conservation through modern methods includes cryopreservation of sperm, artificial insemination, and induction of synthetic diapause hormones to be explored.

(4) Genetic resources should be categorized as most sensitive and sensitive based on their availability in one place or in more than one place, respectively.

(5) The most sensitive genetic resources should be conserved in more than one place by establishing backup stations under the control of main germplasm station.

(6) Establishment of centers for preservation of endangered/local species under *in situ* condition is required.

(7) Use of silkworm genetic resources for nonsericultural use other than cocoon production needs importance.

9. Conclusion

Though silkworm has been domesticated for hundreds of generations, based on available literature, it is speculated that it has not experienced any major reduction of genetic diversity due to phenotypic selection and breeding. But there is concern that bottlenecks may restrict breeding flexibility and slow response to new opportunities, pests, pathogens, and other practices in the future. To broaden the gene pool of silkworm, exotic elite strains were required to be introduced from various countries. The genomes of introduced exotic germplasm will broaden the gene pool; thereby diversity can be maintained. The original genetic composition of genetic resources should be maintained by avoiding genetic drift and selection process. Maintaining adequate population size can prevent the loss of genetic variability due to genetic drift [114]. Study on effects of inbreeding on inbreeding coefficients in silkworm populations is limited. Hence understanding the effects of inbreeding for various traits can be very crucial points in the management of germplasm. As suggested by Doreswamy and Subramanya Gopal [115] during stock maintenance in germplasm centers, rigid selection for more numbers of generations is required to retain original characteristics of the inbred lines and also reduces the deleterious effects of inbreeding.

Conflict of Interests

The authors declare that there is no conflict of interests regarding the publication of this paper.

References

[1] J. Nagaraju and M. R. Goldsmith, "Silkworm genomics—progress and prospects," *Current Science*, vol. 83, no. 4, pp. 415–425, 2002.

[2] Y. Banno, T. Shimada, Z. Kajiura, and H. Sezutsu, "The silkworm—an attractive bioresource supplied by Japan," *Experimental Animals*, vol. 59, no. 2, pp. 139–146, 2010.

[3] D. Charlesworth and T. R. Meagher, "Effects of inbreeding on the genetic diversity of populations," *Philosophical Transactions of the Royal Society B: Biological Sciences*, vol. 358, no. 1434, pp. 1051–1070, 2003.

[4] W. L. Brown, "Genetic diversity and genetic vulnerability—an appraisal," *Economic Botany*, vol. 37, no. 1, pp. 4–12, 1983.

[5] V. R. Rao and T. Hodgkin, "Genetic diversity and conservation and utilization of plant genetic resources," *Plant Cell, Tissue and Organ Culture*, vol. 68, no. 1, pp. 1–19, 2002.

[6] I. A. Matus and P. M. Hayes, "Genetic diversity in three groups of barley germplasm assessed by simple sequence repeats," *Genome*, vol. 45, no. 6, pp. 1095–1106, 2002.

[7] S. A. Mohammadi and B. M. Prasanna, "Analysis of genetic diversity in crop plants—salient statistical tools and considerations," *Crop Science*, vol. 43, no. 4, pp. 1235–1248, 2003.

[8] A. A. Jaradat, M. Shahid, and A. Al-Maskri, "Genetic diversity in the Batini barley landrace from Oman: II. Response to salinity stress," *Crop Science*, vol. 44, no. 3, pp. 997–1007, 2004.

[9] Z. Ahmad, S. U. Ajmal, M. Munir, M. Zubair, and M. S. Masood, "Genetic diversity for morpho-genetic traits in barley germplasm," *Pakistan Journal of Botany*, vol. 40, no. 3, pp. 1217–1224, 2008.

[10] M. R. Goldsmith, "Recent progress in silkworm genetics and genomics," in *Molecular Biology and Genetics of the Lepidoptera*, M. R. Goldsmith and F. Marec, Eds., pp. 25–48, CRC, Boca Raton, Fla, USA, 2009.

[11] P. C. Mahalanobis, "On the generalized distance in statistics," *Proceedings of National Academy of Sciences, India*, vol. 2, no. 1, pp. 49–55, 1936.

[12] C. R. Rao, *Advanced Statistical Methods in Biometrical Research*, John Wiley and Sons, New York, NY, USA, 1952.

[13] M. S. Jolly, R. K. Datta, M. K. R. Noamani et al., "Studies on genetic divergence in mulberry silkworm *Bombyx mori* L," *Sericologia*, vol. 29, no. 4, pp. 545–553, 1989.

[14] G. Subba Rao, S. K. Das, and N. K. Das, "Genetic divergence among fifteen multivoltine genetic stocks of silkworm (*Bombyx mori* L.)," *Indian Journal of Sericulture*, vol. 30, no. 1, pp. 72–74, 1991.

[15] R. Govindan, S. Rangaiah, T. K. Narayana Swamy, M. C. Devaiah, and R. S. Kulkarni, "Genetic divergence among multivoltine genotypes of silkworm (*Bombyx mori* L.)," *Environmental Ecology*, vol. 14, no. 4, pp. 757–759, 1996.

[16] N. B. Pal, S. M. Moorthy, M. Z. Khan, N. K. Das, and K. Mandal, "Analysis of genetic diversity in some multivoltine silkworm genotypes of *Bombyx mori* L," in *Proceedings of the Golden Jubilee National Conference on Sericulture Innovations: Before and Beyond*, Abstract, p. 68, CSRTI, Mysore, India, January 2010.

[17] K. Mandal, S. M. Moorthy, S. Sen, N. K. Das, and C. R. Sahu, "An analysis of genetic variation and diversity in bivoltine silkworm (*Bombyx mori* L.) genotypes," in *Proceedings of the National Symposium on Deccan Biodiversity Co-Existance of Funal Species in Changing Landscapes*, Abstract, p. 38, Osmania University, Hyderabad, December 2010.

[18] P. Mukherjee, S. Mukherjee, and P. Kumaresan, "An analysis of genetic divergence in Indian multivoltine silkworm (*Bombyx mori* L.) germplasm," *Sericologia*, vol. 39, no. 3, pp. 337–347, 1999.

[19] N. B. Pal and S. M. Moorthy, "Assessment of variability in larval and cocoon traits in some genotypes of bivoltine silkworm, *Bombyx mori* L," *International Journal of Research in Biological Sciences*, vol. 1, no. 4, pp. 59–65, 2011.

[20] S. K. Sen, B. P. Nair, S. K. Das et al., "Relationship between the degree of heterosis and genetic divergence in the silkworm, *Bombyx mori* L," *Sericologia*, vol. 36, no. 2, pp. 215–225, 1996.

[21] P. Kumaresan, T. S. Mahadevamurthy, K. Thangavelu, and R. K. Sinha, "Further studies on the genetic divergence of multivoltine silkworm (*Bombyx mori* L.) genotypes based on economic characters," *Entomon*, vol. 28, no. 3, pp. 193–198, 2003.

[22] B. Mohan, N. Balachandran, M. Muthulakshmi et al., "Stratification of silkworm (*Bombyx mori* L.) germplasm for establishing core-set using characterisation data," *International Journal of Tropical Agricultur*, vol. 29, no. 3-4, pp. 325–329, 2011.

[23] G. Subba Rao, S. K. Das, N. K. Das, and S. Nandi, "Genetic divergence among bivoltine races of silkworm (*Bombyx mori*)," *Indian Journal of Agricultural Sciences*, vol. 59, no. 12, pp. 761–765, 1989.

[24] M. Farooq and H. P. Puttaraju, "Genetic divergence in bivoltine silkworm *Bombyx mori* L," in *Proceedings of the National Seminar on Mulberry Sericulture Research in India*, Abstract, p. 170, KSSRDI, Bangalore, India, November 2001.

[25] A. R. Hughes, B. D. Inouye, M. T. J. Johnson, N. Underwood, and M. Vellend, "Ecological consequences of genetic diversity," *Ecology Letters*, vol. 11, no. 6, pp. 609–623, 2008.

[26] N. Mittal and A. K. Dubey, "Microsatellite markers—a new practice of DNA based markers in molecular genetics," *Pharmacognosy Reviews*, vol. 3, no. 6, pp. 235–246, 2008.

[27] B. L. Fisher and M. A. Smith, "A revision of Malagasy species of *Anochetus mayr* and *Odontomachus latreille* (hymenoptera: formicidae)," *PLoS ONE*, vol. 3, no. 5, Article ID e1787, 2008.

[28] B. J. McGill, B. J. Enquist, E. Weiher, and M. Westoby, "Rebuilding community ecology from functional traits," *Trends in Ecology and Evolution*, vol. 21, no. 4, pp. 178–185, 2006.

[29] R. J. Parkash, P. Yadav, and M. Vashisht, "Allozymic variation at ADH locus in some *Drosophila* species," *Perspectives in Cytology and Genetics*, vol. 8, pp. 495–502, 1992.

[30] N. Yoshitake and M. Eguchi, "Distribution of blood esterase types in various strains of the silkworm, *Bombyx mori* L," *Japan Journal of Sericulture*, vol. 34, pp. 95–98, 1965 (Japanese).

[31] N. Yoshitake and M. Akiyama, "Genetic aspect on the esterase activities of the egg in the silkworm *Bombyx mori* L," *Japan Journal of Sericulture*, vol. 34, pp. 327–332, 1965 (Japanese).

[32] M. Eguchi, N. Yoshitake, and H. Kai, "Types and inheritance of blood esterase in the silkworm, *Bombyx mori* L," *Japan Journal of Genetics*, vol. 40, pp. 15–19, 1965.

[33] M. Eguchi and N. Yoshitake, "Interrelation of non specific esterase among various tissues in the silkworm, *Bombyx mori* L," *Japan Journal of Sericulture*, vol. 36, pp. 193–198, 1967.

[34] M. Eguchi, Y. Takahama, M. Ikeda, and S. Horii, "A novel variant of acid phosphatase isozyme from hemolymph of the silkworm, *Bombyx mori*," *Japan Journal of Genetics*, vol. 63, no. 2, pp. 149–157, 1988.

[35] A. Shabalina, "Esterase genetic polymorphism in haemolymph of larvae *Bombyx mori*," *Comptes Rendus de l'Academie Bulgare des Sciences*, vol. 43, pp. 105–110, 1990.

[36] P. Somasundaram, K. Ashok kumar, K. Thangavelu, P. K. Kar, and R. K. Sinha, "Preliminary study on isozyme variation in silkworm germplasm of *Bombyx mori* (L.) and its implication for conservation," *Pertanika Journal of Tropical Agricultural Science*, vol. 27, no. 2, pp. 163–171, 2004.

[37] S. M. Moorthy, S. K. Das, P. R. T. Rao, S. Raje Urs, and A. Sarkar, "Evaluation and selection of potential parents based on selection indices and isozyme variability in silkworm, *Bombyx mori* L," *Internatioanl Journal of Industrial Entomology*, vol. 14, pp. 1–7, 2007.

[38] T. Staykova and D. Grekov, "Stage specificity and polymorphism of haemolymph esterases in races and hybrids of silkworm (*Bombyx mori* L.) kept in Bulgaria," in *Proceedings of the International Workshop on Silk Handcrafts Cottage Industries and Silk Enterprises Development in Africa, Europe, Central Asia and the Near East, & 2nd Executive Meeting of Black, Caspian seas and Central Asia Silk Association (BACSA)*, pp. 667–674, Bursa, Turkey, March 2006.

[39] T. Staykova, "Genetically-determined polymorphism of non-specific esterases and phosphoglucomutase in eight introduced breeds of the silkworm, *Bombyx mori*, raised in Bulgaria," *Journal of Insect Science*, vol. 8, article 18, 2008.

[40] K. Ashok Kumar, P. Somasundaram, A. V. Bhaskara Rao, P. Vara Prasad, C. K. Kamble, and S. Smitha, "Genetic diversity and enzymes among selected silkworm races of *Bombyx mori* (L.)," *International Journal of Science and Nature*, vol. 2, no. 4, pp. 773–777, 2011.

[41] B. B. Patnaik, T. Datta, A. K. Saha, and M. K. Majumdhar, "Isozymic variations in specific and nonspecific esterase and its thermostability in silkworm, *Bombyx mori* L," *Journal of Environmental Biology*, vol. 33, pp. 837–842, 2012.

[42] L. Ronqui, M. Aparecida, and M. C. C. Ruvolo Takasusuk, "Genetic analysis of isoenzymes polymorphisms in silkworm (*Bombyx mori* L.) strains," *Acta Scientiarum Biological Sciences*, vol. 35, no. 2, pp. 249–254, 2013.

[43] T. Staykova, E. Ivanova, D. Grekov, and K. Avramova, "Genetic variability in silkworm (*Bombyx mori* L.) strains with different origin," *Acta Zoologica Bulgarica*, vol. 4, pp. 89–94, 2012.

[44] T. Staykova, E. N. Ivanova, P. Zenov, Y. VaSileva, D. Arkova Pantaleeva, and Z. Petkov, "Acid phosphatase as a marker for differentiation of silkworm (*Bombyx mori*) strains," *Biotechnology and Biotechnological Equipment*, vol. 24, no. 2, pp. 379–384, 2010.

[45] N. Yoshitake, M. Eguchi, and A. Akiyama, "Genetic control of esterase and acid phatase in the silkworm," *Journal of Sericulture Science in Japan*, vol. 35, pp. 1–6, 1966.

[46] J. Glaszmann, B. Kilian, H. Upadhyaya, and R. Varshney, "Accessing genetic diversity for crop improvement," *Current Opinion in Plant Biology*, vol. 13, no. 2, pp. 167–173, 2010.

[47] S. A. Wani, M. A. Bhat, Z. Buhroo, M. A. Ganai, and N. Majid, "Role of molecular markers in silkworm improvement," *International Journal of Recent Scientific Research*, vol. 4, no. 5, pp. 515–523, 2013.

[48] G. M. Nagaraja and J. Nagaraju, "Genome fingerprinting of the silkworm, *Bombyx mori*, using random arbitrary primers," *Electrophoresis*, vol. 16, no. 9, pp. 1633–1638, 1995.

[49] N. Thanananta, P. Saksoong, and S. Peyachoknagul, "RAPD technique in silkworm (*Bombyx mori*): strain differentiation and identification," *Thammasat International Journal of Science and Technology*, vol. 2, no. 2, pp. 47–51, 1997.

[50] K. D. Reddy, J. Nagaraju, and E. G. Abraham, "Genetic characterization of the silkworm *Bombyx mori* by simple sequence repeat (SSR)-anchored PCR," *Heredity*, vol. 83, no. 6, pp. 681–687, 1999.

[51] L. Zhang, Y. Huang, X. Miao, M. Qian, and C. Lu, "Microsatellite markers application on domesticated silkworm and wild silkworm," *Insect Science*, vol. 12, no. 6, pp. 413–419, 2005.

[52] J. H. Huang, S. H. Jia, Y. Zhang et al., "The polymorphism of silkworm, *Bombyx mori* (L.) amylase gene," *Scientia Agricultura Sinica*, vol. 39, no. 11, pp. 2390–2394, 2006.

[53] M. Li, C. Hou, X. Miao, A. Xu, and Y. Huang, "Analyzing genetic relationships in *Bombyx mori* using intersimple sequence repeat amplification," *Journal of Economic Entomology*, vol. 100, no. 1, pp. 202–208, 2007.

[54] S. B. Dalirsefat and S. Z. Mirhoseini, "Assessing genetic diversity in Iranian native silk-worm (*Bombyx mori*) strains and Japanese commercial lines using AFLP markers," *Iran Journal of Biotechnology*, vol. 5, no. 1, pp. 54–63, 2007.

[55] C. X. Hou, M. W. Li, Y. H. Zhang et al., "Analysis of SSR fingerprints in introduced silkworm germplasm resources," *Agricultural Sciences in China*, vol. 6, no. 5, pp. 620–627, 2007.

[56] F. Malik, H. P. Puttaraju, and S. N. Chatterjee, "Assessment of genetic diversity and association of PCR anchored ISSR markers with yield traits in silkworm, *Bombyx mori*," *Sericologia*, vol. 49, no. 1, pp. 1–13, 2009.

[57] J. G. Nagaraju and L. Singh, "Assessment of genetic diversity by DNA profiling and its significance in silkworm, *Bombyx mori*," *Electrophoresis*, vol. 18, no. 9, pp. 1676–1681, 1997.

[58] A. K. Awasthi, P. K. Kar, P. P. Srivastava et al., "Molecular evaluation of bivoltine, polyvoltine and mutant silkworm (*Bombyx mori* L.) with RAPD, ISSR and RFLP-STS markers," *Indian Journal of Biotechnology*, vol. 7, no. 2, pp. 188–194, 2008.

[59] E. Talebi, M. Khademi, and G. Subramanya, "RAPD markers for understanding of the genetic variability among the four silkworm races and their hybrids," *Middle-East Journal of Scientific Research*, vol. 7, no. 5, pp. 789–795, 2011.

[60] E. M. Furdui, L. A. Mărghita, D. Dezmirean, I. F. Pop, C. Coroian, and I. Paşca, "Genetic phylogeny and diversity of some Romanian silkworms based on RAPD technique," *Animal Science and Biotechnologies*, vol. 44, no. 1, pp. 204–208, 2011.

[61] S. Z. Mirhoseini, S. B. Dalirsefat, and M. Pourkheirandish, "Genetic characterization of iranian native *Bombyx mori* strains using amplified fragment length polymorphism markers," *Journal of Economic Entomology*, vol. 100, no. 3, pp. 939–945, 2007.

[62] K. Vijayan, C. V. Nair, and S. Raje Urs, "Assessment of genetic diversity in the tropical mulberry silkworm (*Bombyx mori* L.) with mtDNA-SSCP and SSR markers," *Emirate Journal of Food Agriculture*, vol. 22, no. 2, pp. 71–83, 2010.

[63] K. Y. Kim, E. M. Lee, I. H. Lee et al., "Intronic sequences of the silkworm strains of *Bombyx mori* (Lepidoptera: Bombycidae): high variability and potential for strain identification," *European Journal of Entomology*, vol. 105, no. 1, pp. 73–80, 2008.

[64] R. I. Tunca, T. Staykova, E. Ivanova, M. Kence, and D. Grekov, "Differentiation of silkworm, *Bombyx mori* strains measured by RAPD analyses," in *Proceedings of the Scientific and Technical Reports of the International Conference on Sericulture Challenges in the 21st Century (Serichal 2007) and the 3rd BACSA Meeting*, Abstract, pp. 29–30, Vratza, Bulgaria, September 2007.

[65] T. S. Jagadeesh Kumar, "Molecular dynamics of genomic DNA of silkworm breeds for screening under higher temperature Regimes utilising ISSR-primers," *International Journal of Science, Environment and Technology*, vol. 2, no. 2, pp. 275–285, 2013.

[66] Y. Liu, L. Qin, Y. Li et al., "Comparative genetic diversity and genetic structure of three chinese silkworm species *Bombyx mori* L. (Lepidoptera: Bombycidae), Antheraea pernyi guérin-meneville and samia cynthia ricini donovan (Lepidoptera: Saturniidae)," *Neotropical Entomology*, vol. 39, no. 6, pp. 967–976, 2010.

[67] K. Yukuhiro, H. Sezutsu, T. Tamura et al., "Little gene flow between domestic silkmoth *Bombyx mori* and its wild relative *Bombyx mandarina* in Japan, and possible artificial selection on the CAD gene of *B. mori*," *Genes Genetics Systemics*, vol. 87, pp. 331–340, 2012.

[68] D. S. Falconer and T. F. C. Mackay, *Introduction to Quantitative Genetics*, Longman, Delhi, India, 1996.

[69] Y. Vigouroux, Y. Matsuoka, and J. Doebley, "Directional evolution for microsatellite size in maize," *Molecular Biology and Evolution*, vol. 20, no. 9, pp. 1480–1483, 2003.

[70] J. Diamond, *Guns, Germs, and Streel: The Fates of Human Societies*, Norton, New York, NY, USA, 1997.

[71] M. D. Purugganan and D. Q. Fuller, "The nature of selection during plant domestication," *Nature*, vol. 457, no. 7231, pp. 843–848, 2009.

[72] T. Sakudoh, T. Nakashima, Y. Kuroki et al., "Diversity in copy number and structure of a silkworm morphogenetic gene as a result of domestication," *Genetics*, vol. 187, no. 3, pp. 965–976, 2011.

[73] Q. Xia, Y. Guo, Z. Zhang et al., "Complete resequencing of 40 genomes reveals domestication events and genes in silkworm (*Bombyx mori*)," *Science*, vol. 326, no. 5951, pp. 433–436, 2009.

[74] H. Yu, Y. Shen, G. Yuan et al., "Evidence of selection at melanin synthesis pathway loci during silkworm domestication," *Molecular Biology and Evolution*, vol. 28, no. 6, pp. 1785–1799, 2011.

[75] Y. Guo, Y. Shen, W. Sun, H. Kishino, Z. Xiang, and Z. Zhang, "Nucleotide diversity and selection signature in the domesticated silkworm, *Bombyx mori*, and wild silkworm, *Bombyx mandarina*," *Journal of Insect Science*, vol. 11, pp. 155–165, 2011.

[76] M. D. Loveless and J. L. Hamrick, "Ecological determinants of genetic structure in plant populations," *Annual Review of Ecology, Evolution, and Systematics*, vol. 15, pp. 65–95, 1984.

[77] J. L. Hamrigk and M. J. W. Godt, "Effects of life history traits on genetic diversity in plant species," *Philosophical Transactions of the Royal Society B: Biological Sciences*, vol. 351, no. 1345, pp. 1291–1298, 1996.

[78] K. E. Holsinger, "Reproductive systems and evolution in vascular plants," *Proceedings of the National Academy of Sciences of the United States of America*, vol. 97, no. 13, pp. 7037–7042, 2000.

[79] P. Gepts, "Plant genetic resources conservation and utilization: the accomplishments and future of a societal insurance policy," *Crop Science*, vol. 46, no. 5, pp. 2278–2292, 2006.

[80] E. Bennett, "Wheats of the Mediterranean basin," in *Survey of Crop Genetic Resources in Their Centre of Diversity, First Report*, O. H. Frankel, Ed., pp. 1–8, FAO-IBP, 1973.

[81] T. Gamo, "Recent concepts and trends in silkworm breeding," *Farming Japan*, vol. 10, no. 6, pp. 11–22, 1976.

[82] H. Ohi and A. Yamashita, "On the breeding of the silkworm races J137 and C137," *Bulletin of Sericulture Experiment Station*, vol. 27, pp. 97–139, 1977.

[83] L. L. Ren, L. Mini, and C. Hell, "Stability of double cross hybrid combined with current silkworm varieties for spring and early autumn under normal rearing condition," *Acta Serica Sinica*, vol. 14, no. 1, pp. 42–44, 1988.

[84] R. K. Datta, H. K. Basavaraja, N. Mal Reddy et al., "Evolution of new productive bivoltine hybrids CSR2 × CSR4 and CSR2 × CSR5," *Sericologia*, vol. 40, pp. 151–174, 2000.

[85] M. W. Bruford, D. G. Bradley, and G. Luikart, "DNA markers reveal the complexity of livestock domestication," *Nature Reviews Genetics*, vol. 4, no. 11, pp. 900–910, 2003.

[86] M. A. Toro, J. Fernández, and A. Caballero, "Molecular characterization of breeds and its use in conservation," *Livestock Science*, vol. 120, no. 3, pp. 174–195, 2009.

[87] A. H. D. Brown, "The genetic diversity of germplasm collections," in *Proceedings of the Workshop on the Genetic Evaluation of Plant Genetic Resources*, pp. 9–11, Canada Research Branch, Agriculture Canada, Toronto, Canada, 1988.

[88] M. R. Macnair, V. E. Macnair, and B. E. Martin, "Adaptive speciation in *Mimulus*: an ecological comparison of *M. cupriphilus* with its presumed progenitor, *M. guttatus*," *New Phytologist*, vol. 112, no. 3, pp. 269–279, 1989.

[89] X. Vekemans and C. Lefèbvre, "On the evolution of heavy-metal tolerant populations in *Armeria maritima*: evidence from allozyme variation and reproductive barriers," *Journal of Evolutionary Biology*, vol. 10, no. 2, pp. 175–191, 2001.

[90] S. R. Whitt, L. M. Wilson, M. I. Tenaillon, B. S. Gaut, and E. S. Buckler, "Genetic diversity and selection in the maize starch pathway," *Proceedings of the National Academy of Sciences of the United States of America*, vol. 99, no. 20, pp. 12959–12962, 2002.

[91] A. P. Moller and A. Pomiankowski, "Fluctuating asymmetry and sexual selection," *Genetica*, vol. 89, no. 1–3, pp. 267–279, 1993.

[92] M. G. Bulmer, "The effect of selection on genetic variability," *American Nature*, vol. 105, pp. 201–211, 1971.

[93] C. Pélabon, M. L. Carlson, T. F. Hansen, N. G. Yoccoz, and W. S. Armbruster, "Consequences of inter-population crosses on developmental stability and canalization of floral traits in Dalechampia scandens (Euphorbiaceae)," *Journal of Evolutionary Biology*, vol. 17, no. 1, pp. 19–32, 2004.

[94] A. R. Pradeep, S. N. Chatterjee, and C. V. Nair, "Genetic differentiation induced by selection in an inbred population of the silkworm *Bombyx mori*, revealed by RAPD and ISSR marker systems," *Journal of Applied Genetics*, vol. 46, no. 3, pp. 291–298, 2005.

[95] V. A. Strunnikov, *Control over Reproduction, Sex and Heterosis of the Silkworm*, Harwood Academic Publishers, Luxembourg, 1995.

[96] A. Seidavi, "Estimation of genetic parameters and selection effect on genetic and phenotype trends in silkworm commercial pure lines," *Asian Journal of Animal and Veterinary Advances*, vol. 5, no. 1, pp. 1–12, 2010.

[97] S. Wright, *EVolution and the Genetics of Populations. Volume 1: Genetic and Biometric Foundations*, University of Chicago Press, Chicago, Ill, USA, 1968.

[98] P. Buri, "Gene frequency in small populations of mutant *Drosophila*," *Evolution*, vol. 10, pp. 367–402, 1956.

[99] R. Bijlsma, J. Bundgaard, and A. C. Boerema, "Does inbreeding affect the extinction risk of small populations? Predictions from *Drosophila*," *Journal of Evolutionary Biology*, vol. 13, no. 3, pp. 502–514, 2000.

[100] D. H. Reed, "The effects of population size on population viability: from mutation to environmental catastrophes," in *Conservation Biology: Evolution in Action*, S. P. Carroll and C. W. Fox, Eds., pp. 16–35, Oxford University Press, New York, NY, USA, 2008.

[101] D. Charlesworth and B. Charlesworth, "Quantitative genetics in plants: the effect of the brreeding system on genetic variability," *Evolution*, vol. 49, no. 5, pp. 911–920, 1995.

[102] D. Charlesworth, B. Charlesworth, and C. Strobeck, "Selection for recombination in self-fertilising species," *Genetics*, vol. 93, pp. 237–244, 1979.

[103] J. L. Hamrick and M. J. Godt, "Allozyme diversity in plant species," in *Plant Population Genetics, Breeding, and Genetic Resources*, A. H. D. Brown, M. T. Clegg, A. L. Kahler, and B. S. Weir, Eds., pp. 43–63, Sinauer, Sunderland, Mass, USA, 1990.

[104] R. C. Lacy, "Impacts of inbreeding in natural and captive populations of vertebrates: implications for conservation," *Perspectives in Biology and Medicine*, vol. 36, no. 3, pp. 480–496, 1993.

[105] R. Lande, "Risk of population extinction from fixation of new deleterious mutations," *Evolution*, vol. 48, no. 5, pp. 1460–1469, 1994.

[106] M. C. Whitlock, "Selection, load and inbreeding depression in a large metapopulation," *Genetics*, vol. 160, no. 3, pp. 1191–1202, 2002.

[107] S. Wright, *Evolution and Genetics of Populations. Volume 3: Experimental Results and Evolutionary Deductions*, University of Chicago Press, Chicago, Ill, USA, 1977.

[108] K. Zittlau, *Population genetic analyses of North American caribou (Rangifer tarandus) [Ph.D. dissertation]*, Department of Biological Sciences, University of Alberta, Edmonton, Canada, 2004.

[109] D. Li, Y. Guo, H. Shao et al., "Genetic diversity, molecular phylogeny and selection evidence of the silkworm mitochondria implicated by complete resequencing of 41 genomes," *BMC Evolutionary Biology*, vol. 10, no. 1, article 81, 10 pages, 2010.

[110] M. L. Wang, J. A. Mosjidis, J. B. Morris, Z. B. Chen, N. A. Barkley, and G. A. Pederson, "Evaluation of *Lespedeza* germplasm genetic diversity and its phylogenetic relationship with the genus *Kummerowia*," *Conservation Genetics*, vol. 10, no. 1, pp. 79–85, 2009.

[111] M. Goodman, "Broadening the genetic diversity in maize breeding by use of Exotic germplasm," in *The Genetics and Exploitation of Heterosis in Crops*, G. Coors and S. Pandey, Eds., pp. 130–149, ASA-CSSA-SSSA, Madison, Wis, USA, 1999.

[112] R. Frankham, J. D. Ballou, and D. A. Briscoe, *Introduction to Conservation Genetics*, Cambridge University Press, Cambridge, UK, 2002.

[113] W.-H. Li, "Maintenance of genetic variability under the joint effect of mutation, selection and random drift," *Genetics*, vol. 90, no. 2, pp. 349–383, 1978.

[114] P. S. Guzman and K. R. Lamkey, "Effective population size and genetic variability in the BS11 maize population," *Crop Science*, vol. 40, no. 2, pp. 338–342, 2000.

[115] J. Doreswamy and S. Gopal, "Inbreeding effects on quantitative traits in random mating and selected populations of the mulberry silkworm, *Bombyx mori*," *Journal of Insect Science*, vol. 12, pp. 1–6, 2012.

[116] B. K. Gupta, V. K. Kharoo, and M. Verma, "Genetic divergence in bivoltine strains of silkworm (*Bombyx mori* L.)," *Bioved*, vol. 3, no. 2, pp. 143–146, 1992.

[117] T. K. Narayanaswamy, R. Govindan, S. R. Anantha Narayana, and S. Ramesh, "Genetic divergence among some breeds of silkworm *Bombyx mori* L," *Entomon*, vol. 27, no. 3, pp. 319–321, 2002.

[118] P. Kumaresan, P. R. Koundinya, S. A. Hiremath, and R. K. Sinha, "An analysis of genetic variation and divergence on silk fibre characteristics of multivoltine silkworm (*Bombyx mori* L.) genotypes," *International Journal of Industrial Entomology*, vol. 14, no. 1, pp. 23–32, 2007.

[119] M. Farooq, M. A. Khan, and M. N. Ahmad, "Genetic divergence among bivoltine genotypes of silkworm, *Bombyx mori* L," in *Proceedings of the National Workshop on Seri-Biodiversity Conservation*, pp. 94–98, Central Sericultural Germplasm Resources Centre, Hosur, India, March 2009.

[120] M. S. Nezhad, S. Z. Mirhosseini, S. Gharahveysi, M. Mavvajpour, and A. R. Seidavi, "Analysis of genetic divergence for classification of morphological and larval gain characteristics of peanut cocoon silkworm (*Bombyx mori* L.) germplasm," *American-Eurasian Journal of Agriculture and Environmental Sciences*, vol. 6, no. 5, pp. 600–608, 2009.

[121] D. B. Zanatta, J. P. Bravo, J. F. Barbosa, R. E. F. Munhoz, and M. A. Fernandez, "Evaluation of economically important traits from sixteen parental strains of the silkworm *Bombyx mori* L, (Lepidoptera: Bombycidae)," *Neotropical Entomology*, vol. 38, no. 3, pp. 327–331, 2009.

[122] A. Maqbool and H. U. Dar, "Genetic divergence in some bivoltine silkworm (*Bombyx mori* L.) breeds," 2010, http://dspaces.uok.edu.in/jspui//handle/1/1100.

[123] M. Salehi Nezhad, S. Z. Mirhosseini, S. Gharahveysi, M. Mavvajpour, A. R. Seidavi, and M. Naserani, "Genetic diversity and classification of 51 strains of silkworm *Bombyx mori* (Lepidoptera: Bombycidae) germplasm based on larval phenotypic data using Ward's and UPGMA methods," *African Journal of Biotechnology*, vol. 9, no. 39, pp. 6594–6600, 2010.

[124] E. Talebi and G. Subramanya, "Genetic distance and heterosis through evaluation index in the silkworm, *Bombyx mori* (L.)," *American Journal of Applied Sciences*, vol. 6, no. 12, pp. 1981–1987, 2009.

[125] P. M. Lizardi, "Genetic polymorphism of silk fibroin studied by two-dimensional translation pause fingerprints," *Cell*, vol. 18, no. 2, pp. 581–589, 1979.

[126] T. Egorova, E. Naletova, and Y. Nasirillaev, "Polymorphic system of silkworm haemolymph esterases as a criterion to make programs for parental specimens crossing," *Biochemistry of Insects*, pp. 54–62, 1985 (Russian).

[127] M. Eguchi, Y. Takahama, M. Ikeda, and S. Horii, "A novel variant of acid phosphatase isozyme from hemolymph of the silkworm, *Bombyx mori* L," *Japanese Journal of Genetics*, vol. 63, pp. 149–157, 1988.

[128] S. N. Chatterjee and R. K. Datta, "Hierarchical clustering of 54 races and strains of the mulberry silkworm, *Bombyx mori* L: significance of biochemical parameters," *Theoretical and Applied Genetics*, vol. 85, no. 4, pp. 394–402, 1992.

[129] S. K. Das, P. K. Chinya, S. Patinaik, S. K. Sen, and G. Subba Rao, "Studies on genetic variability of heamolymph esterases in some genetic stocks of mulberry silkworm *Bombyx mori* L," *Perspectives in Cytology and Genetics*, vol. 7, pp. 421–425, 1992.

[130] M. Nei and W. H. LI, "Mathematical model for studying genetic variation in terms of restriction endonucleases," *Proceedings of the National Academy of Sciences of the United States of America*, vol. 76, no. 10, pp. 5269–5527, 1978.

[131] F. C. Yeh, R. Yang, and T. Boyle, "Population genetic analysis POPGENE version 1.31: Microsoftwindow-based freeware for population genetic analysis," Quick User's Guide, A Joint Project Developed by Centre for International Forestry Research and University of Alberta, Alberta, Canada, 1999.

[132] K. Etebari, S. Z. Mirhoseini, and L. Matindoost, "A study on interspecific biodiversity of eight groups of silkworm (*Bombyx mori*) by biochemical markers," *Insect Science*, vol. 12, no. 2, pp. 87–94, 2005.

[133] V. Kumar, S. K. Ashwath, and S. B. Dandin, "Heamolymph protein variability among the silkworm (*Bombyx mori*) breeds and assessment of their genetic relationship," in *Proceedings of the Asia Pacific Congress of Sericulture and Insect Biotechnology (APSERI '06)*, Abstract, p. 54, Sangju, Republic of Korea, October 2006.

[134] S. Bakkappa and G. Subramanya, "Electrophoretic haemolymph protein pattern in a few bivoltine races of the silkworm, *Bombyx mori*," *The Bioscan*, vol. 5, no. 4, pp. 541–544, 2010.

[135] J. Anuradha, S. Somasundaram, S. Vishnupriya, and A. Manjula, "Storage protein-2 as a dependable biochemical index for screening germplasm stocks of the silkworm *Bombyx mori* (L.)," *Albanian Journal of Agriculture Science*, vol. 11, pp. 141–148, 2012.

[136] T. Staykova, "Inter-and intra-population genetic variability of introduced silkworm (*Bombyx mori* L.) strains raised in Bulgaria," *Journal of Biosciences and Biotechnology*, vol. 2, no. 1, pp. 73–77, 2003.

[137] J. Nagaraju, K. D. Reddy, G. M. Nagaraja, and B. N. Sethuraman, "Comparison of multilocus RFLPs and PCR-based marker systems for genetic analysis of the silkworm, *Bombyx mori*," *Heredity*, vol. 86, no. 5, pp. 588–597, 2001.

[138] M. Li, L. Shen, A. Xu et al., "Genetic diversity among silkworm (*Bombyx mori* L., Lep., Bombycidae) germplasms revealed by microsatellites," *Genome*, vol. 48, no. 5, pp. 802–810, 2005.

[139] P. P. Srivastava, K. Vijayan, A. K. Awasthi, P. K. Kar, K. Thangavelu, and B. Saratchandra, "Genetic analysis of silkworms (*Bombyx mori*) through RAPD markers," *Indian Journal of Biotechnology*, vol. 4, no. 3, pp. 389–395, 2005.

[140] B. C. K. Murthy, B. M. Prakash, and H. P. Puttaraju, "Fingerprinting of non-diapausing silkworm, *Bombyx mori*, using random arbitrary primers," *Cytologia*, vol. 71, no. 4, pp. 331–335, 2006.

[141] S. B. Dalirsefat and S. Z. Mirhoseini, "Assessing genetic diversity in Iranian native silkworm (*Bombyx mori* L.) strains and Japanese commercial lines using AFLP markers," *Iranian Journal of Biotechnology*, vol. 5, no. 1, pp. 25–33, 2007.

[142] D. Velu, K. M. Ponnuvel, M. Muthulakshmi, R. K. Sinha, and S. M. H. Qadri, "Analysis of genetic relationship in mutant silkworm strains of *Bombyx mori* using inter simple sequence repeat (ISSR) markers," *Journal of Genetics and Genomics*, vol. 35, no. 5, pp. 291–297, 2008.

[143] D. Eroğlu and S. Cakir Arica, "Molecular genetic analysis of three Turkish local silkworm breeds (Bursa Beyazı, Alaca and Hatay Sarısı) by RAPD-PCR method," *Journal of Applied Biological Sciences*, vol. 3, no. 2, pp. 17–20, 2009.

[144] K. Ashok Kumar, P. Somasundaram, K. M. Ponnuvel, G. K. Srinivasa Babu, S. M. H. Qadri, and C. K. Kamble, "Identification of genetic variations among silkworm races of *Bombyx mori* (L.) through bio-molecular tools," *Indian Journal of Sericulture*, vol. 48, no. 2, pp. 116–125, 2009.

[145] K. Sanjeeva reddy, C. A. Mahalingam, K. A. Murugesh, and S. Mohankumar, "Exploring the genetic variability in *Bombyx mori* L. with molecular marker," *Karnataka Journal of Agricultural Sciences*, vol. 22, pp. 479–483, 2009.

[146] K. A. Murugesh, S. Mohankumar, and C. A. Mahalingam, "Molecular marker analysis on genetic variation in domesticated silkworm," *Trends in Biosciences*, vol. 3, no. 2, pp. 102–105, 2010.

[147] P. P. Srivastava, K. Vijayan, P. K. Kar, and B. Saratchandra, "Diversity and marker association in tropical silkworm breeds of *Bombyx mori* (Lepidoptera: Bombycidae)," *International Journal of Tropical Insect Science*, vol. 31, no. 3, pp. 182–191, 2011.

[148] B. Vlaic, L. A. Mărghitaş, A. Vlaic, and P. Raica, "Analysis of genetic diversity of mulberry silkworm (*Bombyx mori* L.) using RAPD molecular markers," *Animal Science and Biotechnologies*, vol. 69, no. 1-2, pp. 292–296, 2012.

[149] R. Radjabi, A. Sarafrazi, A. Tarang, K. Kamali, and S. Tirgari, "Intraspecific biodiversity of Iranian local races of silkworm *Bombyx mori* by ISSR (Inter-Simple Sequence Repeat) molecular marker," *World Journal of Zoology*, vol. 7, no. 1, pp. 17–22, 2012.

[150] S. M. Moorthy, N. Chandrakanth, S. K. Ashwath, V. Kumar, and B. B. Bindroo, "Genetic diversity analysis using RAPD marker in some silkworm breeds of *Bombyx mori* L," *Annals of Biological Research*, vol. 4, no. 12, pp. 82–88, 2013.

[151] N. Chandrakanth, S. M. Moorthy, P. Anusha et al., "Evaluation of genetic diversity in silkworm (*Bombyx mori* L.) strains using microsatellite markers," *International Journal of Biotechnology and Allied Fields*, vol. 2, no. 3, pp. 73–93, 2014.

Permissions

The contributors of this book come from diverse backgrounds, making this book a truly international effort. This book will bring forth new frontiers with its revolutionizing research information and detailed analysis of the nascent developments around the world.

We would like to thank all the contributing authors for lending their expertise to make the book truly unique. They have played a crucial role in the development of this book. Without their invaluable contributions this book wouldn't have been possible. They have made vital efforts to compile up to date information on the varied aspects of this subject to make this book a valuable addition to the collection of many professionals and students.

This book was conceptualized with the vision of imparting up-to-date information and advanced data in this field. To ensure the same, a matchless editorial board was set up. Every individual on the board went through rigorous rounds of assessment to prove their worth. After which they invested a large part of their time researching and compiling the most relevant data for our readers. Conferences and sessions were held from time to time between the editorial board and the contributing authors to present the data in the most comprehensible form. The editorial team has worked tirelessly to provide valuable and valid information to help people across the globe.

Every chapter published in this book has been scrutinized by our experts. Their significance has been extensively debated. The topics covered herein carry significant findings which will fuel the growth of the discipline. They may even be implemented as practical applications or may be referred to as a beginning point for another development. Chapters in this book were first published by Hindawi Publishing Corporation; hereby published with permission under the Creative Commons Attribution License or equivalent.

The editorial board has been involved in producing this book since its inception. They have spent rigorous hours researching and exploring the diverse topics which have resulted in the successful publishing of this book. They have passed on their knowledge of decades through this book. To expedite this challenging task, the publisher supported the team at every step. A small team of assistant editors was also appointed to further simplify the editing procedure and attain best results for the readers.

Our editorial team has been hand-picked from every corner of the world. Their multi-ethnicity adds dynamic inputs to the discussions which result in innovative outcomes. These outcomes are then further discussed with the researchers and contributors who give their valuable feedback and opinion regarding the same. The feedback is then collaborated with the researches and they are edited in a comprehensive manner to aid the understanding of the subject.

Apart from the editorial board, the designing team has also invested a significant amount of their time in understanding the subject and creating the most relevant covers. They scrutinized every image to scout for the most suitable representation of the subject and create an appropriate cover for the book.

The publishing team has been involved in this book since its early stages. They were actively engaged in every process, be it collecting the data, connecting with the contributors or procuring relevant information. The team has been an ardent support to the editorial, designing and production team. Their endless efforts to recruit the best for this project, has resulted in the accomplishment of this book. They are a veteran in the field of academics and their pool of knowledge is as vast as their experience in printing. Their expertise and guidance has proved useful at every step. Their uncompromising quality standards have made this book an exceptional effort. Their encouragement from time to time has been an inspiration for everyone.

The publisher and the editorial board hope that this book will prove to be a valuable piece of knowledge for researchers, students, practitioners and scholars across the globe.

List of Contributors

Irene Martín-Forés
Laboratory 8, Department of Ecology, Universidad Complutense de Madrid, c. José Antonio Novais, 28040 Madrid, Spain

Berta Martín-López and Carlos Montes
Social-Ecological Systems Laboratory, Department of Ecology, Universidad Autónoma de Madrid, c. Darwin, 28049 Madrid, Spain

F. Semprucci
Dipartimento di Scienze della Terra, della Vita e dell Ambiente (DiSTEVA), Università degli Studi di Urbino "Carlo Bo", Località Crocicchia, 61029 Urbino, Italy

James D'Souza and Bernard Felinov Rodrigues
Department of Botany, Goa University, Taleigao, Goa 403 206, India

Raquel Costa de Luca Rebello and Adriana Hamond Regua-Mangia
Departamento de Ciencias Biologicas, EscolaNacional de Saude Publica Sergio Arouca (Ensp), Fundac ao Oswaldo Cruz (FIOCRUZ), Rua Leopoldo Bulhoes 1480 Manguinhos, 21041-210 Rio de Janeiro, RJ, Brazil

Karen Machado Gomes and Rafael Silva Duarte
Departamento de Microbiologia Medica, Instituto de Microbiologia Paulo de Goes, Universidade Federal do Rio de Janeiro (UFRJ), Rio de Janeiro, RJ, Brazil

Caio Tavora Coelho da Costa Rachid
Departamento de Microbiologia Geral, Instituto de Microbiologia Paulo de Goes, Universidade Federal do Rio de Janeiro (UFRJ), Rio de Janeiro, RJ, Brazil

Julius M. Assam
Tanzania Fisheries Research Institute, P.O. Box 90, Kigoma, Tanzania

Emmanuel A. Sweke
Tanzania Fisheries Research Institute, P.O. Box 90, Kigoma, Tanzania
Graduate School of Fisheries Sciences, Hokkaido University, 3-1-1 Minato-cho, Hakodate 041-8611, Japan

Takashi Matsuishi
Faculty of Fisheries Sciences, Hokkaido University, 3-1-1 Minato-cho, Hakodate 041-8611, Japan

Abdillahi I. Chande
Tanzania Fisheries Research Institute, P.O. Box 90, Kigoma, Tanzania
Marine Parks and Reserves, P.O. Box 7565, Dar es Salaam, Tanzania

Muposhi Victor Kurauwone, Utete Beven, Kupika Olga and Tarakini Tawanda
Department of Wildlife & Safari Management, Chinhoyi University of Technology, P. Bag 7724, Chinhoyi, Zimbabwe

Muvengwi Justice
Department of Environmental Science, Bindura University of Science Education, P. Bag 1020, Bindura, Zimbabwe

Chiutsi Simon
Department of Travel & Recreation Management, Chinhoyi University of Technology, P. Bag 7724, Chinhoyi, Zimbabwe

Adrian C. Newton, Elena Cantarello, Natalia Tejedor and Gillian Myers
Centre for Conservation Ecology and Environmental Science, School of Applied Sciences, Bournemouth University, Talbot Campus, Poole, Dorset BH12 5BB, UK

Juan Li and Dajun Wang
Center for Nature and Society, College of Life Sciences, Peking University, Beijing 100871, China

Zhi Lu
Center for Nature and Society, College of Life Sciences, Peking University, Beijing 100871, China
Shan Shui Conservation Center, Beijing 100871, China

George B. Schaller
Panthera and Wildlife Conservation Society, 8 West 40th Street, 18th Floor, New York, NY 10018, USA

Thomas M. McCarthy
Snow Leopard Program, Panthera, 8 West 40th Street, 18th Floor, New York, NY 10018, USA

Zhala Jiagong
Shan Shui Conservation Center, Beijing 100871, China

Ping Cai
Wildlife Conservation and Management Bureau, Qinghai Forestry Department, Xining 810008, Qinghai, China

Lamao Basang
Sanjiangyuan National Nature Reserve, Qinghai Forestry Department, Xining 810008, Qinghai, China

Nikunj B. Gajera, Arun Kumar Roy Mahato and V. Vijay Kumar
Gujarat Institute of Desert Ecology, Mundra Road, P.O. Box 83, Bhuj, Kachchh, Gujarat 370001, India

C. Holyoake, R. Donaldson and K. Warren
Conservation Medicine Program, College of Veterinary Medicine, School of Veterinary and Life Sciences, Murdoch University, Perth, WA 6150, Australia

G. Flacke
Conservation Medicine Program, College of Veterinary Medicine, School of Veterinary and Life Sciences, Murdoch University, Perth, WA 6150, Australia
School of Animal Biology, University of Western Australia, 35 Stirling Highway, Crawley, WA 6009, Australia

P. Becker
Center for Species Survival, Smithsonian Conservation Biology Institute, National Zoological Park, Front Royal, VA 22630, USA
Centre for Wildlife Management, University of Pretoria, Pretoria 0002, South Africa

D. Cooper
Ezemvelo KZN Wildlife, Queen Elizabeth Park, Pietermaritzburg 3202, South Africa

M. Szykman Gunther
Center for Species Survival, Smithsonian Conservation Biology Institute, National Zoological Park, Front Royal, VA 22630, USA
Department of Wildlife, Humboldt State University, Arcata, CA 95521, USA

I. Robertson
Veterinary Epidemiology Programme, College of Veterinary Medicine, School of Veterinary and Life Sciences, Murdoch University, Perth, WA 6150, Australia

Patience Zisadza-Gandiwa and Edson Gandiwa
Scientific Services, Gonarezhou National Park, Parks and Wildlife Management Authority, Private Bag 7003, Chiredzi, Zimbabwe

Cheryl T. Mabika and Olga L. Kupika
Department ofWildlife and Safari Management, Chinhoyi University of Technology, Private Bag 7724, Chinhoyi, Zimbabwe

Chrispen Murungweni
Department of Animal Production and Technology, Chinhoyi University of Technology, Private Bag 7724, Chinhoyi, Zimbabwe

Periasamy Alagesan and Baluchamy Ramanathan
Post Graduate and Research Department of Zoology, Yadava College, Madurai, Tamil Nadu 625 014, India

Federico Morelli
DiSTeVA, University of Urbino, Scientific Campus, 61029 Urbino, Italy

Thaís X. Melo and Elvio S. F. Medeiros
Grupo Ecologia de Rios do Semiarido, Universidade Estadual da Paraíba-UEPB,
Rua Horacio Trajano de Oliveira, S/N, Cristo Redentor, 58020-540 Joao Pessoa, PB, Brazil

A. N. Petrov and E. L. Nevrova
Institute of Biology of the Southern Seas, National Academy of Sciences of Ukraine, Sevastopol 99011, Ukraine

H. Tynsong
Ministry of Environment and Forest, North Eastern Regional Office, Shillong 793021, Meghalaya, India

M. Dkhar
Union Christian Collage Umiam, Ri Bhoi, Shillong 793122, Meghalaya, India

B. K. Tiwari
Department of Environmental Studies, North-Eastern Hill University, Shillong 793022, Meghalaya, India

John Carter
Kiesha's Preserve, Paris, ID 83261, USA

Allison Jones
Wild Utah Project, Salt Lake City, UT 84101, USA

Mary O'Brien
Grand Canyon Trust, Flagstaff, AZ 86001, USA

Jonathan Ratner
Western Watersheds Project, Pinedale, WY 82941, USA

George Wuerthner
Foundation for Deep Ecology, Bend, OR 97708, USA

M. K. Rajesh, K. Samsudeen, P. Rejusha, Shafeeq Rahman and Anitha Karun
Division of Crop Improvement, Central Plantation Crops Research Institute, Kasaragod Kerala, 671124, India

C. Manjula
Division of Crop Improvement, Central Plantation Crops Research Institute, Kasaragod Kerala, 671124, India
Nehru Arts and Science College, Kanhangad Kerala, 671314, India

Tewodros Kumssa and Afework Bekele
Department of Zoological Sciences, Addis Ababa University, P.O. Box 1176, Addis Ababa, Ethiopia

Roghayeh Najafzadeh and Kazem Arzani
Department of Horticultural Sciences, Tarbiat Modares University (TMU), P.O. Box 14115-336, Tehran, Iran

Naser Bouzari
Horticultural Section, Stone Fruit Research Group, Seed and Plant Improvement Research Institute of Karaj (SPII), P.O. Box 31585-4119, Karaj, Iran

Ali Saei
Genomics Section, Agricultural Biotechnology Research Institute of Iran (ABRII), P.O. Box 85135-487, Isfahan, Iran

Monnanda Somaiah Nalini
Department of Studies in Botany, University of Mysore, Manasagangotri, Mysore, Karnataka 570 006, India

Ningaraju Sunayana and Harischandra Sripathy Prakash
Department of Studies in Biotechnology, University of Mysore, Manasagangotri, Mysore, Karnataka 570 006, India

Bharat Bhusan Bindroo and Shunmugam Manthira Moorthy
Central Sericultural Research and Training Institute, Srirampura, Mysore, Karnataka 570 008, India

www.ingramcontent.com/pod-product-compliance
Lightning Source LLC
Chambersburg PA
CBHW070154240326
41458CB00126B/4834